Nutrition and Fish Health

FOOD PRODUCTS PRESS
Aquaculture
Carl David Webster, PhD
Senior Editor

Nutrition and Fish Health by Chhorn Lim and Carl D. Webster

Additional Titles of Related Interest:

Introduction to the General Principles of Aquaculture by Hans Ackefors, Jay V. Huner, and Mark Konikoff

Freshwater Crayfish Aquaculture in North America, Europe, and Australia: Families Astacide, Cambaridae, and Parastacidae edited by Jay V. Huner

Economics of Aquaculture by Curtis M. Jolly and Howard A. Clonts

Nutrition and Fish Health

Chhorn Lim
Carl D. Webster
Editors

CRC Press is an imprint of the
Taylor & Francis Group, an informa business

CRC Press
Taylor & Francis Group
6000 Broken Sound Parkway NW, Suite 300
Boca Raton, FL 33487-2742

First issued in paperback 2019

© 2001 by Taylor & Francis Group, LLC
CRC Press is an imprint of Taylor & Francis Group, an Informa business

No claim to original U.S. Government works

ISBN-13: 978-1-56022-887-5 (hbk)
ISBN-13: 978-0-367-39707-4 (pbk)

This book contains information obtained from authentic and highly regarded sources. Reasonable efforts have been made to publish reliable data and information, but the author and publisher cannot assume responsibility for the validity of all materials or the consequences of their use. The authors and publishers have attempted to trace the copyright holders of all material reproduced in this publication and apologize to copyright holders if permission to publish in this form has not been obtained. If any copyright material has not been acknowledged please write and let us know so we may rectify in any future reprint.

Except as permitted under U.S. Copyright Law, no part of this book may be reprinted, reproduced, transmitted, or utilized in any form by any electronic, mechanical, or other means, now known or hereafter invented, including photocopying, microfilming, and recording, or in any information storage or retrieval system, without written permission from the publishers.

For permission to photocopy or use material electronically from this work, please access www.copyright.com (http://www.copyright.com/) or contact the Copyright Clearance Center, Inc. (CCC), 222 Rosewood Drive, Danvers, MA 01923, 978-750-8400. CCC is a not-for-profit organization that provides licenses and registration for a variety of users. For organizations that have been granted a photocopy license by the CCC, a separate system of payment has been arranged.

Trademark Notice: Product or corporate names may be trademarks or registered trademarks, and are used only for identification and explanation without intent to infringe.

Library of Congress Cataloging-in-Publication Data

Nutrition and fish health / Chhorn Lim, Carl D. Webster, editors.
p. cm.
Includes bibliographical references (p.).
ISBN 1-56022-887-3 (hardcover : alk. paper)
1. Fishes—Diseases. 2. Fishes—Nutrition. I. Lim, Chhorn. II. Webster, Carl. D.

SH171.N88 2001
639.8—dc21

00-049478

Visit the Taylor & Francis Web site at
http://www.taylorandfrancis.com

and the CRC Press Web site at
http://www.crcpress.com

To my wife, Brenda, and our children,
Chheang Chhun, Chhorn Jr., and Brendan.

Chhorn Lim

To my wife, Caroline, our daughter, NancyAnn,
and our other "children," Darwin, Poppins,
Michael, KC, and Barley.

Carl D. Webster

CONTENTS

ABOUT THE EDITORS

Chhorn Lim, PhD, has more than 25 years of experience in aquaculture nutrition and feed development research. Currently, he is the Nutrition Scientist at the Fish Diseases and Parasites Research Laboratory, U.S. Department of Agriculture, Agricultural Research Services at Auburn, Alabama, where he conducts research on the interactions between nutrition, immune system function, and disease resistance. He also serves as Affiliate Professor of the Department of Fisheries and Allied Aquaculture, Auburn University, and an Affiliate Researcher of the University of Hawaii Institute of Marine Biology in Kaneohe, Hawaii. Dr. Lim has performed several long- and short-term consultancies and has received several honors and awards in recognition of his contributions and outstanding achievements. He is a member of several professional organizations and an editorial board member of the *Journal of Applied Aquaculture.* He has authored or co-authored more than 75 publications and abstracts, including book chapters and a book of which he is the senior editor.

Carl D. Webster, PhD, has more than 10 years of experience in aquaculture nutrition and diet development research. Currently, he is Principal Investigator for Aquaculture at the Aquaculture Research Center, Kentucky State University, where he conducts research on nutrition requirements and practical diet formulations for fish and crustacean species that are currently or potentially cultured. He is also Associate Professor of the Department of Math and Sciences at Kentucky State University and is Adjunct Professor of the Department of Animal Sciences at the University of Kentucky. Dr. Webster has been elected twice to serve as Secretary/Treasurer of the U.S. Chapter of the World Aquaculture Society (now known as the U.S. Aquaculture Society, a Chapter of the World Aquaculture Society), and was elected President of the U.S. Chapter of the World Aquaculture Society. He is a member of several professional organizations and is editor of the *Journal of Applied Aquaculture.* He has authored or co-authored more than 60 publications in refereed, peer-reviewed journals, numerous lay publications, and several book chapters.

CONTRIBUTORS

Shannon K. Balfry, PhD, West Vancouver Laboratory, West Vancouver, British Columbia, Canada.

Joyce J. Evans, PhD, USDA-ARS Aquatic Animal Health Research Laboratory, Auburn, Alabama.

Ann L. Gannam, PhD, U.S. Fish and Wildlife Service, Abernathy Fish Technology Center, Longview, Washington.

Delbert M. Gatlin III, PhD, Department of Wildlife and Fisheries Sciences and Faculty of Nutrition, Texas A&M University, College Station, Texas.

Joseph M. Groff, VMD, PhD, Department of Pathology, Microbiology, and Immunology, School of Veterinary Medicine, University of California, Davis, California.

Ronald W. Hardy, PhD, Hagerman Fish Culture Experiment Station, University of Idaho, Hagerman, Idaho.

David A. Higgs, PhD, Department of Fisheries and Oceans, West Vancouver Laboratory, West Vancouver, British Columbia, Canada.

Myung Y. Kim, PhD, Department of Fisheries and Allied Aquacultures, Auburn University, Auburn, Alabama.

Phillip H. Klesius, PhD, USDA-ARS Aquatic Animal Health Research Laboratory, Auburn, Alabama.

Scott E. LaPatra, PhD, Clear Springs Foods, Inc., Research Division, Buhl, Idaho.

Meng H. Li, PhD, Thad Cochran National Warmwater Aquaculture Center, Mississippi State University, Stoneville, Mississippi.

Donald V. Lightner, PhD, Department of Veterinary Science and Microbiology, University of Arizona, Tucson, Arizona.

Rebecca Lochmann, PhD, Department of Aquaculture and Fisheries, University of Arkansas at Pine Bluff, Pine Bluff, Arkansas.

Richard T. Lovell, PhD, Department of Fisheries and Allied Aquacultures, Auburn University, Auburn, Alabama.

Bruce B. Manning, PhD, Thad Cochran National Warmwater Aquaculture Center, Mississippi State University, Stoneville, Mississippi.

Veronica O. Okwoche, PhD, Department of Fisheries and Allied Aquacultures, Auburn University, Auburn, Alabama.

Harold Phillips, MSc, Department of Aquaculture and Fisheries, University of Arkansas at Pine Bluff, Pine Bluff, Arkansas.

John A. Plumb, PhD, Southeastern Cooperative Fish Disease Project, Department of Fisheries and Allied Aquacultures, Alabama Agriculture Experiment Station, Auburn University, Auburn, Alabama.

Robert C. Reigh, PhD, Aquaculture Research Station, Louisiana Agricultural Experiment Station, Louisiana State University, Baton Rouge, Louisiana.

Edwin H. Robinson, PhD, Thad Cochran National Warmwater Aquaculture Center, Mississippi State University, Stoneville, Mississippi.

Robin M. Schrock, MSc, U.S. Geological Survey, Biological Resources Division, Columbia River Research Laboratory, Western Fisheries Research Center, Cook, Washington.

Wendy M. Sealey, PhD, Department of Wildlife and Fisheries Sciences and Faculty of Nutrition, Texas A&M University, College Station, Texas.

Craig A. Shoemaker, PhD, USDA-ARS, Aquatic Animal Health Research Laboratory, Auburn, Alabama.

Joseph R. Tomasso, PhD, Department of Aquaculture, Fisheries, and Wildlife, Clemson University, Clemson, South Carolina.

Charles R. Weirich, PhD, Aquaculture Research Station, Louisiana Agricultural Experiment Station, Louisiana State University, Baton Rouge, Louisiana.

Foreword

I am honored to write some words of introduction to this important book titled *Nutrition and Fish Health*. It spans the topics of nutrition, the interrelationship between nutrition and immune function, and disease resistance of economically important fish and shellfish. *Nutrition and Fish Health* is a compilation of chapters written by scientists who have been actively involved in the field of nutrition and aquatic animal health. The editors, Dr. Chhorn Lim and Dr. Carl D. Webster, are leading authorities in fish and shellfish nutrition. Further, during the past several years, Dr. Lim's research efforts have been in the area of nutrition as it relates to fish health. This book is the first such effort that brings together information on the nutritional well-being of aquatic animals and the interrelationship to infectious disease, resistance, and the ability of their immune systems to control diseases.

Aquaculture production has greatly increased in recent years. Fish have become less dependent on natural food and more on prepared diets. Therefore, the availability of least cost, nutritionally balanced diets is critical to the success of the aquaculture industry. Given the important relationship between nutrition and infectious diseases, poor nutrition or poor feeding practices may lead to a reduced immune system response and lower the ability of fish and shellfish to resist disease. This book begins with up-to-date overviews of the diseases of warmwater fish, economically important diseases of salmonids, and shrimp viral diseases. Marine fish and larval fish nutrition are also discussed. Subsequent articles detail feeding management to disease resistance, vitamin and mineral requirements, immunostimulants and aflatoxins to better understand the interrelationship of nutrition and fish health. I believe this book will serve as an excellent reference for students, fish nutritionists, fish health specialists, and aquaculturists.

Yolanda J. Brady, PhD
Associate Professor of Aquatic Animal Health
Department of Fisheries
and Allied Aquacultures
Auburn University
Auburn, AL

Preface

Aquaculture constitutes a vital and rapidly growing segment of agriculture worldwide. New technological advances and increased demand for fish as a source of animal protein are expected to accelerate the industry's growth in the near future. As the industry continues to expand, the culture methods have become more intensive for the purpose of producing higher yields per unit area. Infectious disease is the major cause of economic loss in intensive culture operations. In the United States, current methods for disease treatment are limited to a number of government-approved antibiotics or chemotherapeutics that are marginally effective. Some of the problems arising from the use of the drugs, either by means of medicated diets or water treatments, are high cost, creation of antibiotic-resistant pathogens, required withdrawal period, and environmental contamination. For these reasons, aquaculturists are interested in developing cost-effective preventive measures that can prevent the outbreak or reduce the severity of epizootics. One such preventive measure is the development of various nutritional strategies that might lessen or eliminate diseases, something which is currently being examined by the aquaculture industry.

The significance of nutrition as a key factor in maintaining the health of humans and other animal species, including fish, has been recognized for many years. Earlier research on the relationships between nutrition, immune response, and disease resistance has focused on humans and other terrestrial animals. In the past two decades, however, attempts to conduct similar studies utilizing fish have met with limited success due to an incomplete understanding of the immune response in fish. Evidence from either unintentional or intentional infection of fish occurring in some of these nutritional investigations appears to indicate that most, if not all, dietary nutrients have an influence on immune function and disease resistance. A deficiency or excess of any nutrient could have a profound effect on the infection and survival of fish, largely through its effects on host defense mechanisms. Other factors such as nutrient bioavailability and interactions, the presence of immunostimulants and toxins, and feeding management also influence fish health. This clearly demonstrates the potential role that nutrition can play in improving immune response and disease resistance in fish.

This book should be useful for nutritionists, disease specialists, feed formulators, students, extension specialists, and aquaculturists. It begins with an overview of diseases affecting warm-water and cold-water fish and shrimp viral diseases. Nutritional strategies affecting the health of marine and baitfish and nutritional deficiencies in commercial aquaculture settings are provided. Immune system functions and disease resistance in fish are presented to provide the readers with a better understanding of the effects of dietary nutrients (ascorbic acid, iron, and other minerals, and lipids and fatty acids); additives and contaminants (immunostimulants and mycotoxins); and feed allowance as they relate to fish health. The impacts of dietary lipids and environment on the stress tolerance of fish and altering environmental tolerances of fish through dietary modifications are also presented. The use of vaccines as a management strategy to prevent diseases of aquatic species is also included to provide a balanced description of disease preventive measures.

The information contained in this book is by no means complete. Moreover, it is apparent from this book that fish immunonutrition is a relatively young discipline. It is hoped, however, that this book will fulfill its intended purpose to provide a better understanding of and put into perspective the complex interrelationship between nutrition and fish health, as well as to accelerate research in this area. It is also hoped that future aquaculture diets will be formulated to provide not only optimum growth and feed efficiency but also improved fish health.

Chhorn Lim
Carl D. Webster

Acknowledgments

The editors gratefully acknowledge the contributions made by the chapter authors and Katherine Tave, who assisted in the editing of this book. The preparation of this book has involved the cooperative efforts of many people, to whom we are extremely appreciative. Our gratitude is also extended to our families for their enormous patience and support.

Chapter 1

Overview of Warm-Water Fish Diseases

John A. Plumb

INTRODUCTION

Infectious diseases are among the most often-cited problems facing warm-water aquaculture (USDA 1997), but diseases are natural events in all animal groups, and fish are no different in this regard. It has been estimated that about 10 percent of all cultured fish die because of infectious diseases; however, in some instances this estimate may be conservative. Some channel catfish operations report that nearly 75 percent of newly hatched swim-up fry die before they reach market size (unpublished). Regardless of the method used to measure the effect of infectious diseases, it is obvious that they cost the warm-water aquaculture industry millions of dollars annually in dead fish, reduced production, expenditures for chemotherapeutics, vaccines, and human resources to institute health maintenance practices.

During the past 30 years, a dramatic increase in the number of fish disease cases has been recorded at diagnostic laboratories throughout the southeastern United States (Mitchell 1997). In the early 1970s, there were 600 to 700 disease cases per year in this geographical region; however, during the 1980s the number grew to 5,000 to 6,000 per year. The increase was due in large part to expansion of aquaculture, primarily the channel catfish industry, but intensification of the production of hybrid striped bass and other species has also contributed to the increase. There are several other reasons for the increase, not the least of which is the nature of aquaculture itself: it emphasizes increasing fish production in a given body of water and enhances stressful environmental conditions that increase susceptibility of fish to disease (Plumb 1999). During the 1990s the number of reported disease incidences has not increased, due largely to better aquaculture management.

Another reason for the increase in reported fish disease cases is the greater availability of trained biologists and veterinarians who have acquired skills in fish health and a greater abundance of laboratories that regard fish health as a primary mission. There is also a greater concern for fish health in the industry itself, as well as an increased awareness of environmental conditions and their relationship to aquatic animal health. In addition, research techniques, diagnostic tools adapted from veterinary and human medicine, and molecular biology have been instrumental in expanding the scientific base of aquatic animal health.

Infectious diseases are generally seasonal, especially in warm-water aquaculture (see Figure 1.1). Peak disease incidence is in spring from March to June and in September and October, periods when water temperatures are between 20 and 28°C, which is optimum for many fish pathogens. It appears that warm summer and cool winter temperatures are not conducive to a high incidence of infectious fish diseases, although they do occur during these periods. Also, disease incidence increases in spring because of the abundance of juvenile fish, which are more susceptible to disease than older fish; the immunity of older fish is reduced as result of overwintering; and the resistance of adult fish is compromised by spawning activities. Elevated disease incidence in autumn may also be associated with moving and stocking juveniles into production ponds.

Mortality patterns of infectious diseases range from acute, with a high percentage of fish dying on a daily basis, to subacute and chronic, when the mortality extends over several weeks (see Figure 1.2). Most infectious diseases of warm-water fish are subacute to chronic, and only a few produce acute mortalities.

INFECTIOUS DISEASE AGENTS

Viruses, bacteria, water molds (fungi), and parasites cause diseases of freshwater fish in warm-water aquaculture. Some of these agents are obligate pathogens (require a host for survival in nature), but many are facultative organisms found free-living in aquaculture waters. These facultative agents are opportunistic and cause problems when the host's resistance is compromised. The following discussion includes only the most common pathogens that affect warm-water aquaculture. One should realize that less common pathogens also cause mortality of fish, and with the continued expansion and intensification of warm-water aquaculture, new diseases and pathogens are very likely to be discovered.

Viruses

The most serious virus disease in warm-water aquaculture in the United States is channel catfish virus disease (CCVD), caused by channel catfish virus (CCV), of juvenile channel catfish during their first summer of life. Channel catfish virus is a herpesvirus that attacks swim-up fry to 10 cm fingerlings when water temperatures are 25°C or above. Under certain optimum conditions, mortality due to CCV can be acute and reach up to 90 percent, especially in densely populated tanks or ponds. CCVD is often exacerbated by secondary columnaris (bacteria) infections on the skin and

FIGURE 1.1. Seasonal Occurrence of Fish Disease Cases in the Southeastern United States from the Early 1970s Through the Mid-1990s

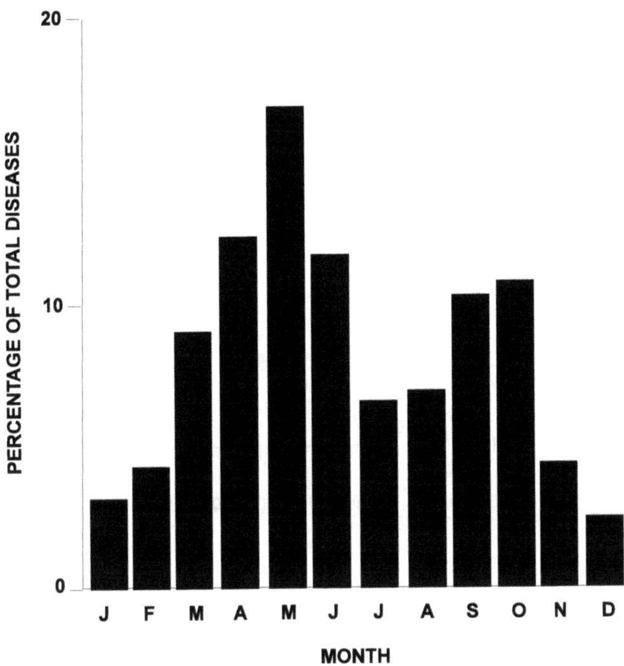

Source: Mitchell 1997.

fins. There are other viruses of cultured warm-water fishes, but generally they are of little consequence.

Bacteria

Bacteria cause more infectious disease problems than any other group of pathogens in warm-water aquaculture. Motile aeromonad septicemia (MAS), caused by *Aeromonas hydrophila* and related species, is a common disease in freshwater fishes, particularly during spring. These organisms are facultative opportunists and usually cause infection following environmental stressors or are associated with skin injury, fish transport, other trauma, and even other parasites. MAS-infected fish have hemorrhaged, inflamed, and extensively necrotic lesions in the skin and muscle, but the disease may also become systemic in a wide range of species, particularly channel catfish.

FIGURE 1.2. Hypothetical Mortality Curves of Fish Suffering from Various Kinds of Diseases

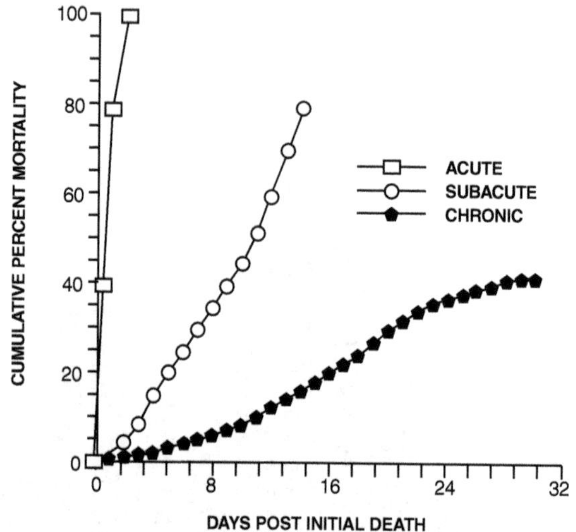

Enteric septicemia of catfish (ESC), caused by *Edwardsiella ictaluri,* is a serious disease that almost exclusively affects cultured channel catfish. The bacterium produces small white spots on the skin, which then become inflamed and necrotic. A severe hyperemia often develops on the skin on the lower jaw and abdomen. An open necrotic lesion may occur in the cranium between the eyes in chronic infections. In systemic infections the liver is mottled, the spleen is dark red, and the kidney is pale. The viscera are generally hyperemic, and bloody fluid may accumulate in the body cavity. ESC is most serious in spring and fall when water temperatures are 18 to 28°C; fewer outbreaks occur in summer and winter. All age groups are susceptible, but most economic ESC-associated losses occur during the grow-out stages because of the value of the fish. Mortalities can be acute but more often are subacute to chronic.

Columnaris is nearly as serious as ESC in channel catfish but also infects numerous other warm-water fishes. This disease, caused by *Flavobacterium columnare* (formerly *Flexibacter columnaris*), manifests itself as pale necrotic lesions on the skin, fins, and gills. Although systemic infections of *F. columnare* occur, most damage to fish results from injury to the skin and gills. Columnaris is usually associated with stressors and injuries associated with seining, handling, and fish transport. Columnaris can develop quickly in susceptible fish populations and cause acute mortality.

Streptococcus spp. infections have been known to affect fish for about 40 years but only recently have been considered serious pathogens of cultured warm-water fish in the United States. While streptococci are known to infect a variety of fish species, their presence in intensively cultured tilapia and striped bass has enhanced their visibility. Several species of *Streptococcus* can cause disease of fish, but *S. iniae* is currently of most concern. Streptococci cause systemic infections in cultured tilapia and striped bass in closed, intensive, recirculating culture systems in which environmental conditions are marginal. Channel catfish are not seriously affected by streptococci. Poor water quality, traumatic injury to the skin, and less serious parasites (e.g., *Trichodina*) are predisposing factors to *Streptococcus* infections. Affected fish may develop mild inflammation of the skin, but more often the abdomen is enlarged; the eyes are opaque, inflamed, and protruding; the body cavity may contain a red, cloudy exudate; and the body may form a U-shaped configuration just prior to death. Mortality is usually chronic. *Streptococcus iniae* is also of concern because of its potential to infect humans: infections are reported to occur in open cuts, abrasions, or sores acquired on hands during cleaning or handling of infected fish.

Mycobacteriosis is an infrequent infection in certain warm-water fish species. *Mycobacterium marinum* is the most commonly implicated species and most seriously affects striped bass in closed recirculating facilities where infections are chronic and protracted. Fish develop darker than normal pigmentation; skin inflammation occurs; and the eyes become opaque, hemorrhagic, and protruding. The liver and spleen develop dense granulomas, which give the organs a pale sandpaper-like texture. Because of its chronic nature, mycobacteriosis mortalities are usually chronic. *Mycobacterium marinum* is also infectious to humans: it causes hard granulomatous lesions on the skin of the hands and wrists. Humans usually contract mycobacteriosis through contamination of open wounds during handling of infected fish.

Water Molds (Fungi)

Water molds, also known as fungi, usually cause secondary infections on fish as a result of physical injury or environmental stressors, especially rapid temperature reduction. The major pathogen is *Saprolegnia parasitica,* which has been considered a fungus in the past, but its flagellated spores make it a water mold. It produces white to brownish cottony growths on any body surface, including fins, scales, spines, operculum, eyes, mouth, and gills. Many fish pathologists do not consider water molds serious fish pathogens because of their nearly exclusive secondary nature, but once fish become infected with this opportunist they are unlikely to recover. All age groups of all fish species are susceptible to water mold. Dead eggs are particularly susceptible to water mold, and these can infect healthy eggs during incubation, especially when water temperatures are below the optimum for the species in question. Saprolegniasis is particularly troublesome to fish during cool and cold weather and is the principal culprit in "winter saprolegniasis" of cultured catfish.

Parasites

Many types and species of parasites infect warm-water fish, some of which are highly pathogenic, while others are little more than nuisances. Nevertheless, parasites are important in the overall consideration of warm-water fish health. Parasites of fish include protozoa, helminthic worms, and crustacea (Mitchum 1995).

Many species of protozoa, sporozoans, and other single-cell parasites affect warm-water fishes. Most of these occur on the gills and skin, but some parasitize internal tissues and the digestive tract, these generally being less serious. The better-known external protozoa are *Ichthyobodo necatrix, Trichodina* spp., and *Ichthyophthirius multifiliis* in freshwater fishes (Mitchum 1995).

Ichthyobodo necatrix is a teardrop-shaped protozoan about the size of a red blood cell and it has two flagella. On wet-mount slides the attached cell flickers like a candle. Skin lesions are pale but sometimes the epithelium is destroyed, while gills become pale and swollen and produce excessive mucus. This parasite is more prevalent during cool temperatures, at which time it can cause high mortality in many fish species, especially channel catfish.

Trichodina spp. are saucer-shaped protozoa that possess cilia along the margin and a denticular ring near the center. This parasite glides on the surface of gills and skin, where it causes mucus production and occasionally ulcerative lesions. *Trichodina* is seldom the primary cause of mortality, but when present in very high numbers it can do so; more often it irritates the skin, which allows secondary infections of opportunistic bacteria. This appeared to be the case in one instance in which moderate *Trichodina* spp. irritated the skin of tilapia in a closed-culture system, which led to a *Streptococcus* infection. When the *Trichodina* were removed with formalin the bacterial infection disappeared.

Ichthyophthirius multifiliis (Ich), an obligate pathogen, is a serious parasite of a variety of warm-water fishes, especially cultured channel catfish. Its complex life cycle includes the adult (trophozoite) stage, which appears as numerous white, pinhead-sized spots embedded in the skin that give the fish a sandpaper-like surface. Upon maturity, these cells drop off of the fish, attach to a substrate, where they form a pseudocyst, and undergo division to produce the motile infectious cell (tomite). The tomite then swims to a new fish host within three days under optimum conditions (or it dies), where it attaches and embeds in the epithelium to develop the adult trophozoite. Ich often causes acute mortality when water temperatures are 16 to 23°C, but at temperatures above 25°C the parasite does not multiply, and the disease progresses slowly at 16°C or less.

Helminthic parasites include monogenetic and digenetic trematodes, nematodes, and cestodes. Monogenetic trematodes (flatworms) have simple life cycles and attach to the skin or gills. Most fish have some flatworms, but they create a health problem only when present in large numbers, at which time they cause mucus production and irritation. Generally, a given species of trematode affects a particular species of fish.

Digenetic trematodes have a complex life cycle that includes the larval stages in the eye, muscle, or visceral tissue of fish; the adult worm lives in the digestive tract of fish-eating birds, with a crustacean or snail serving as an intermediate host between bird and fish. The two most frequently encountered digenetic trematodes are the visible larvae of the white grub, *Postodisplostomum minimum,* in visceral tissue and the yellow grub, *Clinostomum marginatum,* encysted in the muscle of a variety of fish species. A third trematode is the eye fluke larvae of *Diplostomulum spathaecum* that parasitizes the eye of catfishes and centrarchids. Larval trematodes seldom cause great injury to fish; however, they do affect their appearance and culinary quality, and the eye fluke causes blindness. None are host specific.

Adult nematodes (roundworms) can occur in the eye, the viscera, or the lumen of the intestine. Adult cestodes (tapeworms) are found in the intestine of fish, while larvae may be encysted in visceral tissue. Generally, neither of these groups of parasites is a serious threat to fish health.

Parasitic Crustacea

Parasitic crustacea infest the gills and skin of fish, where they are often visible without magnification. These non-host-specific adults attach to the skin by modified appendages or use sucking mouth parts to penetrate the fish's epithelium. *Learnea* is the most common parasitic crustacean. The head of the sticklike adult is embedded in the skin and two attached egg sacks are affixed to the body. Juveniles are found on the gills, sometimes in large numbers. *Ergasilus* is another crustacean that attaches to the gills by modified anterior appendages where, if present in large numbers, it can cause serious injury. *Argulus* is a large crustacean that moves about on the skin of fish and uses piercing mouth parts to injure the fish skin.

DISEASE CONTROL

There are two basic approaches to fish diseases in aquaculture; the first is to do nothing and let the disease take its course; however, this attitude is not very common. The second is to take steps to prevent diseases from occurring or to reduce their effect when they do occur, bearing in mind that they are natural events and it is impractical to try to eliminate all disease organisms. The most important approach to disease control is through "best management practices" to prevent diseases from occurring or to reduce their effects by maximizing environmental conditions and creating an environment that is best suited for optimum survival, growth, feed conversion ratio, and overall production of fish (Plumb 1999). These include keeping water quality at an optimum level (high oxygen and low ammonia, carbon dioxide, organic loads, etc.); using moderate stocking densities; providing the highest quality feeds at proper feeding rates; reducing stress and trauma when handling or moving fish; and utilizing the most genetically improved broodstock available.

Another "best management practice" is to use U.S. Food and Drug Administration (FDA)-approved drugs in a legal and judicious manner when diseases do occur (Plumb 1999). Drugs provide a tool for disease control in aquaculture, but they cannot be used indiscriminately as "cure-alls" to overcome all of the ills of aquaculture, many of which result from poor management. That many drugs are not as efficacious as desired discourages their use by many aquaculturists. Successful chemotherapy depends on initiating proper drug application in the early stages of disease.

The list of legal drugs available to aquaculture is short. Terramycin and Romet are registered antibiotics that are incorporated into the feed for some bacterial infections. Formalin is FDA registered as a bath for external parasites on all fish species. Copper sulfate and potassium permanganate are used to treat some external bacterial and parasitic infections, but these chemicals are approved by the Environmental Protection Agency (EPA) only for treating algae and oxidizing organic material, respectively, in waters. These chemicals are permitted in waters containing fish as long as the guidelines set forth by the EPA for algae control or organic oxidation are not exceeded. Potassium permanganate and formalin are used as prophylaxes to reduce external parasites; sodium chloride (salt) is used as a stress mediator when fish are handled. Parasites do not develop resistance to these chemicals, but Terramycin and Romet (antimicrobials) should not be used in any manner other than that recommended because bacteria may develop resistance to them.

Although research continues on other drugs for aquaculture, the registration process is long, arduous, and expensive and the possibility of FDA approval is not guaranteed. In view of this, it is unlikely that any new therapeutics will be forthcoming in the near future. However, in the event that new drugs are registered, it is most likely that their use will be by prescription through a veterinarian.

Vaccination to prevent some bacterial diseases is becoming a potential tool in warm-water aquaculture. Recent experimental vaccination in warm-water aquaculture indicates an effective means of preventing enteric septicemia *(Edwardsiella ictaluri)* of catfish, and possibly other diseases. With continued emphasis on fish vaccination research, this tool may have a broader application to warm-water aquaculture in the future.

REFERENCES

Mitchell, A. J. 1997. Fish disease summaries for the southeastern United States. *Aquaculture Magazine* 23(1):87-93.

Mitchum, D. L. 1995. *Parasites of Fishes in Wyoming.* Cheyenne, WY: Wyoming Game and Fish Department.

Plumb, J. A. 1999. *Health Maintenance and Principle Microbial Diseases of Cultured Fishes.* Ames, IA: Iowa State University Press.

USDA. 1997. *Reference of 1996 U. S. Catfish Health and Production Practices. Catfish '97.* Ft. Collins, CO: United States Department of Agriculture.

Chapter 2

An Overview of the Economically Important Diseases of Salmonids

Joseph M. Groff
Scott E. LaPatra

INTRODUCTION

Prior to any discussion of disease in an individual organism or population, a conceptual framework of disease needs to be established by definition of the pertinent terminology. Therefore, *disease* can be defined as any definitive morbid condition or process that has a characteristic set of symptoms or qualities. The various aspects of disease include the cause, or *etiology,* the developmental process, or *pathogenesis,* the biochemical and morphological alterations of the cells and tissues, and the functional significance or clinical consequence of these alterations. The etiology of disease may be *intrinsic* (genetic) or *extrinsic* (acquired); the latter includes *infectious, environmental, toxic,* and *nutritional* etiologies. *Neoplastic* disease may have an extrinsic and/or intrinsic component, whereas a disease with an uncertain or unknown etiology is referred to as *idiopathic* disease. Concerning etiology, disease may be due to a single etiology, such as a highly virulent infectious agent, or may be multifactorial. For example, a primary and/or secondary infectious disease may occur in fish exposed to poor water quality or low concentrations of a toxin. Disease can be further classified according to the progression and severity of the condition. *Acute disease* has a rapid onset and progression, whereas *chronic disease* has a slow progression and long duration. Disease that is neither acute nor chronic may be classified as *subacute* or *subchronic,* whereas disease that has an extremely rapid progression can be considered *peracute. Clinical disease* is apparent and characterized by observations and/or the results of tests, whereas *subclinical disease* is not apparent or does not result in clinical manifestations and is difficult to characterize. Subclinical disease may progress to clinical disease.

Infection is often used synonymously with disease but is more correctly defined as the invasion and colonization of the tissues by microbial pathogens and the consequent response of the host to this event. A *pathogen* is any organism capable of causing disease, whereas *pathogenicity* is the ability of an organism to produce disease. Pathogenicity of an infectious agent is dependent on the contagious and invasive properties of the pathogen and the ability of the pathogen to resist defense mechanisms of the host that will vary with a particular strain. Infection that results in apparent symptoms, i.e., disease, is often referred to as *clinical infection* but is more correctly characterized as a clinical disease due to an infectious etiology. In contrast, a *subclinical infection* is synonymous with *asymptomatic infection* and does not result in disease. Therefore, the detection or presence of any infectious agent does not imply the presence of disease. Asymptomatic infections may progress to clinical disease or may remain subclinical, although the host may function as a reservoir of infection to other members of the population—this is referred to as the *carrier state*. Furthermore, exposure to infectious agents is a normal and continual event that does not necessarily result in infection or clinical disease during the life span of any individual organism. The manifestation of clinical disease in a population, or *epizootiology*, is dependent on a complex interaction among the host, environment, and pathogen. For example, the ability of infectious hematopoietic necrosis virus (IHNV) to cause disease in salmonids is dependent on the status of the host, including species, age, and life stage; water quality parameters such as temperature; and the strain of IHNV (LaPatra 1998). Pathogens are normal components of the aquatic ecosystem that have coexisted and evolved with the host in the natural environment and generally do not result in serious disease within a wild population. However, the propagation of fish, especially in intensive culture operations, generally provides conditions that affect the complex interaction of the host and pathogen. These conditions often exacerbate the manifestation of disease in a cultured population but do not create the host-pathogen interaction.

This chapter discusses the infectious diseases of salmonids that are considered economically important due to their regional or international impact on commercial salmonid operations. However, it must be emphasized that any disease condition, regardless of etiology, can have an adverse economic impact in any individual facility or operation. For example, fish maintained in a facility with marginal or poor water quality may be further compromised by an external parasitic infection, such as *Ichthyophthirius multifiliis*, that can result in morbidity and variable mortality. The economic impact of morbidity can often be significant in these situations, due to the reduced feed conversion and consequent reduced growth rate in the population. Morbidity can also result in downgrading and rejection of the product, which is often a significant loss in commercial salmonid operations. Reviews of the various diseases that affect salmonids, but not included in this discussion, have previously been summarized and should be consulted as necessary (Wolf 1988; Austin and Austin 1993; Inglis, Roberts, and Bromage 1993; Stoskopf 1993; Thoesen 1994; Noga 1995; Kent and Poppe 1998). These texts also review proper diagnostic methods and techniques that are essential and cannot be overemphasized in a discussion of disease. Proper diagnosis includes a review of the history; evaluation of the husbandry conditions, including the water quality and nutrition; a complete necropsy examination of multiple fish that includes a gross and microscopic examination; and ancillary laboratory tests that are sensitive and specific for a definitive determination of the etiology or etiologies. An example of an improper diagnostic effort that is not uncommon but results in an incorrect diagnosis is an examination that is limited to evaluation of cutaneous and branchial wet-mount preparations. This limited examination may reveal an external parasitic infection that may not be the primary cause of the disease, although this cannot be determined without a complete diagnostic effort. The consequences of an incorrect diagnosis are obvious and may result in an avoidable economic loss due to the additional morbidity and mortality in the population and the potential recurrence of disease.

BACTERIAL DISEASES

Motile Aeromonad Septicemia

Motile aeromonad septicemia (MAS) is a common disease of fish, including salmonids, and other aquatic animals that inhabit freshwater, but MAS can also occur in brackish water (Hazen et al. 1978). The motile aeromonads are a heterogeneous group of ubiquitous, mesophilic, Gram-negative bacteria that are also a normal component of the microbial flora of fish (Trust and Sparrow 1974; Hazen et al. 1978; Ugajin 1979; LeBlanc et al. 1981). The motile taxon has been divided into three species—*Aeromonas hydrophila, A. sobria,* and *A. caviae*—although the heterogeneity of the group has resulted in difficult separation and incomplete taxonomic placement of these bacteria (Austin and Austin 1993). Therefore, motile aeromonads that do not conform to the characteristic biochemical phenotype of the designated species are not uncommon isolates from fish (Wakabayashi et al. 1981; Austin and Austin 1993).

Epizootiology

Transmission of the motile aeromonads is horizontal and can occur by direct contact or indirectly through the water. The latter may occur following contamination of the environment with pathogenic strains of bacteria shed by diseased or carrier fish (Bullock, Conroy, and Snieszko 1971; Wolke 1975; Schäperclaus 1991). Motile aeromonads may exhibit a chemotactic response to fish mucus (Hazen et al. 1982) that facilitates transmission of the bacteria. Bacteria may also be transmitted by protozoas, copepods, monogenean flatworms, leeches, snails, amphibians, birds, and nonsalmonid fish (Bullock, Conroy, and Snieszko 1971; Schäperclaus 1991), although transmission of the bacteria via any fomite should be considered a likely source of infection. Contaminated feeds have also been implicated as a source of infection (King and Shotts 1988).

The significance of these bacteria as primary pathogens remains controversial, although it is generally not disputed that the motile aeromonads can result in significant disease conditions (Austin and Austin 1993; Roberts 1993). However, bacterial virulence is highly variable and dependent on the bacterial strain and resistance of the host (Austin and Austin 1993; Roberts 1993; Thune, Stanley, and Cooper 1993). Clinical disease generally occurs in fish that are predisposed to infection due to poor conditions, including inadequate nutrition, adverse environmental conditions, or other stress-mediated conditions such as increased density, capture/transport, and spawning (Bullock, Conroy, and Snieszko 1971; Schäperclaus 1991; Austin and Austin 1993; Roberts 1993). Environmental conditions that predispose fish to infection include low dissolved oxygen, increased organic loads and nitrogenous waste concentrations, and increased water temperature (Groberg et al. 1978; Hazen et al. 1978; Nieto et al. 1985). Preexistent infectious diseases due to parasites, viruses, or other bacteria also increase the susceptibility of fish to motile aeromonad infections (Bullock, Conroy, and Snieszko 1971; Wolke 1975; Schäperclaus 1991; Austin and Austin 1993). However, improved husbandry has generally prevented the occurrence of MAS in freshwater salmonid operations, although isolated epizootics may occur with poor husbandry. In this context, conditions that predispose fish to infection need to be recognized and corrected or eliminated for successful treatment and prevention of disease recurrence (Roberts 1993).

Clinical disease may occur at any time of the year but generally has a higher prevalence in the spring and early summer at temperatures greater than 10°C, due to the concomitant increase in temperature and nutrient level of the water, which provides ideal conditions for an increased bacterial load. The increased metabolic requirements of juvenile fish increase both the susceptibility to infection and the mortality following infection in this age class, although disease is not restricted to a specific age class. For example, an increased susceptibility in rainbow trout, *Oncorhynchus mykiss,* has been associated with sexual maturation and reproductive activity (Roberts 1993).

Pathogenesis

Ingestion of the bacteria results in an anterior enteritis followed by a primary bacteremia and bacterial colonization of the internal viscera and integument (Schäperclaus 1991). The integument has also been suggested as a primary site of infection, especially in fish with preexistent cutaneous lesions or a compromised integument (Schäperclaus 1991; Plumb 1994). Properties of the motile aeromonads associated with virulence include adherence to the extracellular matrix of the host tissues and the avoidance of phagocytosis and/or intracellular killing by the production of bacterial cytotoxins (Austin and Austin 1993; Thune, Stanley, and Cooper 1993). Clinical disease is characterized by hemorrhage and extensive tissue damage that is typical of a Gram-negative septicemia, due to the production of various exotoxins or extracellular products (ECP) that have enzymatic activity, such as hemolysin, gelatinase, elastase, caseinase, lipase, various proteases, and a dermonecrotic factor (Austin and Austin 1993; Thune, Stanley, and Cooper 1993). The production of acetylcholinesterase by *A. hydrophila* has also been reported and may have a significant role in the pathogenesis of disease and consequent mortality in the host (Nieto et al. 1991). However, tissue damage may also be partially due to the host inflammatory response that results in the release of lysozymes and toxic free radicals. Additional factors associated with bacterial virulence include the presence of a bacterial surface-array matrix, or S-layer, and a bacterial endotoxin or lipopolysaccharide (LPS) that constitutes the bacterial somatic or O-antigen (Austin and Austin 1993; Roberts 1993; Thune, Stanley, and Cooper 1993). However, the relative contribution of these various bacterial properties and toxins to bacterial virulence and the pathogenesis of disease is incomplete and requires further investigation.

Clinical Disease

Fish with MAS may present with various clinical symptoms dependent on the pathogenicity of the bacterial strain, relative susceptibility of the host, and chronicity of the disease (Bullock, Conroy, and Snieszko 1971; Wolke 1975; Schäperclaus 1991; Austin and Austin 1993; Roberts 1993; Plumb 1994). Clinical signs and lesions are nonspecific but characteristic of a Gram-negative septicemia and include anorexia, lethargy, dark coloration, and a progressive loss of equilibrium. External lesions include a diffuse hyperemia of the integument and fins; petechial and ecchymotic hemorrhage of the integument and fins, especially at the base of the fins; pallor of the gills due to anemia, although the gills may be bright red due to hyperemia or mottled a dark red due to hemorrhage; prolapse with swelling and hemorrhage of the vent; corneal opacity due to corneal edema; exophthalmos; ophthalmitis; subcutaneous edema; scale loss; necrosis of the fins; and focal to multifocal erosions/ulcerations of the integument that may occur early in the progression of the disease. Initially, the cutaneous lesions are generally convex, opaque white to pale red lesions that progress to shallow, concave ulcerations of variable size that have hemorrhagic, necrotic centers and are delineated by a rim of white, necrotic tissue. Further progression of the ulcerative lesions may result in necrosis or deep abscesses of the subcutaneous muscle, whereas chronic ulcers may be delineated by a peripheral zone of melanin pigment. Opportunistic bacterial, parasitic, and fungal infections of the cutaneous ulcers are common.

Internal lesions in acute disease include diffuse hyperemia with petechial and ecchymotic hemorrhage of the coelomic membranes and viscera; a variable amount of serous or serosanguinous ascitic fluid that may be purulent and malodorous; a flaccid, hyperemic intestine that is hemorrhagic and devoid of ingesta but often contains a white to yellow, mucoid material; and enlargement of the viscera, especially the liver, kidney, and spleen. The enlarged kidney and spleen are often friable and bright red but may have a pale, mottled appearance due to multiple areas of liquefactive necrosis secondary to bacterial colonization and microabscess formation. Severe infections may result in focally extensive to diffuse, liquefactive necrosis with exudation of a viscous fluid from the cut surface of the affected organ, especially the kidney. The liver may be congested and hemorrhagic or may be a pale brownish yellow due to hepatic lipidosis or a dark greenish brown due to bile stasis. Likewise, the gall bladder may be prominently distended with a dark green, viscous bile fluid. Fibrinous adhesions within the coelomic cavity may occur with resolution of the ascites.

The microscopic lesions in MAS are also nonspecific and characteristic of a Gram-negative septicemia (Wolke 1975; Roberts 1993). There is generalized edema with hyperemia and hemorrhage of the capillaries and small vessels, regardless of location. Inflammation of the affected tissues is generally characterized by an infiltration of mixed inflammatory cells, although the periphery of the cutaneous ulcers may contain a predominance of granulocytes. Inflammation of the integument results in spongiosis followed by epidermal necrosis and scale loss with subsequent ulceration and progressive involvement of the dermis and subcutaneous musculature. Focal necrosis of the internal viscera that may be liquefactive corresponds to the gross lesions and is often associated with the presence of individual bacteria or bacterial colonies. Coagulative necrosis of the splenic connective tissue may also be a feature of the disease (Bullock, Conroy, and Snieszko 1971). An exudative enteritis characterized by increased mucus production occurs early in the progression of the disease, followed by necrosis and sloughing of the intestinal mucosa (Schäperclaus 1991). However, interpretation of intestinal lesions following a prolonged postmortem interval is difficult, if not impossible, due to rapid autolysis, even if the specimen has been refrigerated or kept on ice prior to fixation. Melanomacrophage centers in the kidney and spleen may contain bacteria and are generally increased in size and distribution. Likewise, histiocytes and circulating macrophages may contain intracellular melanin, lipofuscin, and/or hemosiderin pigments secondary to hemolysis and rupture of the melanomacrophage centers (Roberts 1989). Lipid hepatopathy and bile stasis are also nonspecific but common findings in moribund fish that are anorexic.

The chronic form of MAS may be the more common presentation in mature fish. Fish with chronic disease may be lethargic and generally present with ulcerations of the integument without concurrent lesions or septicemia. Therefore, it can be argued that reference to the chronic disease as motile aeromonad septicemia may not be appropriate (Plumb 1994). Regardless, fish with chronic disease are often indistinguishable from fish with chronic furunculosis.

Diagnosis

A diagnosis of MAS is based on the presence of typical gross and microscopic lesions in affected fish, followed by isolation and identification of the bacterium. Bacterial cultures should be obtained from the kidney and other affected tissues of multiple fish, since the ubiquitous, opportunistic, motile aeromonads are common contaminants in bacterial cultures that may readily overgrow and obscure a more fastidious primary bacterial pathogen such as *A. salmonicida*. Overinterpretation of bacterial cultures should also be avoided, since the motile aeromonads are components of the external and internal microflora of the host. In this context, bacterial cultures of external lesions are generally difficult to interpret, due to the expected isolation of motile aeromonads from the external surfaces. Bacterial isolation can be achieved using standard enriched media such as blood agar or tryptic soy agar (TSA) or selective media such as Rimler-Shotts medium prior to incubation at 20 to 25°C for 24 to 48 hours (Austin and Austin 1993). Incubation on nonselective nutrient media at 25°C generally results in white to tan, round, raised, 2 to 3 mm bacterial colonies. The bacteria are motile by a single polar flagellum, Gram-negative, fermentative bacilli or coccobacilli that are approximately 0.8 to 1.0 × 1.0 to 3.5 μm and resistant to the pteridine vibriostat O/129. Phenotypic characterization of the motile aeromonads has previously been summarized (Austin and Austin 1993). Automated systems can be used for bacterial identification (Teska, Shotts, and Hsu 1989), although rapid identification systems should be used with caution due to the potential for false results (Toranzo et al. 1986). Immunological tests such as agglutination assays and fluorescent antibody tests (FAT) have been used for rapid confirmation of the bacteria (Austin and Austin 1993), although the sensitivity and specificity of these methods may be compromised due to the heterogeneity of the taxon.

Treatment and Prevention

Treatment and prevention of MAS is dependent on improvement of husbandry conditions to decrease stress and maintain water quality. Likewise, husbandry practices that eliminate the potential transmission of the bacteria should be incorporated into the routine management of the facility. The use of antimicrobials such as oxytetracyline, sulfadimethoxine-ormetoprim, or nifurpirinol administered in the feed may be warranted depending on the severity of the disease, although bacterial resistance to various antimicrobials is not uncommon (Austin and Austin 1993; Noga 1995). The fluoroquinolones, such as enrofloxacin, have also been successfully used for treatment of MAS. However, the extra-label use of nitrofurans and fluoroquinolones in food animals is currently not permitted in the United States (Payne et al. 1999). Likewise, moribund fish are often anorexic, which precludes the use of medicated feeds. Vaccines are not currently available, and development of an effective vaccine remains a challenge due to the heterogeneity of the motile aeromonads, although immersion or injectable autogenous vaccines can be easily developed for endemic bacterial strains that cause disease in a facility (Austin and Austin 1993).

Furunculosis

Furunculosis is one of the most important diseases of both freshwater and marine salmonids. The etiologic agent, *A. salmonicida,* is a nonmotile, Gram-negative bacterium that is currently composed of three subspecies—*A. salmonicida salmonicida, A. salmonicida masoucida,* and *A. salmonicida achromogenes*—although there have been proposals to reclassify the subspecies as previously reviewed (Austin and Austin 1993; Munro and Hastings 1993). The typical subspecies—*A. salmonicida salmonicida*—is generally associated with the septicemic disease referred to as furunculosis, although the atypical subspecies—*A. salmonicida achromogenes* and *A. salmonicida masoucida*—may also result in septicemia or cutaneous lesions without septicemia. Regardless, the species is a relatively homogeneous group of bacteria in contrast to the motile aeromonads. The recent publication by Bernoth et al. (1997) summarizes the various aspects and current knowledge of furunculosis.

Epizootiology

A. salmonicida is an obligate pathogen of fish, although the bacterium may survive in water, sediment, and decaying flesh for prolonged periods of several weeks to months (Dubois-Darnaudpeys 1977; McCarthy 1980; Rose, Ellis, and Munro 1989, 1990). The primary reservoir of infection is generally considered to be salmonid and various nonsalmonid fish (McCarthy 1980; Treasurer and Cox 1991; Frerichs, Millar, and McManus 1992; Austin and Austin 1993). The mode of transmission remains controversial but may be indirect via the environment or direct from infected or carrier fish that shed bacteria from cutaneous lesions or in the feces (Bullock and Stuckey 1975a; Dubois-Darnaudpeys 1977; McCarthy 1980; Smith et al. 1982; Rose, Ellis, and Munro 1989). Carrier fish that shed bacteria in the feces may present with clinical disease following stress such as increased water temperature (McCarthy 1980), and fish infected in freshwater retain the infection following transfer to seawater (Smith et al. 1982). Transmission of the bacterium may also occur via fomites, parasites, and possibly contaminated feed (McCarthy 1980; King and Shotts 1988). The bacterium has also been isolated from the surfaces of the sea louse, *Lepeophtheirus salmonis* (Nese and Enger 1993). Lesions of the integument, including lesions due to ectoparasite infections, may also predispose fish to infection (McCarthy 1980; Munro and Hastings 1993). In contrast, vertical transmission via the ova has not been demonstrated as a mode of transmission (Bullock and Stuckey 1975a). All ages and species of salmonids are susceptible to furunculosis, although Atlantic salmon, *Salmo salar,* are most susceptible to the disease, whereas rainbow trout are most resistant. Clinical disease occurs most often in the spring, with an increase in water temperature, although the disease can occur at 6°C (Evelyn et al. 1998).

Pathogenesis

Bacterial entry may occur via the oropharynx, gills, rectum, nares, and lateral line, although the intestine and integument are considered the primary sites of infection (McCarthy 1980; Hodgkinson, Bucke, and Austin 1987). Primary infection of the intestinal tract results in bacteremia and hematogenous spread of the bacteria, with subsequent localization in the viscera and integument. Similar to other bacterial infections in fish, the pathogenicity of *A. salmonicida* is variable and dependent on the bacterial strain and relative susceptibility of the host. Properties of the bacteria associated with virulence and pathogenesis of disease include the bacterial lipopolysaccharide (LPS) endotoxin and the presence of a bacterial extracellular layer, referred to as the additional layer or A-layer (Austin and Austin 1993; Munro and Hastings 1993). The A-layer has been associated with bacterial adherence to host tissues and survival within phagocytes. Bacterial virulence has also been associated with exotoxins or extracellular products (ECP), including hemolysin, phospholipase, leukocidin, and various proteases (Austin and Austin 1993). Leukopenia with a limited or absent host cellular inflammatory response is characteristic of the disease and has been attributed to a bacterial leukocidin (Ellis 1991). In fish with cutaneous ulcers, mortality may ultimately be due to the continual loss of electrolytes and organic plasma solutes and the consequent inability to maintain normal osmoregulation (Munro and Hastings 1993).

Clinical Disease

Three forms of clinical disease have been described for furunculosis—the peracute, acute, and subacute to chronic forms. Peracute disease is generally restricted to juvenile fish that present without apparent gross lesions, except for a possible dark coloration, mild exophthalmos, and hemorrhage at the base of the fins prior to mortality (McCarthy and Roberts 1980). Endothelial necrosis of the atrium has also been reported in fry with peracute disease. The acute form of furunculosis is most common and generally affects subadult and adult fish (McCarthy 1975b). Fish with acute disease present with clinical symptoms that are typical of a bacterial septicemia prior to death in two to three days. These symptoms include a dark coloration, lethargy, anorexia, and hemorrhage at the base of the fins. Finally, the subacute to chronic form of the disease is the least common form and generally occurs in adult fish (McCarthy and Roberts 1980). Fish with subacute to chronic disease may present with lethargy; exophthalmos; generalized edema; hemorrhage of the gills or pallor of the gills secondary to anemia; and hemorrhage of the fins, nares, and vent (Ferguson and McCarthy 1978; McCarthy and Roberts 1980; Roberts 1989). The occurrence of cutaneous lesions is most common in the subacute to chronic form of the disease but is not a consistent finding. These lesions may be large, dark red, raised vesicles—the classical *furuncle*—that contain a sanguinous or serosanguinous to hemorrhagic fluid prior to ulceration of the vesicle that subsequently results in a large, hemorrhagic ulcer. As previously mentioned, cutaneous lesions occur secondary to systemic infection, following hematogenous localization of the bacteria within the integument and subcutis.

Internal gross and microscopic lesions in the acute to chronic disease are also typical of a bacterial septicemia and include diffuse hyperemia and congestion with hemorrhage of the viscera and coelomic membranes, often with a pale or mottled appearance of the affected viscera (Ferguson and McCarthy 1978; McCarthy and Roberts 1980). The kidney, spleen, and liver may be enlarged with a variable amount of necrosis, although liquefactive necrosis of the affected viscera is generally less common and severe compared to motile aeromonad septicemia, due to the limited or absent cellular inflammatory response (Ellis 1991). Bacterial colonies are often present in the affected organs, especially the gills, spleen, kidney, liver, and myocardium. A catarrhal, necrotizing enteritis may also occur in fish with subacute to chronic disease. Degranulation of the gastric submucosal and branchial eosinophilic granular cells may be a diagnostic feature of the disease (Vallejo and Ellis 1989). As previously mentioned, fish with *A. salmonicida* infections are anemic, with a marked leukopenia due to bacterial leukocidin (Ellis 1991). Affected fish may also be hypoglycemic due to hemodilution and/or loss of glucose in fish with cutaneous ulcers, although bacterial utilization of glucose may also contribute to the hypoglycemia.

Diagnosis

A preliminary diagnosis of furunculosis is based on the gross and microscopic findings, including the presence of bacterial colonies in affected tissues. Bacterial colonies may be observed by cytological examination of Gram-stained tissue imprints and/or histological examination of tissues. However, a definitive diagnosis requires positive identification of the bacterium. Bacterial isolation is generally more successful if cultures of the kidney and other affected viscera from multiple fish are obtained immediately following the manifestation of clinical symptoms (Noga 1995). Enriched media such as brain-heart infusion agar (BHIA), tryptic soy agar (TSA), or 5 percent blood agar, should be used for primary isolation (Austin and Austin 1993); BHIA is generally the preferred medium. Cultures should be incubated at 20 to 22°C and assessed daily for bacterial growth that occurs after two to three days at room temperature; growth does not occur at temperatures greater than 37°C (Austin and Austin 1993).

Bacterial colonies are small, circular, gray to tan-brown, and easily moved along the surface of the agar with an inoculating loop (Shotts and Teska 1989). Colonies of the typical subspecies produce a characteristic brown, diffusible pigment that is more easily observed on clear agar such as BHIA or TSA, whereas the atypical subspecies either do not produce pigment or require several days of growth for production of pigment (Austin and Austin 1993). The bacteria are Gram-negative, nonmotile, oxidase- and catalase-positive, facultative anaerobes that are approximately 0.3 to 1.0 × 1.0 to 3.5 μm with bipolar-staining and resistant to the pteridine vibriostat O/129 (Austin and Austin 1993). Definitive identification is obtained by phenotypic profiles, although the atypical subspecies do not conform to the biochemical phenotype of the typical subspecies (Shotts and Teska 1989; Austin and Austin 1993). Phenotypic profiles of the typical and atypical subspecies of *A. salmonicida* have previously been summarized (Austin and Austin 1993).

Immunological tests have been used as alternative methods of bacterial identification, due to the lack of sensitivity and delay using standard diagnostic techniques. Immunological methods include the enzyme-linked immunosorbent assay (ELISA) (Yoshimizu et al. 1993, Hiney, Kilmartin, and Smith 1994), indirect fluorescent antibody test (IFAT) (Sakai, Atsuta, and Kobayashi 1986), latex bead agglutination test (McCarthy 1975a), and slide agglutination test, although the latter is not recommended due to the auto-agglutination properties of the bacterium (Rabb, Cornick, and McDermott 1964). However, these techniques are generally not routinely performed in diagnostic laboratories, although the latex bead agglutination test is commercially available (Microtek Ltd., Victoria, British Columbia*). In contrast, the polymerase-chain reaction (PCR) method has also been used for rapid detection of bacterial nucleic acid in fish tissues (Gustafson, Thomas, and Trust 1992; Hiney et al. 1992; Miyata, Inglis, and Aoki 1996; Hoie, Heum, and Thoresen 1997) and has often been used in combination with bacterial culture and identification as the preferred diagnostic method in several laboratories. However, bacterial isolation is required for the assessment of antimicrobial sensitivity.

Treatment and Prevention

If possible, fish with furunculosis should be removed from the facility prior to the initiation of treatment. Various oral or injectable antimicrobials such as amoxicillin, oxytetracycline, furazolidone, oxolinic acid, flumequine, and potentiated sulfonamides have been used successfully for the treatment of furunculosis (Austin and Austin 1993; Noga 1995; Evelyn et

---

*Use of trade or manufacturer's name does not imply endorsement.

al. 1998). However, bacterial resistance, often with multiple resistance to these various antimicrobials, is not uncommon (Hastings and McKay 1987; Tsoumas, Alderman, and Rodgers 1989; Barnes et al. 1990a; Austin and Austin 1993; Munro and Hastings 1993) and has apparently increased in direct proportion to the use of antimicrobials (Aoki et al. 1983). More recently, effectiveness of the fluoroquinolones—sarafloxacin and enrofloxacin—against *A. salmonicida* has been demonstrated (Barnes et al. 1990b; Barnes et al. 1991; Inglis and Richards 1991; Martinsen et al. 1991). However, as previously mentioned, the extra-label use of nitrofurans and fluoroquinolones in food animals is currently not permitted in the United States (Payne et al. 1999). Furthermore, mixed infections composed of multiple *A. salmonicida* strains may also occur and result in differential sensitivity to any particular antimicrobial. Therefore, sensitivities of multiple isolates need to be determined prior to the selection and administration of antimicrobials. Otherwise, recurrence of disease is not uncommon following the completion of antimicrobial therapy (Hastings and McKay 1987).

Prevention of furunculosis is essential, since the disease is difficult to control. Preventive measures include fallowing netpen sites and other facilities for six weeks to several months prior to stocking, disinfection of the system prior to stocking, elimination or reduction of stress, segregation of age classes, use of water not exposed to wild or cultured fish, elimination or avoidance of infected fish, and vaccination (Munro and Hastings 1993). Infected fish can be avoided by surface disinfection of eggs and quarantine of stocks for detection of asymptomatic carriers prior to movement or introduction of fish. Detection of asymptomatic carriers can be achieved by a stress test. Briefly, fish are injected with an immunosuppressant agent such as prednisolone acetate and held at 18°C for three weeks prior to evaluation (Bullock and Stuckey 1975a; McCarthy and Roberts 1980). The kidney and lower intestine are then cultured for bacteria (Rose, Ellis, and Munro 1989) or evaluated using a more sensitive technique such as PCR for the detection of asymptomatic carriers. Finally, various immersion and injectable vaccine preparations are commercially available for prevention of furunculosis, although the most effective protection is generally provided by intraperitoneal injection (Evelyn et al. 1998). Combination vaccines against furunculosis and other bacterial diseases, especially vibriosis, are generally used in vaccination programs.

Vibriosis

Vibriosis is an important disease of fish, including salmonids, and other aquatic organisms maintained in marine and brackish environments (Frerichs and Roberts 1989). In many respects, vibriosis may be considered the marine equivalent of MAS or furunculosis and has often been referred to as saltwater furunculosis. The most common species associated with disease in salmonids is *Vibrio anguillarum,* although inclusion of this species within the genus *Listonella* has previously been proposed (MacDonnell and Colwell 1985). Regardless, the species is a heterogeneous group of bacteria, although only a few strains are associated with disease as determined by the lipopolysaccharide somatic or O-antigen (Kitao, Aoki, and Muroga 1984; Tajima, Ezura, and Kimura 1985; Sorensen and Larsen 1986). In contrast, *V. ordalii* is a more homogeneous group of bacteria that has been identified as the cause of vibriosis in salmonids from New Zealand and the Pacific Northwest region of North America (Schiewe, Trust, and Crosa 1981; Wards et al. 1991).

Epizootiology

Vibrios are ubiquitous, facultative pathogens that are also a normal component of the external and internal microflora of fish (Noga 1995). An increase in environmental loads of bacteria has been associated with an increase in organic content and salinity of the water. Bacterial transmission is horizontal, either indirectly via the water or directly from fish to fish (Kanno, Nakai, and Muroga 1989), and can be transmitted by feeding contaminated fish products. Conditions that predispose fish to disease include an increase in temperature, organic load, salinity, and density, among various other stressors. Vibriosis is most common in spring and summer, due to the increase in temperature, and is generally most severe at 15 to 21°C (Evelyn et al. 1998). All species and ages of salmonids are susceptible to the disease, and severe mortality may occur in unvaccinated populations, although smolts during the first summer in seawater are extremely susceptible to the disease (Evelyn et al. 1998).

Pathogenesis

The integument and intestine are considered the primary sites of infection (Kanno, Nakai, and Muroga 1989). Primary infection is followed by a bacteremia and subsequent colonization of the internal viscera, although bacteremia in vibriosis due to *V. ordalii* generally occurs later in the course of the disease. Bacterial virulence is variable among strains, and highly pathogenic bacteria may cause disease in the absence of obvious stressful conditions (Evelyn et al. 1998). Properties of the bacteria that apparently confer virulence among isolates include resistance to the normal bacteriocidal properties of the host serum and an increased ability to sequester iron (Trust et al. 1981). Virulent bacteria also produce various extracellular products similar to the aeromonads, such as hemolysin and various proteases (Austin and Austin 1993; Hjeltnes and Roberts 1993; Thune, Stanley, and Cooper 1993). However, the relative contribution of these bacterial properties and products to bacterial virulence and the pathogenesis of disease is incomplete and requires further investigation.

Clinical Disease

Three forms of vibriosis have previously been described—the peracute, acute, and subacute to chronic forms—that are similar to the various forms of disease that can occur with MAS and furunculosis (Hjeltnes and Roberts 1993). Peracute disease is generally restricted to juvenile fish that present with anorexia and dark coloration prior to rapid mortality. Histologically, fish with peracute disease may have a vacuolar cardiomyopathy with renal and splenic necrosis.

Fish with acute disease generally present with symptoms that are characteristic of a bacterial septicemia such as anemia; hemorrhage of the integument and the base of the fins; necrosis of the fins; coelomic distension due to ascites; enlargement of the kidney and spleen; and hemorrhage and necrosis of the internal viscera, especially the kidney and spleen. Liquefactive necrosis of the kidney may occur in severe infections. Large, multifocal, coalescent, hemorrhagic foci or hematomas in the liver that have been referred to as *peliosis hepatis* are a characteristic lesion of vibriosis but may also occur with other diseases, such as infectious salmon anemia (Evelyn et al. 1998). Affected fish may also have necrosis of the intestinal mucosa that often results in yellow, mucoid intestinal contents. Fish with acute vibriosis may also present with fluctuant, subcutaneous vesicles or cavitations that contain a serosanguinous or hemorrhagic fluid prior to ulceration of the lesion.

Histologically, there is necrosis of the internal viscera, especially the liver, heart, kidney, and spleen, often with depletion of the hematopoietic tissue. The necrotic lesions that occur with *V. ordalii* infections are generally focal, in contrast to the focally extensive to diffuse coalescent lesions that are common with *V. anguillarum* infections. Vibriosis due to *V. anguillarum* is generally characterized by a diffuse distribution of individual bacteria throughout the vasculature and internal viscera. In contrast, the presence of bacterial microcolonies in the gills and necrotic lesions of the viscera are common in fish with vibriosis due to *V. ordalii* (Ransom et al. 1984)

The clinical presentation and gross lesions in fish with the subacute to chronic form of vibrosis are nonspecific and similar to fish with any subacute to chronic, Gram-negative bacterial infection. Fish with subacute to chronic vibriosis may have anemia, exophthalmos, corneal edema with corneal ulceration, and intracoelomic hemorrhage that often results in the formation of fibrous adhesions. Granulomatous inflammation of the skeletal muscle can result in partial or complete loss of the carcass, although this may not be apparent until the fish are processed. Histologically, there is often an increased deposition of hemosiderin within the renal and splenic melanomacrophage centers.

Diagnosis

A preliminary diagnosis of vibriosis is based on gross and microscopic findings, whereas a definitive diagnosis requires bacterial isolation and identification. Bacteria can generally be observed on Gram-stained imprints of the affected tissues. Bacterial cultures should be obtained from the kidney and other affected tissues of multiple fish and incubated at 20 to 22°C. Since the bacteria are ubiquitous and may be normal components of the external and internal microflora, cautious interpretation of bacterial cultures is recommended to avoid overinterpretation of the results. Bacteria are easily isolated on marine agar or enriched agar, such as TSA supplemented with 1.5 percent sodium chloride (NaCl), although supplementation with salts may not be required for pathogenic strains of bacteria. Colonies are generally small, raised, convex, and cream to tan; growth of *V. ordalii* is slower than that of *V. anguillarum*. The bacteria are Gram-negative, motile, short (0.5 × 2.0 μm), curved bacilli that are catalase- and oxidase-positive and are sensitive to the pteridine vibriostat O/129. The phenotypic biochemical profiles of both species have been previously summarized (Austin and Austin 1993; Hjeltnes and Roberts 1993). Identification can be performed using rapid identification systems such as API-20E (Analtytab, Plainview, New York; Grisez, Ceusters, and Ollevier 1991) or by serological methods such as slide agglutination using commercially available antisera (Microtek-Bayer, Sidney, British Columbia). However, immunofluorescence assays for the detection of bacteria in tissue are not useful (Evelyn et al. 1998).

Treatment and Prevention

Various oral and injectable antimicrobials such as oxytetracycline, potentiated sulfonamides, oxolinic acid, and florfenicol have been successfully used for the treatment of vibriosis (Hjeltnes and Roberts 1993; Evelyn et al. 1998). Treatment with antimicrobials administered in the feed is the most practical method but needs to be initiated early in the course of the disease, prior to the onset of anorexia in affected fish. Bacterial resistance to the common antimicrobials is also a concern that requires the evaluation of antimicrobial sensitivity, prior to the administration of antimicrobials. Therefore, control and prevention is accomplished by avoidance and attention to proper management practices that eliminate stress. Vaccination should be an essential component of any husbandry program in marine salmonid operations. Vaccines for vibriosis are available from a variety of manufacturers and are generally provided as multivalent vaccines for protection against other common diseases, such as furunculosis and cold-water vibriosis. Maximum protection is achieved by vaccination of immunocompetent fish that are at least 5 to 10 g prior to shipment or movement (Evelyn et al. 1998). Vaccines can be administered by immersion, although injection generally provides the most efficacious results. Revaccination of fish after introduction to netpens is often practiced, although this may cause additional stress to fish (Evelyn et al. 1998).

Cold-Water Vibriosis

Cold-water vibriosis (or Hitra disease) is a problem in marine salmonid operations of the North Atlantic regions of Europe and North America (Egidius et al. 1981; Bruno, Hastings, and Ellis 1986; O'Halloran and Henry 1993). The disease is caused by *V. salmonicida* (Egidius et al. 1986), which is a ubiquitous bacterium in water and sediment (Enger, Husevag, and Goksoyr 1989). The bacterium is psychrophilic, and severe disease generally occurs at temperatures of 2 to 5°C during the winter months, although disease may occur throughout the year. Environmental stress and nutrition may be additional factors that contribute to the pathogenesis of disease. Disease may also occur in nonsalmonids such as Atlantic cod, *Gadus morhua* (Jorgensen et al. 1989).

The role of bacterial toxins in bacterial virulence and the pathogenesis of disease is not currently known, although the clinical presentation and lesions associated with the disease are characteristic of a bacterial septicemia. Affected fish can present with peracute, acute, or subacute to chronic forms of the disease similar to furunculosis or vibriosis (Hjeltnes and Roberts 1993). Generally, fish with cold-water vibriosis present with anorexia, depression, disorientation, dark coloration, exophthalmos, coelomic distension due to ascites, swelling and prolapse of the vent, pallor of the gills, and hyperemia/petechial hemorrhage of the integument, including the base of the fins. Internally, there is often hemorrhage of the viscera, including the swim bladder, which often contains a serosanguinous fluid, whereas the liver may have a yellow discoloration (Hjeltnes and Roberts 1993; Evelyn et al. 1998). Microscopic lesions are nonspecific and characteristic of a bacterial septicemia but often include hemorrhage and necrosis of the intestine with sloughing of the mucosa and severe necrosis of the skeletal and cardiac muscle. Bacteria are also present in the affected tissues, especially the muscle, heart, kidney, spleen, and liver, and can be more easily identified using Giemsa-stained tissue sections (Fjolstad and Heyeraas 1985). Fish with the chronic form of the disease may have cutaneous ulcers and a pseudomembranous coelomitis and epicarditis.

Diagnosis

A preliminary diagnosis of cold-water vibriosis is based on gross and microscopic findings. This includes a direct cytological examination of ascitic fluid or fluid from the swim bladder that often contains motile bacilli. However, a definitive diagnosis is based on bacterial isolation and identification. Cultures can be obtained from the ascitic or swim bladder fluids, kidney, and other affected viscera. The use of enriched media such as TSA or blood agar supplemented with 1.5 to 2.0 percent NaCl prior to incubation at 15°C is recommended for bacterial isolation. Small, translucent colonies generally appear in three to five days at 12 to 16°C. A definitive diagnosis is confirmed with biochemical profiles, as previously summarized, that include sensitivity to novobiocin and the pteridine vibriostat O/129 (Holm et al. 1985). A diagnosis of cold-water vibriosis can also be confirmed using immunological techniques such as FAT, agglutination assays, or immunohistochemistry (Evensen, Espelid, and Hastein 1991; Evelyn et al. 1998).

Treatment and Prevention

Fish with the disease should be isolated if possible, since increased environmental concentrations of the bacterium have been associated with outbreaks of the disease (Enger, Husevag, and Goksoyr 1989). Treatment with various antimicrobials, as previously mentioned for vibriosis, is often successful, although clinical intervention should occur early in the course of the disease. In this context, routine culture and isolation of the opportunistic bacterial pathogens in the facility, including *V. salmonicida,* is recommended for determination of antimicrobial sensitivities prior to the onset of clinical disease (Evelyn et al. 1998). Although *V. salmonicida* is not considered highly pathogenic, strategies for prevention of the disease include the elimination of stress and vaccination of fish. Vaccines for cold-water vibriosis are available from a variety of manufacturers and are generally provided as multivalent vaccines for additional protection against furunculosis and vibriosis.

Winter Ulcers

Winter ulcers is a serious disease of Atlantic salmon that is caused by *Vibrio vulnificus* and *V. wodanis*. Mortality generally occurs in smolts during the winter, following introduction of fish to seawater, although fish of all ages may be affected throughout the year (Evelyn et al. 1998). This may result in an additional economic impact due to the rejection or downgrading of market-sized fish (Evelyn et al. 1998). A decreased ability of the host to maintain normal osmoregulation during cold temperatures may be involved in the pathogenesis of the disease (Evelyn et al. 1998).

Fish with the disease may be lethargic and congregate at the sides of the enclosure, although the characteristic feature of the disease is ulceration of the integument, which typically occurs on the flanks. The ulcers can be variable in size but are often several centimeters in diameter and delineated by a hemorrhagic border. Resolution of the lesions results in a margin composed of white granulation tissue. Pathogenesis of the cutaneous lesions may be due to the formation of capillary thrombi that develop at decreased temperatures (Salte et al. 1994). Microscopically, the cutaneous ulcers are characterized by a moderate to severe infiltration of mixed inflammatory cells, with degeneration and necrosis of the integument and subcutaneous tissues secondary to bacterial colonization.

A preliminary diagnosis of winter ulcers is based on the presence of cutaneous ulcers, although this is a nonspecific finding that may occur with various infectious diseases. Therefore, bacterial isolation and identification is required for a definitive diagnosis. Treatment with medicated feeds is generally successful, although the disease may recur following completion of the antimicrobial therapy (Evelyn et al. 1998). Commercial vaccines are not currently available for prevention of winter ulcers, although autogenous vaccines can be developed for individual facilities as necessary.

Columnaris Disease

Members of the family Cytophagaceae (order Cytophagales) are an extremely heterogeneous and diverse group of chromogenic (yellow-pigmented) bacteria that are difficult to isolate and identify using routine methods (Austin and Austin 1993; Wakabayashi 1993). Representatives of this diverse group have previously been classified as *Myxobacterium, Chondococcus, Myxococcus, Sporocytophaga, Cytophaga, Flexibacter,* and *Flavobacterium,* although these various classifications have resulted in confusion, which underscores the need for reclassification of these bacteria. Therefore, the recent proposal to reclassify several of these bacteria within the genus *Flavobacterium* (Bernardet et al. 1996) will be adapted for consistency in this discussion.

Columnaris disease is a common and often serious disease of freshwater fish, including salmonids, and is caused by the bacterium *Flavobacterium columnare* (Austin and Austin 1993; Wakabayashi 1993; Bernardet et al. 1996). The species and disease names are derived from the characteristic and diagnostic feature of the bacterium to form columns in wet-mount preparations of infected tissue.

Epizootiology

F. columnare is ubiquitous in freshwater environments and may be isolated from the integument and internal organs of healthy fish (Austin and Austin 1993; Wakabayashi 1993). The bacterium is an opportunistic pathogen that may survive for prolonged periods in water. Bacterial survival has been associated with an optimal calcium carbonate hardness of approximately 70 ppm (Chowdhury and Wakabayashi 1988) and an increased organic content of the water (Fijan 1968). In contrast, a calcium carbonate hardness less than 50 ppm and a pH of 6.0 or less decreases survival of the bacteria (Fijan 1968). The bacterium can be a primary pathogen but is often an opportunistic pathogen that is associated with poor water quality and stressful conditions (Austin and Austin 1993; Wakabayashi 1993). Environmental conditions that predispose fish to infection include increased organic loads, increased ammonia and nitrite concentrations, and decreased dissolved oxygen concentrations. Stressful conditions that increase the susceptibility of fish to disease include trauma, handling, and crowding. Preexistent lesions or external bacterial and parasitic infections of the integument further increase the susceptibility of fish, although this is not a prerequisite for infection (Wakabayashi 1993). Transmission of the bacterium is indirect via the environment, including cohabitation with carrier fish, or direct via contact with infected fish (Austin and Austin 1993; Wakabayashi 1993). Contaminated fomites are also a likely source of infection, including nets. Additionally, netting can result in cutaneous lesions that further predispose fish to infection.

Disease may occur at various water temperatures, although the incidence of disease is increased from May to October when ambient water temperatures exceed 15°C (Austin and Austin 1993; Wakabayashi 1993). Disease at lower temperatures is generally due to highly virulent strains of the bacteria that are adapted to colder temperatures (Frerichs and Roberts 1989; Dalsgaard 1993). Juvenile fish have an increased susceptibility to the disease, with an increased mortality following infection, whereas mature fish are generally less susceptible but may be carriers of the bacterium.

Pathogenesis

Columnaris disease is characterized as an acute to chronic infection of the gills and integument, including the fins (Wolke 1975). Bacterial virulence and pathogenicity is strain-dependent and has been correlated with hemagglutination activity of the bacterium and adhesion to host tissues (Dalsgaard 1993; Thune, Stanley, and Cooper 1993). Bacterial adhesion is mediated by a slime layer composed of a capsular acid-mucopolysaccharide that may function as an adhesin during gliding. Several extracellular proteases that result in proteolysis of connective tissue, such as gelatinase, caseinase, and chondroitinase, have been identified. Infection results only in a limited or absent host inflammatory response that may be inhibited by bacterial proteases.

Progression of the disease is dependent on water temperature and may be extremely rapid at higher temperatures, with an incubation period of less than 24 hours and mortality after two to three days (Bullock, Conroy, and Snieszko 1971; Schäperclaus 1991; Austin and Austin 1993). Acute disease with or without external lesions may result in systemic infections and mortality without internal lesions, especially in juvenile fish (Austin and Austin 1993; Dalsgaard 1993; Thune, Stanley, and Cooper 1993; Wakabayashi 1993). Mortality is most common in fish with gill lesions, due to hypoxia/anoxia and loss of normal osmoregulatory function, although the latter may also occur in fish with cutaneous lesions (Bullock, Conroy, and Snieszko 1971; Frerichs and Roberts 1989).

Clinical Disease

The characteristic features of columnaris disease are the external lesions of the gills, oropharynx, and integument (Bullock, Conroy, and Snieszko 1971; Wolke 1975; Austin and Austin 1993; Wakabayashi 1993). Lesions may be localized or multifocal, with concurrent involvement of various tissues. Integument lesions may occur in various locations, including the head, jaws, and fins, and are initially opaque white to gray-white or blue, punctate foci or plaques, often with peripheral erythema that correspond to foci of epithelial proliferation due to bacterial colonization. The lesions may become more extensive and confluent, with eventual sloughing of the epidermis and the development of shallow ulcerations. Extensive, confluent lesions often involve the dorsal midline posterior to the dorsal fin, with extension along the lateral flanks, which have been referred to as saddle lesions or saddleback lesions, due to their characteristic appearance (Wolke 1975; Wakabayashi 1993). The ulcers may be delineated by a zone of white necrotic tissue and/or have an orange to yellow pigmentation due to massive colonization and proliferation of the yellow-pigmented bacteria. Deep ulcers with exposure of the subcutaneous muscle may occur in severe cases. Lesions of the oropharynx are often covered by an increased amount of a yellow, mucoid material. Opportunistic fungal infections of the external lesions are not uncommon, although, in contrast to aeromonad infections, secondary bacterial infections are less common, especially early in the progression of the disease, possibly due to bacterial production of bacteriocins (Wakabayashi 1993). Histologically, the cellular inflammatory response is minimal or absent (Dalsgaard 1993; Thune, Stanley, and Cooper 1993).

The gross appearance of the branchial lesions is a characteristic feature of columnaris disease that can be used as a presumptive diagnosis. These lesions are typically brownish green to orange-yellow foci of necrosis that initially involve the margin of the hemibranch, with progressive extension toward the gill arch (Wolke 1975; Austin and Austin 1993; Wakabayashi 1993). Histologically, there is vascular congestion and hemorrhage with epithelial hyperplasia and fusion of the secondary lamellae, followed by epithelial necrosis and sloughing. Progressive necrosis often results in exposure of the connective tissue stroma of the primary gill filaments. As previously mentioned, mortality is most common in fish with severe gill lesions.

Diagnosis

A preliminary diagnosis of columnaris disease is dependent on the observation of long, slender bacilli that are approximately 0.5 to 1.0 × 4 to 10 μm and have a characteristic gliding motion in wet-mount preparations. As previously mentioned, the bacteria generally aggregate into columns during examination of wet-mount preparations. Although a definitive diagnosis is based on bacterial isolation and identification, observation of the bacterium in wet-mounts is generally sufficient for a diagnosis, since the bacterium requires special media for isolation. The various media used for bacterial isolation have previously been reviewed (Austin and Austin 1993). Specifically, a nonenriched medium with high moisture content, such as *Cytophaga* agar or tryptone-yeast extract (TYE), is the preferred medium, although TSA may also be used for bacterial isolation. Otherwise, a 1:10 dilution of nutrient broth in 1 percent agar can be used as a substitute medium. A selective *Cytophaga* agar (SCA) that contains antibiotics can be used for bacterial isolation in mixed infections or to prevent contamination of the cultures. Incubation at 15 to 18°C results in bacterial colonies that are flat, smooth or mucoid to rhizoid, and spreading, with a pale yellow to yellowish orange pigmentation (Austin and Austin 1993; Wakabayashi 1993). Rhizoid strains generally adhere to the agar and are commonly associated with localized infections in salmonids. The species is a homogeneous, strictly aerobic group of bacteria that is identified by phenotypic profiles, which include production of flexirubin pigment (Austin and Austin 1993; Wakabayashi 1993). Agglutination tests have also been used for definitive diagnosis of the bacterium (Morrison et al. 1981).

Treatment and Prevention

The most practical treatment of columnaris disease generally involves bath administration of salt or various chemical disinfectants and antimicrobials such as potassium permanganate, copper sulfate, or oxytetracycline, although the use of oral or injectable antimicrobials is recommended for severe cases of the disease (Austin and Austin 1993; Wakabayashi 1993; Noga 1995). Bath or oral administration of oxytetracycline is generally effective, although bath administration should be avoided in hard water. Prevention of the disease requires improvement of environmental conditions and the elimination of stress. Exposure to wild fish that may potentially carry the bacterium or water exposed to wild fish should be avoided. Recently, a killed vaccine for columnaris disease (Fryvacc 1, Aqua Health Ltd., Charlottetown, Prince Edward Island, Canada) has been licensed for use in the United States and Canada.

Marine Columnaris

As the name implies, marine columnaris is similar to columnaris disease in freshwater fish, although marine columnaris may be a nonspecific condition that includes several diseases with separate etiologies but without distinct characteristics. Specifically, the causative agents of these various diseases include *Flexibacter maritimus, Cytophaga* sp., and other nonclassified myxobacterial-like organisms (Kent et al. 1988; Frelier et al. 1994; Handlinger, Soltani, and Percival 1997; Evelyn et al. 1998). However, the identification and taxonomic placement of these various marine bacteria is incomplete and requires further investigation (Austin and Austin 1993). In this context, the recent proposal for placement of similar freshwater bacteria within the genus *Flavobacterium* did not include the pathogenic species of marine bacteria (Bernardet et al. 1996).

Epizootiology and Pathogenesis

The diseases collectively referred to as marine columnaris have been reported in rainbow trout and Atlantic salmon from Tasmania and the Pacific Northwest region of North America (Kent et al. 1988; Frelier et al. 1994; Handlinger, Soltani, and Percival 1997; Evelyn et al. 1998). Disease in Atlantic salmon smolts due to *Cytophaga* sp. generally occurs within one to three weeks following introduction to seawater and persists for three to four weeks, although the disease may be less severe in smolts introduced during late spring and summer (Kent et al. 1988). A separate condition in Atlantic salmon due to a myxobacterial-like organism, generally occurs in smolts during the first summer in seawater and often results in high mortality (Frelier et al. 1994). Lesions are restricted to the mouth and oral cavity and initially involve the teeth. This observation suggests that development of the disease may be related to preexistent lesions of the mouth that may occur secondary to stress, feeding on crustaceans or hard pelleted feed, and/or biting of nets (Evelyn et al. 1998). Severe disease may be associated with high salinity (Evelyn et al. 1998). The disease has also been reported in Arctic char, *Salvelinus alpinus,* raised in netpens (Evelyn et al. 1998).

Clinical Disease

Fish with marine columnaris due to *F. maritimus* present with lesions similar to freshwater salmonids with columnaris disease (Handlinger, Soltani, and Percival 1997). The characteristic features of the disease due to *Cytophaga* sp. in Atlantic salmon smolts are the large, opaque white, cutaneous lesions that generally involve the posterior flanks and caudal peduncle (Kent et al. 1988). Severe lesions may be hemorrhagic and ulcerative with involvement of the subcutaneous skeletal muscle, although there is no involvement of the gills and internal viscera. A hypernatremia may also occur in fish with severe lesions, suggestive of compromised osmoregulation. Fraying of the fins and erosion of the caudal fin may occur in Pacific salmon, but the disease is generally less severe in these species (Evelyn et al. 1998).

A separate condition in Atlantic salmon with a myxobacterial-like infection of the oral cavity results in lethargy, anorexia, and emaciation that is often accompanied by flashing or shaking of the head (Frelier et al. 1994). The oral cavity contains focal to multifocal, yellow-pigmented lesions that are often ulcerated, depending on the chronicity of the disease. Pigmentation of the lesions is due to extensive bacterial colonization. There may also be involvement of the branchial arches and proximal esophagus, with erosion of the jaws in severe disease. Histologically, the lesions are associated with an infiltration of mixed inflammatory cells with extensive bacterial colonization (Frelier et al. 1994).

Diagnosis

A preliminary diagnosis of these various diseases is based on the gross and microscopic findings, including direct observation of the filamentous bacteria in wet-mount preparations. A definitive diagnosis is dependent on bacterial isolation and identification, although this requires specialized media such as marine agar or *Cytophaga* agar made with 50 to 70 percent sterile seawater (Austin and Austin 1993; Evelyn et al. 1998). Serial dilutions in 50 percent sterile seawater prior to subculture on solid media at 15°C has also been recommended for isolation of the bacteria (Evelyn et al. 1998). Bacterial colonies are small, yellow to yellow-green, diffuse, and rhizoid on *Cytophaga* agar but smooth and yellow-orange on marine agar. The bacteria are Gram-negative, long, slender, filamentous, motile or nonmotile, and approximately 0.5 to 0.7 × 4 to 20 μm, depending on the species. However, a routine diagnosis is generally based on wet-mount findings, due to the difficult isolation and identification of these various bacteria.

Treatment and Prevention

Treatment of these various diseases is similar to columnaris disease, although bath or immersion administration of antimicrobials in netpen facilities is not practical. Therefore, treatment is generally restricted to the use of medicated feeds, such as potentiated sulfonamides, and is recommended in severe infections (Evelyn et al. 1998). Prevention of the disease is preferred and is achieved by elimination and avoidance of stress and other conditions that predispose fish to disease. Prevention of traumatic lesions that often occur during transport and handling has been associated with a decreased prevalence of disease (Evelyn et al. 1998). Vaccines are not currently available for protection against these various disease conditions.

Cold-Water Disease

Epizootiology and Pathogenesis

Cold-water disease (CWD) or peduncle disease is caused by *Flavobacterium psychrophilum* and is often a serious problem in salmonid culture operations. The bacterium is ubiquitous in freshwater environments and may be endemic in freshwater salmonid facilities but can also occur in the marine environment. The bacterium has been isolated from the external surfaces of salmonids and may therefore be a normal component of the microbial flora (Holt, Rohovec, and Fryer 1993). In this context, cutaneous lesions may predispose fish to infection and subsequent disease (Holt, Rohovec, and Fryer 1993). All species and ages of salmonids are susceptible to the disease, although juvenile fish may have an increased susceptibility. Transmission of the bacterium is horizontal and can occur either directly from fish to fish or indirectly through the water (Holt, Rohovec, and Fryer 1993). Although the bacterium has been isolated from the reproductive tissues and surface of the ova, there is no direct evidence for vertical transmission. Clinical disease is associated with virulence of the bacterial strain and water temperature. Factors associated with bacterial virulence have not been adequately investigated but may be similar to other members of the genus, such as *F. columnare* (Dalsgaard 1993; Holt, Rohovec, and Fryer 1993). Disease is generally most severe at 15°C but can occur at temperatures as low as 4°C, whereas mortality is decreased at temperatures greater then 15°C (Holt, Rohovec, and Fryer 1993). Therefore, the incidence of disease is generally increased in the spring as temperatures approach 15°C.

Clinical Disease

The clinical symptoms and lesions of fish with cold-water disease are similar to fish with columnaris disease (Holt, Rohovec, and Fryer 1993). Affected fish may have a dark coloration with cutaneous erosions and ulcerations. Epithelial erosions of the yolk sac can also occur in sac fry. However, in contrast to columnaris disease, fish with cold-water disease often have systemic infections, with bacterial colonization of the gills, heart, spleen, and kidney, that may occur in the absence of cutaneous lesions. The affected viscera may be hemorrhagic, although inflammation is mild or absent. The bacterium apparently has an affinity for the mineralized connective tissue of the posterior cranium and vertebral column that results in musculoskeletal and spinal deformities secondary to a necrotizing osteochondritis (Kent, Groff, et al. 1989; Ostland, McGrogan, and Ferguson 1997). Fish that survive the disease may also present with neurological disease and/or deformities several months postinfection.

Diagnosis

A presumptive diagnosis of cold-water disease is based on the characteristic gross lesions and observation of the typical bacteria in wet-mount preparations. Histological examination is also presumptive, although the use of silver stains such as Steiner or Warthin-Starry highlights the bacteria. Bacterial isolation and identification is required for a definitive diagnosis, although this may not be necessary for routine cases. Similar to *F. columnare,* isolation of *F. psychrophilum* requires special media such as *Cytophaga* agar, TYE, or dilute nutrient agar (Austin and Austin 1993; Holt, Rohovec, and Fryer 1993), although TSA has also been used for isolation. A selective *Cytophaga* agar (SCA) that contains antibiotics can also be used for bacterial isolation in mixed infections or to prevent contamination of the cultures. In contrast to *F. columnare,* the pathogen may be isolated from the affected internal viscera, including the brain (Kent, Groff, et al. 1989). Bacterial colonies are yellow with thin, spreading margins. The strictly aerobic bacteria are slender, filamentous, Gram-negative bacilli that are approximately 0.75 × 1.5 to 7.5 μm with a characteristic flexing or gliding motility. Phenotypic information has previously been summarized (Bernardet and Grimont 1989; Bernardet and Kerouault 1989; Austin and Austin 1993).

Treatment and Prevention

Treatment for cold-water disease is the same as for columnaris disease. Bath administration of potassium permanganate or oxytetracycline is generally effective (Austin and Austin 1993; Holt, Rohovec, and Fryer 1993), although oral or injectable antibiotics such as oxytetracycline are recommended for treatment of severe or systemic disease. Prevention of the disease involves the elimination of stress and incorporation of management strategies to improve water quality and prevent cutaneous lesions. The use of shallow enclosures with reduced water flow has been suggested as a method to reduce cutaneous lesions in sac fry. Exposure to wild fish that may carry the bacterium or water contaminated with the bacterium should be avoided. Disinfection of eggs is also recommended as a routine procedure in salmonid operations. Likewise, fish that carry the bacterium in the reproductive tissues and eggs should be rejected for use as broodstock. Vaccines are currently being developed and should be commercially available in the near future.

Bacterial Gill Disease

Bacterial gill disease (BGD) is caused by *Flavobacterium branchiophilum* and generally results in chronic morbidity with low mortality in freshwater salmonid facilities, although mortality may approach 25 percent (Speare et al. 1991b). However, the primary economic impact of the disease is due to the chronic morbidity that results in reduced feed conversion and, consequently, reduced growth rate.

Epizootiology

Horizontal transmission of the bacterium is indirect via the water (Ferguson et al. 1991), although the bacterium is not considered a normal component of the external microflora, since it is difficult to isolate from healthy fish (Heo, Kasai, and Wakabayashi 1990). The bacterium is considered a primary pathogen (Ferguson et al. 1991), although disease is generally associated with poor environmental quality such as increased turbidity, ammonia concentrations, and density or decreased dissolved oxygen (DO) concentrations (Turnbull 1993). Epizootics may occur at temperatures of 20°C or less (Turnbull 1993). The disease may also affect other freshwater fish that may potentially carry and transmit the bacterium.

Pathogenesis and Clinical Disease

Differential pathogenicity may occur among bacterial strains, although the factors and mechanisms of bacterial virulence have not been determined and require further investigation. As the name implies, bacterial gill disease is restricted to the gills and results in a chronic proliferation or hyperplasia of the secondary lamellar epithelium (Ferguson et al. 1991; Turnbull 1993). Clinical signs of affected fish include anorexia, lethargy, dyspnea, coughing, and flared opercula. The gills generally have a swollen and pale appearance on gross examination but may be hyperemic. An increased mucous exudate is not an uncommon finding in affected fish and may trail from the opercular chamber. Bacteria initially colonize the proximal aspects of the secondary lamellae, with progressive involvement of the basal lamellar epithelium (Speare et al. 1991b). Bacterial colonization results in hyperplasia of the lamellar epithelium, with fusion of adjacent lamellae. Histologically, fusion results in the appearance of interlamellar cystic spaces, often with obliteration of the interlamellar space (Speare et al. 1991a). Fusion of adjacent primary filaments may occur with severe disease. Morbidity and mortality are most probably due to compromised branchial respiration and osmoregulation.

Diagnosis

A preliminary diagnosis of BGD is based on the characteristic gross and microscopic lesions of the gills and the presence of typical bacteria in wet-mount preparations of the gills. A definitive diagnosis requires bacterial isolation and identification, although positive gross and microscopic findings are generally sufficient for a diagnosis. As with other members of this genus such as *F. columnare,* the bacterium is difficult to isolate and requires dilute nutrient media such as *Cytophaga* agar. Bacterial colonies are typically yellow, round, smooth, transparent, and 0.5 to 1.0 mm after two to five days of incubation (Turnbull 1993). The bacteria are Gram-negative, nonmotile, filamentous bacilli that are approximately 0.5 to 1.0 × 4 to 10 μm. The phenotypic profile has previously been summarized (Turnbull 1993). Immunological tests such as FAT (Huh and Wakabayashi 1987; Heo, Kasai, and Wakabayashi 1990) have been recommended for bacterial identification and diagnosis, since the fastidious nature of the bacterium often results in difficult isolation and contamination of the cultures (Turnbull 1993).

Treatment and Prevention

Treatment for BGD is similar to that for columnaris disease or cold-water disease and includes bath administration of salt or various disinfectants and antimicrobials (Austin and Austin 1993; Turnbull 1993). General strategies for prevention of bacterial gill disease include improvement of environmental quality and reduction of stress. Exposure to other species of freshwater fish should be avoided, since these species may carry the bacterium. Commercial vaccines are currently not available for prevention of bacterial gill disease.

Yersiniosis

Yersiniosis, or enteric redmouth disease (ERM), is an important disease of freshwater and marine salmonids but can also affect nonsalmonid species in these environments. The causative agent of yersiniosis is the enteric bacterium *Yersinia ruckeri,* which is a ubiquitous and cosmopolitan pathogen but may also be a normal component of the intestinal microflora, since approximately 10 percent of any population may be carriers of the bacterium (Busch and Lingg 1975). The species is composed of a variety of strains and is currently divided into six serovars, although the most common and pathogenic is the Type I or Hagerman serovar (Austin and Austin 1993; Stevenson, Flett, and Raymond 1993). However, other serovars can result in severe disease.

Epizootiology

Although the bacterium is ubiquitous in the environment, the reservoir of infection is carrier salmonid and nonsalmonid fish (Hunter, Knittel, and Fryer 1980). Rainbow trout are extremely susceptible, although all species and ages of salmonids can be affected by the disease, especially market-sized fish. Transmission of the bacterium is horizontal and can occur either directly from fish to fish or indirectly through the water. The latter commonly occurs following contamination of the environment with pathogenic strains of bacteria shed in the feces by carrier fish or by fish with clinical disease. Bacterial shedding is cyclical and dependent on environmental factors, including exposure to increased temperatures of 15 to 18°C, which may account for the variable prevalence of the bacterium within the environment and captive populations (Bruno and Munro 1989). Clinical disease generally occurs in fish, including carrier fish, that are predisposed to infection due to stress such as increased temperatures and poor water quality (Stevenson, Flett, and Raymond 1993). A majority of fish (75 percent) that survive the disease become carriers of the bacterium (Busch and Lingg 1975). However, transmission of the bacteria via piscivorous birds and mammals or any fomite should also be considered a potential source of infection, whereas vertical transmission does not occur with *Y. ruckeri* (Stevenson, Flett, and Raymond 1993).

Pathogenesis and Clinical Disease

Although fish with yersiniosis present with symptoms and lesions typical of a septicemia, the bacterial virulence factors associated with the disease have generally not been determined (Stevenson, Flett, and Raymond 1993).

Clinical disease with mortality may occur in one week, following infection at 15°C. Initially, there is a low mortality in the affected population that tends to increase and continue for an extended period with progression of the disease. Affected fish with acute disease generally present with anorexia, lethargy, dark coloration, cutaneous hemorrhage, and lesions typical of a bacterial septicemia. The latter include petechial hemorrhage of the internal viscera; enlargement of the spleen and kidney; and yellow, catarrhal to mucoid, fecal pseudocasts due to necrosis of the intestinal mucosa. Hemorrhage of the mouth and jaws is a characteristic lesion of yersiniosis in freshwater salmonids that has resulted in the descriptive designation—*enteric redmouth disease*. However, this lesion is not a characteristic feature of the disease in marine salmonids. Histologically, there is hemorrhagic necrosis of the gills, skeletal muscle, and internal viscera that is often associated with bacterial colonization. There may also be a purulent epicarditis and meningitis (Evelyn et al. 1998). Affected fish may be hypoproteinemic and anemic with a leukocytosis. Fish with subacute to chronic disease may have exophthalmos, hyphema, and distension of the coelomic cavity due to ascites. Hyphema may result in blindness that results in a dark coloration of the affected fish.

Diagnosis

A preliminary diagnosis is based on the characteristic gross and microscopic lesions, whereas a definitive diagnosis requires bacterial isolation and identification. Bacterial cultures from fish with clinical disease should be obtained from the kidney and other affected tissues. The bacterium is often difficult to isolate using standard techniques and enriched agar. Therefore, primary cultures using enriched broth, such as tryptic-soy broth, that are incubated for two days at 18°C prior to subculture on enriched agar, such as TSA, BHIA, or blood agar, may be a more successful procedure (Stevenson, Flett, and Raymond 1993). A selective medium has been used for identification of North American isolates but may not be useful for other isolates (Waltman and Shotts 1984; Rodgers 1992; Austin and Austin 1993). Cultures of the lower intestine have been used as a method to detect carrier fish (Busch and Lingg 1975), although the sensitivity of this technique is compromised due to the mixed bacterial flora of the intestine.

Incubation at 25°C on enriched agar produces round, raised, shiny, tannish white, 2 to 3 mm bacterial colonies (Ross, Rucker, and Ewing 1966). The bacterium is a Gram-negative, slightly curved bacillus that is approximately 1.0 × 3 2.0 to 3.0 µm and motile by peritrichous flagella (Austin and Austin 1993). However, filamentous and nonmotile strains are not uncommon isolates (Stevenson, Flett, and Raymond 1993). The species is phenotypically homogeneous and can be identified using standard techniques, including rapid identification methods such as API 20E, although the latter method may identify *Y. ruckeri* as *Hafnia alvei* (Austin and Austin

1993). Phenotypic biochemical profiles have previously been summarized (Ewing et al. 1978; Austin and Austin 1993). The bacterium can also be identified using immunological techniques including IFAT, ELISA, latex bead agglutination, and immunohistochemical techniques that have been used for the detection of carrier fish (Austin and Austin 1993).

Treatment and Prevention

Treatment for yersiniosis with oxytetracycline, potentiated sulfonamides, and oxolinic acid in the feed has been successful (Rodgers and Austin 1982; Austin and Austin 1993), although antimicrobial resistance is a concern with any pathogenic bacterium. Avoidance of the bacterium is preferred, but difficult, due to bacterial shedding by carrier fish (Austin and Austin 1993). Ideally, fish should not be exposed to potential carrier fish or water exposed to carrier fish. Stress should be reduced and eliminated if possible. Although vertical transmission has not been demonstrated, surface disinfection of eggs is a standard procedure that should be a routine practice in any salmonid operation. However, vaccination against yersiniosis is the best strategy for prevention of disease. Vaccines are commercially available from various manufacturers, although autogenous bacterins can be developed for endemic strains of the bacterium.

Bacterial Kidney Disease

Epizootiology

The causative agent of bacterial kidney disease (BKD) is the obligate, intracellular pathogen *Renibacterium salmoninarum,* which does not survive for prolonged periods in the environment (Evelyn 1993). Horizontal transmission can occur in freshwater and marine environments via cohabitation with infected fish, ingestion, or exposure to contaminated water. Localization of the bacterium within the ova also results in vertical transmission that is not prevented with surface disinfection of the ova (Evelyn, Ketcheson, and Prosperi-Porta 1984; Evelyn, Prosperi-Porta, and Ketcheson 1986). All species of freshwater and marine salmonids are susceptible to infection, although Atlantic salmon are generally more resistant to infection and clinical disease (Evelyn 1993). The disease has a cosmopolitan distribution but has not been reported in Australia, New Zealand, or Russia (Evelyn 1993). Adult and subadult fish greater than six months of age are generally most affected by the disease, although younger fish are also susceptible to infection and clinical disease. Clinical disease is generally associated with stress such as spawning and transfer of smolts to seawater (Fryer and Sanders 1981). Transfer of infected fish may result in clinical disease immediately following movement and relocation, although clinical disease generally occurs during the winter and spring following transfer (Evelyn et al. 1998).

Clinical Disease

Infection often results in significant morbidity and mortality within an affected population. Affected fish may be dark, lethargic, and anorexic with pallor of the gills due to anemia, exophthalmos often with an ophthalmitis, hemorrhage of the vent and base of the fins, and distension of the coelomic cavity due to ascites (Evelyn 1993; Evelyn et al. 1998). The disease is often referred to as *spawning rash* in fish during sexual maturation, due to the multiple cutaneous vesicles of the flanks that contain a serous or serosanguinous fluid. Rupture of the vesicles results in multifocal, cutaneous ulcerations.

Internally, infection results in a chronic, granulomatous inflammation of the liver, heart, spleen, and kidney but may also involve the eye, brain, and skeletal muscle (Wolke 1975; Bruno 1986; Evelyn et al. 1998). The affected viscera, especially the kidney, are enlarged and contain multifocal to diffuse, miliary to focally extensive, white, nodular lesions representative of granulomas. The granulomas are generally discrete and well-encapsulated in species that are more resistant to infection and clinical disease, such as Atlantic salmon, whereas the inflammatory foci are less discrete in more susceptible species, such as coho salmon, *O. kisutch,* and chinook salmon, *O. tshawytscha.* The viscera may be invested by an opaque, white to gray pseudomembrane of variable thickness. Hemorrhage of the liver, intestine, visceral fat, and skeletal muscle is not an uncommon finding. Caseous necrosis of the granulomas is common and may result in the formation of cystic lesions (cavitations) of the skeletal muscle. Fish with localized infections of the meninx without an associated systemic involvement may swim in a spiral or whirling pattern (Speare, Ostland, and Ferguson 1993). Acute disease may occur following stress and is often characterized by a diffuse inflammation without the discrete, nodular lesions that are typical of chronic disease.

Diagnosis

A preliminary diagnosis of BKD depends on an evaluation of the history and the presence of characteristic gross lesions, which is confirmed by examination of tissue imprints and/or histological sections stained with Gram or periodic acid-Shiff (PAS) reagents. Tissue imprints or histological sections from affected fish generally reveal numerous Gram-positive and PAS-positive, intracellular bacilli that are 0.5×3 1.0 to 2.0 μm and often occur in aggregates (Evelyn 1993; Evelyn et al. 1998). Culture and isolation of the bacterium is not recommended for diagnosis of BKD, since the organism is fastidious, exhibits slow growth, and requires enriched media, such as KDM-2 or charcoal agar, that are not commercially available (Evelyn 1993;

Evelyn et al. 1998). The requisite prolonged incubation of three to six weeks at 15 to 18°C can also result in contamination of the cultures with opportunistic bacterial and fungal organisms.

Since the species is immunologically homogeneous, a definitive diagnosis can also be obtained with immunological methods such as FAT (Bullock and Stuckey 1975b; Bullock, Griffin, and Stuckey 1980) or ELISA (Pascho and Mulcahy 1987). An ELISA kit (Diagxotics, Wilton, Connecticut) and antisera to *R. salmoninarum* (Kirkegaard and Perry, Gaithersburg, Maryland) are commercially available. However, the PCR technique for detection of bacterial nucleic acid may become the routine method in diagnostic laboratories (Brown et al. 1995).

Treatment and Prevention

Treatment of fish with antibiotics is generally not practical or efficacious. Parenteral administration of erythromycin in the feed may result in clinical improvement but may not eliminate infection (Evelyn 1993). Fish with clinical disease may also refuse medicated feed secondary to anorexia and/or due to the reduced palatability of the feed. Furthermore, erythromycin is expensive and not approved for use in food fish. Due to these constraints, oxytetracycline has been used as an alternative antibiotic in Canada, although the efficacy of this antibiotic for BKD has not been established.

Control and prevention of BKD is therefore preferred, but difficult, due to the chronic, insidious nature of the disease. Avoidance of BKD is achieved by elimination of infected fish and the use of specific pathogen-free (SPF) broodstock (Elliott, Pascho, and Bullock 1989; Evelyn 1993). Ideally, female broodstock are administered intramuscular (IM) or intraperitoneal (IP) injections of 10 to 20 mg/kg erythromycin at 9 to 57 days prior to spawning to reduce or eliminate infection and prevent vertical transmission (Evelyn, Ketcheson, and Prosperi-Porta 1986; Lee and Evelyn 1994). Best results have been obtained with injections at 12 to 20 (Armstrong et al. 1989) or 15 to 40 (Moffitt 1991) days prior to spawning. This strategy not only reduces or eliminates infection in the adult but also prevents infection of the ova that persists to the alevin stage of development. Regardless, ovarian fluids from female broodstock should also be tested using immunological techniques to confirm the pathogen-free status of broodstock. Specifically, the FAT is less sensitive than ELISA, although a membrane filtration technique has been developed to increase the sensitivity of the FAT (Elliott and Barila 1987). However, ELISA is also more practical for screening a large number of samples. As previously mentioned, the PCR technique may become the preferred diagnostic procedure for detection of clinical disease and carrier fish (Brown et al. 1994; Pascho, Chase, and McKibben 1998). Regardless, surface disinfection of eggs using 100 ppm iodophore for 15 minutes should be a routine procedure in salmonid operations. Progeny should not be exposed to potential carriers and should not be raised in water that may be contaminated with the pathogen. Additional measures include segregation of fish by brood and year class and the elimination of raw fish products in the feed. Recently, a live heterologous vaccine for BKD (Renogen, Aqua Health Ltd., Charlottetown, Prince Edward Island, Canada) that reduces infection, inflammation, and shedding of the bacterium has been licensed for use in Canada.

Piscirickettsiosis

Piscirickettsiosis (or salmonid rickettsial septicemia) is caused by the obligate, intracellular rickettsial organism *Piscirickettsia salmonis* (Cvitanich, Garate, and Smith 1991; Fryer et al. 1992). The disease often has a serious economic impact in pen-raised salmonid operations and has been reported in Chile, Norway, Scotland, Ireland, and Canada, as previously summarized (Evelyn et al. 1998).

Epizootiology

Piscirickettsiosis is generally a primary disease in marine salmonids but may also occur with other conditions such as BKD and *Nucleospora salmonis* infections (Evelyn et al. 1998). Species susceptible to the disease include rainbow trout, pink salmon *(O. gorbuscha),* coho salmon, chinook salmon, and Atlantic salmon. The reservoir of infection has not been identified, although ectoparasites such as sea lice, *Caligus* sp., and isopods, *Ceratothoa gaudichaudii,* may be vectors of transmission (Evelyn et al. 1998). Detection of the pathogen within crustaceans and mollusks suggests that these invertebrates may also be reservoirs of infection (Cvitanich, Garate, and Smith 1991). Regardless, transmission of the pathogen is most probabaly horizontal from fish to fish and has been detected in the feces of infected salmonids (Cvitanich, Garate, and Smith 1991; Evelyn et al. 1998). The pathogen can survive in seawater but not freshwater, although infection has been reported in rainbow trout and coho salmon maintained in freshwater and can be transmitted by cohabitation in freshwater (Lannan and Fryer 1994; Almendras et al. 1997). Anecdotal observations further suggest that vertical transmission of the pathogen is a possibility that may explain the occurrence of infection in freshwater (Evelyn et al. 1998). Conditions that have been associated with clinical disease include inclement weather, rapid changes in temperature, and various husbandry practices, such as grading and net changes (Evelyn et al. 1998). Fish with other disease conditions such as BKD and *N. salmonis* infections may have an increased susceptibility to piscirickettsiosis (Evelyn et al. 1998).

Pathogenesis and Clinical Disease

The portals of entry of the pathogen are the gills and alimentary tract (Almendras et al. 1997), although the mechanisms and factors involved in the pathogenesis of disease have not been determined. Fish with clinical disease generally present with symptoms of a septicemia, including dark coloration, lethargy, anorexia, pallor of the gills due to anemia, distension of the coelomic cavity due to ascites, and congregation at the surface (Branson and Nieto Diaz-Munoz 1991; Cvitanich, Garate, and Smith 1991; Brocklebank et al. 1993; Evelyn et al. 1998). Atlantic salmon with the disease commonly swim on their sides or flash, a symptom that may be related to infection of the central nervous system. Externally, there may be petechial hemorrhage with small, white foci or raised nodules of the integument. Cutaneous ulceration is a common manifestation in rainbow trout and coho salmon but not in Atlantic salmon.

Internal lesions include pallor with petechial hemorrhage of the internal viscera, enlargement of the kidney and spleen that may have a gray discoloration, pseudomembranous epicarditis, and petechial hemorrhage of the skeletal muscle. However, the characteristic lesions in fish with piscirickettsiosis are the multifocal, coalescent, whitish yellow, hepatic granulomas. Rupture of the granulomas may result in single or multiple cavitations of the liver. Histologically, there is a necrotizing, granulomatous inflammation of the liver, spleen, and kidney, with an associated vascular thrombosis (Branson and Nieto Diaz-Munoz 1991; Cvitanich, Garate, and Smith 1991; Brocklebank et al. 1993; Evelyn et al. 1998). Aggregates of the pathogen may occur within hepatocytes, macrophages, and melanomacrophages, including macrophages contained within the vascular thrombi, and may appear as spherical, basophilic to amphophilic, intracytoplasmic inclusions with routine hematoxylin and eosin stains.

Diagnosis

A preliminary diagnosis of piscirickettsiosis is based on the presence of the characteristic gross and microscopic lesions. Identification of the pathogen by cytological examination of stained tissue imprints and/or histological examination of tissue sections is often sufficient for a definitive diagnosis. Tissue imprints stained with Gram, Giemsa, or methylene-blue reagents will reveal pairs or aggregates of intracytoplasmic, basophilic, coccoid organisms that are approximately 0.5 to 1.5 µm (Evelyn et al. 1998). Tissue sections stained with methylene-blue generally provide better results than sections stained with hematoxylin and eosin.

Isolation of the organism in cell culture using CHSE-214 cells is required for a definitive diagnosis (Fryer et al. 1990; Cvitanich, Garate, and Smith 1991), although this may not be necessary, as previously mentioned. An IFAT can also be used for detection and confirmation of the organism in tissue samples or cell cultures (Lannan, Ewing, and Fryer 1991). Detection of pathogen-specific nucleic acid using the PCR technique is highly specific and sensitive (Mauel, Giovannoni, and Fryer 1996) and has become the routine diagnostic method in several laboratories. However, rapid-field diagnostic procedures are often preferred and can be performed with a commercially available ELISA kit (Microtek-Bayer, Sidney, British Columbia, Canada).

Treatment and Prevention

Treatment of fish with antibiotics such as oxytetracycline, flumequine, and oxolonic acid in the feed has resulted in limited success, which may be due to the intracellular nature of the pathogen and the development of antibiotic resistance (Evelyn et al. 1998). Fluoroquinolone injections have also been used for treatment, although the extra-label use of these antibiotics in food animals is not permitted in the United States, as previously mentioned. Therefore, control and prevention of the disease is essential, although difficult, due to the insidious, intracellular nature of the pathogen and the incomplete information concerning the epizootiology of disease. Regardless, measures to control and prevent the disease are similar to strategies used for prevention of BKD and include fallowing farms prior to reuse, segregation of year classes at a single site, control of ectoparasites and other primary infectious diseases such as bacterial kidney disease, culture at reduced densities, and surface disinfection of eggs (Evelyn et al. 1998). Likewise, the commercially available ELISA kit previously mentioned has been used for the identification and rejection of infected broodstock. Unfortunately, a vaccine for protection against piscirickettsiosis is not currently available.

VIRAL DISEASES

Infectious Hematopoietic Necrosis

Epizootiology

Infectious hematopoietic necrosis (IHN) is caused by the rhabdovirus *infectious hematopoietic necrosis virus* (IHNV), and may be the most important viral disease of salmonids in freshwater. However, it also affects wild and cultured fish in the marine environment. The virus is endemic to the Pacific Northwest region of North America but has achieved a more cosmopolitan distribution, probably due to the international shipment of fish and ova. The virus has an affinity for all wild and captive salmonid species of economic importance, although brown trout, *Salmo trutta,* brook trout, *Salvelinus fontinalis,* cutthroat trout, *O. clarki,* and coho salmon are less susceptible to the disease (Wolf 1988). Various viral strains have been identified by electrophoretic separation and cross-reaction with monoclonal antibodies that can be categorized by geographic location and specificity for various host species (Leong et al. 1981; Winton et al. 1988; Ristow and Arnzen 1989; LaPatra 1998). The latter characteristic explains the variable pathogenicity among viral strains, relative to the particular species of salmonid.

The primary reservoir of infection is carrier salmonids (Wolf 1988), although nonsalmonids (Traxler, Kent, and Poppe 1998) and aquatic invertebrates (Mulcahy, Klaybor, and Batts 1990) may also function as viral reservoirs. Viral transmission via a direct or an indirect route can occur in freshwater and seawater (Traxler, Roome, and Kent 1993). In this context, fish with clinical disease have high titers of virus in the feces, urine, and mucus that will facilitate viral transmission during epizootics. Likewise, high titers of virus may occur in reproductive fluids, which subsequently results in contamination of the water during spawning activity. The virus is stable in freshwater and can remain infectious for several months but is less stable in seawater. Vertical transmission has not been conclusively demonstrated, although anecdotal evidence suggests that this may also be a mode of transmission (Meyers et al. 1990; Traxler et al. 1996).

Pathogenesis

The portals of viral entry are the gills and the alimentary tract, via ingestion (Drolet, Rohovec, and Leong 1994). Severe disease generally occurs in juvenile and subadult fish less than two years of age, with the highest mortality in fish less than six months of age (Wolf 1988). The pathogenicity of IHNV is influenced by various factors, including temperature (LaPatra 1998). Severe disease with the highest mortality occurs at 10°C, with an incubation period of 5 to 14 days at this temperature. Clinical disease generally does not occur at temperatures greater than 15°C, although mortality due to IHNV has been reported in rainbow trout fry held at 3 to 18°C (Hetrick, Fryer, and Knittel 1979). In contrast, a significantly reduced mortality occurred in rainbow trout fry and juvenile sockeye salmon, *O. nerka,* that were held at temperatures greater than 15.5°C prior to infection or moved to a higher temperature within 24 hours after infection (Amend 1970, 1976). The results of these studies may have varied due to differences in experimental design, size of the fish, or strain of IHNV (LaPatra 1998). Subadult and adult fish may also be affected, although disease in these fish is generally chronic with a low mortality that occurs over a period of several months.

Clinical Disease

It is important to recognize that fish with acute IHN present with symptoms and lesions that are similar to bacterial septicemia (Traxler, Kent, and Poppe 1998). The clinical symptoms and lesions have previously been summarized (Yasutake 1970, 1975, 1978; Wolf 1988). Fish with clinical disease may present with dark coloration, lethargy and/or erratic swimming and flashing, exophthalmos, hemorrhage at the base of the fins, pallor of the gills due to anemia, and coelomic distension due to ascites. Initial mortality within the population may occur in the larger, precocious fish. Internally, there is generally petechial and ecchymotic hemorrhage of the viscera, with pallor of the liver. The intestine is generally devoid of ingesta but may contain a pale yellowish white to white, catarrhal to mucoid material that may trail from the fish. This latter feature is not pathognomonic for IHNV but has also been described in fish with viral hemorrhagic septicemia and infectious pancreatic necrosis. Cytological examination of spleen and kidney imprints and, to a lesser extent, blood smears often reveal necrotic hematopoietic cells, which have been referred to as necrobiotic bodies. Further evaluation of the blood generally reveals anemia with a leukopenia and a reduced osmolarity (Amend and Smith 1974). Emaciation may be the only symptom in fish with chronic disease, especially subadult and adult fish. Fish that survive the disease or fish with chronic infection may exhibit musculoskeletal and spinal deformities. A neurotropic form of IHNV that occurred during the subsidence of a typical IHNV epizootic and was associated with high concentrations of virus in the brain has also been reported in rainbow trout with a mean weight of 13 g (LaPatra et al. 1995).

Histologically, the characteristic finding is necrosis of the hematopoietic tissue of the spleen and kidney that may be mild to severe and focal to diffuse, depending on the severity and chronicity of the disease. There may also be necrosis of the pancreas, liver, and intestinal submucosa, whereas necrosis of the branchial tissue may occur in subadult and adult fish. Necrosis of the eosinophilic granular cell component of the intestinal submucosa has been considered a highly diagnostic feature of disease in juvenile fish that are three to four months of age, although this is not a specific or consistent finding. Intranuclear and intracytoplasmic inclusions of the exocrine and endocrine pancreatic cells have been described for IHN but are also not a consistent or diagnostic finding.

Diagnosis

A preliminary diagnosis of IHN is dependent on the presence of the characteristic gross and microscopic lesions, whereas a definitive diagnosis requires viral isolation. The latter is achieved by inoculation of tissue homogenates on CHSE-214 or EPC cells prior to incubation at 15°C (Wolf 1988). Tissues used for viral isolation should include the kidney, spleen, and brain, although high viral titers are also found in the mucus and feces during clinical disease. Tissues may be frozen and stored at 4°C prior to isolation if necessary. Cytopathic effect (CPE) generally occurs within seven days postinoculation and results in a typical margination of the chromatin and rounding of the cells that retract from the center of the plaque. Confirmation can be achieved with immunological techniques using polyclonal antibodies such as virus neutralization assays, FAT, or ELISA, since the various viral strains are serologically homogeneous (Winton 1991). The evaluation of anti-IHNV serum antibody titers has been employed to determine previous exposure or infection (LaPatra 1996). The diagnosis of IHN by analysis of tissue samples and cell culture isolates using the PCR method (Arakawa et al.

1990) is currently not a routine diagnostic procedure and is less sensitive than the standard diagnostic techniques.

Prevention

Regulation of temperature may be used for control of IHN in juvenile fish and incubation of eggs, although this may not be practical in various situations (Amend 1970). Likewise, certain strains of IHNV that affect salmonids at warmer temperatures may not be controlled with increased temperatures. Prevention of IHN is by selection and use of pathogen-free broodstock, surface disinfection of ova using iodophores, elimination of infected smolts prior to transfer to seawater, disinfection of facilities, and use of pathogen-free water (Winton 1991; Traxler, Kent, and Poppe 1998). Tissues used for detection of asymptomatic viral infection in adults, especially potential broodstock, should include the kidney, spleen, and postspawning reproductive fluids. Wild fish or water exposed to wild fish that may be carriers of the virus should be avoided. Standard killed vaccines for protection against IHNV are commercially available, although highly efficacious results have been obtained in experimental trials with DNA vaccines that should be available for commercial use in the near future (Anderson et al. 1996).

Viral Hemorrhagic Septicemia

Viral hemorrhagic septicemia (VHS, or Egtved virus) is caused by the rhabdovirus *viral hemorrhagic septicemia virus* (VHSV). VHS is primarily a disease of freshwater rainbow trout and brown trout, although other salmonid and nonsalmonid species in freshwater and marine environments are susceptible to natural or experimental disease. Although the disease has historically been restricted to Europe, the virus has been isolated from steelhead trout *(O. mykiss),* coho salmon, chinook salmon, Pacific cod *(Gadus macrocephalus),* and Pacific herring *(Clupea pallasi),* from the Pacific Northwest region of North America. Three major viral serotypes have been demonstrated for VHSV (Wolf 1988).

Epizootiology and Pathogenesis

The reservoir of infection is probably salmonid and nonsalmonid species (Wolf 1988). Survivors of the disease may become carriers of the virus, although the carrier state may be difficult to detect prior to sexual maturity. The virus is transmitted horizontally and can be shed in the urine and reproductive fluids but not in the feces. However, shedding of the virus occurs only at low temperatures. Vertical transmission of the virus has not been demonstrated. The virus is stable in water for one week at 14°C and survives drying for up to one week at 4°C. All ages of fish are susceptible to the disease, although the most severe disease occurs in juvenile fish (Wolf 1988). Clinical disease can occur at temperatures of 3 to 12°C, with severe mortality at 8 to 10°C, but generally does not occur at temperatures greater than 15°C. The incubation period for the disease at 3 to 12°C is one to two weeks but is prolonged at lower temperatures.

Clinical Disease

Clinical symptoms and lesions are dependent on the severity of the disease but are generally similar to IHN (Yasutake and Rasmussen 1968; Yasutake 1970, 1975). However, clinical symptoms in fish greater than six months of age may be mild or absent. Severe disease, which has also been referred to as acute disease, generally results in high mortality. Affected fish may present with dark coloration; lethargy; leukopenia; and hemorrhage of the gills, base of the fins, and internal viscera. In contrast, chronic disease is characterized by moderate mortality. Fish with chronic disease may present with dark coloration, exophthalmos, coelomic distension due to ascites, pallor of the gills and internal viscera due to anemia, and hemorrhage of the internal viscera. In addition, a nervous phase of the disease has been described that results in darting motions and spiral swimming in affected fish, although mortality is low or absent. Histological lesions are similar to IHN and include necrosis of the renal and splenic hematopoietic tissue and hepatic degeneration. Fish with chronic disease generally have a prominent accumulation of hemosiderin within melanomacrophages.

Diagnosis

A definitive diagnosis of VHS requires viral isolation, which is achieved by inoculation of tissue homogenates on BF-2, CHSE-214, or EPC cell lines prior to incubation at 10 to 15°C (Wolf 1988). Tissues used for viral isolation include the kidney and spleen, although the pyloric cecae, brain, and postspawning reproductive tissue and fluids should also be used for detection of asymptomatic carriers. Initially, cytopathic effect appears as foci of rounded cells that progressively enlarge prior to lysis (Wolf 1988). Confirmation of VHS can be achieved with various immunological techniques, although polyvalent antisera should be used for detection, since there is weak cross-reaction among the three major serotypes (Wolf 1988).

Treatment and Prevention

Strategies employed for the prevention of VHS are similar to those for IHN and include the use of certified pathogen-free stocks, surface disinfection of ova using iodophores, and avoidance of wild fish that may carry the virus and of water supplies that may be contaminated with the virus. A vaccine for protection against VHSV is not currently available, although highly efficacious results have been obtained in experimental trials with DNA vaccines.

Infectious Pancreatic Necrosis

Infectious pancreatic necrosis (IPN) is a serious disease of freshwater and marine salmonids and may be the most important infectious disease of salmonids in Norway. The causative agent—*infectious pancreatic necrosis virus* (IPNV)—is a member of the birnaviruses, which are an extremely heterogeneous family of viruses that can be isolated from various freshwater and marine fish and invertebrates from various geographic locations (Wolf 1988). Due to this heterogeneity, classification of the taxon is complicated, although several major and minor serotypes have been associated with disease in salmonids (Caswell-Reno et al. 1989; Hill and Way 1995).

Epizootiology and Pathogenesis

IPN has been reported from North America, Chile, Europe, and Southeast Asia (Wolf 1988; Traxler, Kent, and Poppe 1998). The virus can cause disease in various species of salmonids, although rainbow trout, brook trout, brown trout, and Atlantic salmon have an increased susceptibility to IPN. The reservoir of infection is primarily salmonids that survive infection, although various IPN-like birnaviruses have been associated with disease in nonsalmonids or have been isolated from various aquatic invertebrates (Wolf 1988). Transmission is horizontal, either directly from contact with infected fish or indirectly through the water. Carrier fish can shed the virus indefinitely in the feces and urine, although viral shedding may be intermittent or may not occur in carrier fish (Billi and Wolf 1969). Vertical transmission has also been demonstrated and may be a cause of mortality in embryos (Wolf, Quimby, and Bradford 1963; Bullock et al. 1976; Fijan and Giorgetti 1978; Wolf 1988).

All ages of fish are susceptible to infection, although clinical disease generally occurs in fish less than six months of age (Wolf 1988; Traxler, Kent, and Poppe 1998). Disease in netpen facilities can occur within a year following transfer of smolts to seawater but is most common several weeks to a few months following transfer (Smail et al. 1992; Jarp et al. 1995). Mortality is most severe at 10 to 14°C, with reduced mortality at temperatures above and below this range. Incubation period is three to ten days at these temperatures, with a peak mortality at 12 to 18 days postinfection, although this is dependent on the age of the fish. Stress in carrier fish may result in clinical disease.

Clinical Disease

A rapid, severe mortality in juvenile fish is highly suggestive of IPN (Wolf 1988; Traxler, Kent, and Poppe 1998). Affected fish often present with lethargy and erratic, spiral, or whirling swimming patterns; anorexia; dark coloration; coelomic distension due to ascites; exophthalmos; opaque white feces that trail from the fish; and pallor of the gills and ventral integument. Internally, fry often have pallor of the internal viscera, whereas petechial hemorrhage of the viscera and fat is more common in older fish. The anterior gastrointestinal tract generally contains a translucent to opaque white, catarrhal exudate, whereas a white, mucoid material is commonly present in the posterior intestine; these findings are highly suggestive of IPN (McKnight and Roberts 1976). Survivor fish may be stunted secondary to pancreatic fibrosis.

The characteristic microscopic lesion of IPN is necrosis of the pancreatic exocrine cells that may contain basophilic, intracytoplasmic inclusions (Yasutake, Parisot, and Klontz 1965; Yasutake 1970). Degeneration and necrosis with karyolysis of individual epithelial cells of the pyloric ceca, which have been referred to as McKnight cells, are also considered diagnostic features of the disease (McKnight and Roberts 1976). There may also be degeneration and necrosis of the renal tubular epithelium, hematopoietic tissue, intestinal mucosa, and hepatocytes.

Diagnosis

The primary differential for IPN in marine salmonids is salmon pancreas disease. A definitive diagnosis of IPN requires isolation and confirmation of the virus with neutralization assays or FAT (Wolf 1988; Traxler, Kent, and Poppe 1998). Tissues recommended for viral isolation in clinical disease include the kidney and pyloric ceca, although reproductive tissues should also be included for detection of asymptomatic carriers; homogenates are inoculated on RTG-2 or CHSE-214 cells at 10 to 20°C (Wolf 1988). Serial dilutions are recommended for quantification of viral titers to prevent toxicity and differentiate carrier fish from fish with clinical disease (Wolf 1988). Tissues can be frozen and stored at 4°C prior to isolation if necessary. The addition of 2 percent bovine serum albumin to fluid samples will stabilize the virus prior to freezing (Yu, MacDonald, and Moore 1982). The virus generally results in rapid lysis of cell cultures (Wolf 1988). Immunohistochemical techniques have also been used to differentiate IPN from salmon pancreas disease.

Treatment and Prevention

Prevention of IPN is by avoidance, although this may be difficult due to the indefinite shedding of the virus by carrier fish. Strategies for prevention of IPN are similar to the stamping-out program used for eradication of infectious salmon anemia and include attention to stress reduction and improvement of environmental quality; use of certified pathogen-free fish; segregation of stocks; avoidance of carrier fish, including wild fish and infected broodstock; surface disinfection of ova; use of pathogen-free water, especially for juvenile fish less than six months of age; and culture of juvenile fish at temperatures less than 10°C (Wolf 1988; Jarp et al. 1995). The virus is highly stable and can survive in water for five days at 15°C and for ten days at 4°C or drying for one month at 10°C (Toranzo et al. 1983). The virus can be inactivated with iodophores, chlorine, formalin, and highly alkaline solutions (pH 12.5) but is generally resistant to ultraviolet irradiation (Wolf 1988). Development of an effective vaccine has been difficult due to the heterogeneity of the various major and minor serotypes, although effective protection has been achieved with a multivalent recombinant vaccine that has been licensed in Norway (Traxler, Kent, and Poppe 1998).

Salmon Pancreas Disease

Salmon pancreas disease (SPD) has been reported in Atlantic salmon raised in netpens from Scotland, Ireland, Norway, and the Pacific Northwest region of North America (Ferguson et al. 1986; McVicar 1987; McLoughlin, Rowley, and Doherty 1998; Traxler, Kent, and Poppe 1998). The causative agent of the disease has not been conclusively determined but is generally considered to be a togavirus (Nelson et al. 1995; McLoughlin et al. 1996). A vitamin E/selenium deficiency has been proposed as a cause of the disease (Ferguson et al. 1986), although the deficiency that occurs in affected fish is probably a consequence of the disease (Bell, McVicar, and Cowey 1987).

The disease occurs in marine salmonids from 6 to 12 weeks to several years of age, following transfer to seawater (McVicar 1987). Mortality is generally low, although survivor fish may have reduced growth and an increased susceptibility to other diseases. Fish with SPD generally present with anorexia, lethargy, and emaciation with internal hemorrhage or atrophy of the pancreatic tissue and associated fat. Histologically, the characteristic lesion of SPD is a necrotizing inflammation of the pancreas, although a ventricular coagulative myocardial necrosis has also been associated with the disease (Ferguson et al. 1986). Resolution of the inflammation often results in pancreatic fibrosis.

A definitive diagnosis is dependent on the presence of characteristic microscopic lesions and/or viral isolation. Viral isolation is achieved by cocultivation of kidney tissue, using CHSE-214 cells at 15°C for 28 days, although several passages may be required prior to detection of CPE (Nelson et al. 1995). Reduction of stress during acute disease may improve recovery of affected fish. A vaccine is not currently available, although development of natural immunity in survivor fish suggests that a vaccine may be commercially viable.

Infectious Salmon Anemia

Infectious salmon anemia (ISA) may be considered an emergent viral disease that has a serious economic impact on Atlantic salmon operations in Norway and, more recently, Atlantic Canada, but has also been reported from Scotland (Traxler, Kent, and Poppe 1998). The disease in Canada has also been referred to as hemorrhagic kidney disease (HKD). A viral agent that is consistent with an orthomyxovirus has recently been determined as the causative agent of ISA in Norway (Dannevig, Falk, and Namork 1995) and HKD in Canada (Mullins, Groman, and Waldowska 1998), although it has not been determined whether the Norwegian and Canadian isolates are similar or separate viral strains.

The disease primarily affects Atlantic salmon raised in seawater, but the disease has also occurred in freshwater facilities that have been supplemented with seawater (Thorud and Djupvik 1988). The virus can also be experimentally transmitted in freshwater. The reservoir of infection has not been definitively determined but is most likely infected salmonids that shed the virus in the cutaneous mucus, feces, and urine (Nylund and Jakobsen 1995).

Nonsalmonids may also be carriers of the virus, whereas sea lice may also function as a vector for transmission of the virus (Nylund, Wallace, and Hovland 1993; Nylund et al. 1994). Clinical disease is associated with a rapid increase in temperature and generally occurs in the spring but may also occur in autumn (Traxler, Kent, and Poppe 1998). Disease may last for several months, with a cumulative mortality that ranges from 15 to 100 percent, although the time course may be shorter at temperatures greater than 10 to 12°C. Severe mortalities of 5 percent per day have been reported with HKD.

The portal of viral entry is probably the gills (Totland, Hjeltnes, and Flood 1996). Affected fish with ISA may have anorexia, lethargy or listlessness, exophthalmos, cutaneous edema and hemorrhage, coelomic distension, and pallor of the gills, often due to severe anemia (Evensen, Thorud, and Olsen 1991). Internal lesions include pallor of the heart, congestion of the internal viscera, and hemorrhage of the visceral fat and capsular surfaces of the viscera. The liver may be severely congested, with a dark red to black coloration. Fish with HKD have coelomic distension and enlargement of the kidney (Byrne et al. 1998). The characteristic microscopic lesion in fish with ISA is a multifocal, hemorrhagic, necrotizing hepatitis that may be focally diffuse to coalescent, although the perivascular parenchyma may be intact or less severely affected (Evensen, Thorud, and Olsen 1991). In contrast, microscopic features of HKD include severe splenic congestion; severe, diffuse renal hemorrhage; necrosis of the renal tubules; and hepatic necrosis (Byrne et al. 1998).

A preliminary diagnosis is based on the presence of characteristic clinical signs and lesions, especially severe anemia that is confirmed by isolation of the virus. Isolation of the ISA virus is achieved using SHK-1 or CHSE-214 cells and is confirmed with serological techniques such as IFAT (Falk and Dannevig 1995). However, detection of viral nucleic acid using PCR may be more sensitive and specific and less labor intensive (Mjaaland et al. 1997).

An aggressive eradication (or stamping-out) program was initially implemented in Norway and has been successful in reducing the prevalence of disease. The strategies for eradication include mandatory disease control programs on smolt farms, disinfection of seawater used in freshwater facilities, disinfection of water and facilities used for processing, isolation of infected sites, and fallowing of farms after processing. A killed, autogenous vaccine (Aqua Health Ltd, Charlottetown, Prince Edward Island, Canada) has recently been licensed for commercial use in Canada.

PARASITIC DISEASES

Diseases due to various parasites such as the helminth parasites (nematodes, trematodes, monogenean flatworms, cestodes, and acanthocephalans), protozoan parasites (amoebae, ciliates, flagellates), myxosporeans, microsporidians, crustaceans (copepods and isopods), and mollusks (mussels) have been reported in fish, including salmonids. However, the parasitic diseases generally do not have a significant economic impact and/or can be easily managed and controlled in commercial salmonid operations. For example, the common cutaneous and branchial parasites of freshwater salmonids, such as the ciliated and flagellated protozoans and the monogenean flatworms, can result in morbidity and mortality in any individual facility but are often secondary to poor management and husbandry conditions, such as poor water quality. A brief review of the various parasitic diseases of salmonids not included in this discussion can be found in Noga (1995) or Kent and Poppe (1998).

Branchial Amoebiasis

Amoebal infections of the gills may result in serious disease conditions in freshwater and marine salmonids. The causative agent of branchial amoebiasis in marine salmonids is *Paramoeba pemaquidensis,* whereas the cochliopodid amoebae are associated with disease conditions in freshwater salmonids. Disease in freshwater salmonids has often been referred to as *nodular gill disease,* due to the branchial nodules that may be grossly apparent in affected fish. Severe disease in marine salmonids due to *P. pemaquidensis* has been reported in rainbow trout and Atlantic salmon from Tasmania (Roubal, Lester, and Foster 1989; Munday et al. 1990, 1993) and in coho salmon from the western United States (Kent, Sawyer, and Hedrick 1988). Amoebal infections in freshwater rainbow trout, similar to recent cases of branchial amoebiasis in rainbow trout from the western United States (LaPatra and Groff, unpublished observations), have been reported from North America and Europe (Daoust and Ferguson 1985).

The causative amoebae are opportunistic pathogens that are ubiquitous in the environment. Risk factors associated with disease have not been adequately determined, although increased density, reduced water flow, fouling of netpens, preexistent branchial lesions and disease conditions, and smoltification may increase the susceptibility of marine salmonids to the disease (Kent, Sawyer, and Hedrick 1988; Kent 1998b). Infection and subsequent disease due to *P. pemaquidensis* are generally most severe at temperatures greater than 15°C in late summer and fall, following transfer of fish from freshwater to seawater (Kent 1998b). Fish with severe disease are generally lethargic and congregate at the surface. Grossly, these fish may have flared opercula and multifocal to diffuse pallor of the gills that often contain an excessive mucous exudate (Kent 1998b). As previously mentioned, branchial nodules may occur in freshwater rainbow trout with branchial amoebiasis (Daoust and Ferguson 1985), although this is not a consistent feature of the disease. Histological examination of affected gills reveals a variable epithelial hyperplasia of the secondary lamellae, with fusion of adjacent lamellae in severe infections, which can be incomplete, resulting in the formation of enclosed interlamellar spaces (Kent, Sawyer, and Hedrick 1988; Munday et al. 1990, 1993). The amoebae are not invasive and generally occur on the surface of the secondary lamellae but may occasionally be found within the interlamellar spaces.

Diagnosis of branchial amoebiasis is achieved by examination of branchial wet-mount preparations and histological sections. Direct examination of branchial scrapings or tissue preparations generally reveals the floating and transitional forms of *P. pemaquidensis* that are typically 20 to 30 μm with several digitiform pseudopodia (Kent 1998b). Transition to the locomotive form of the amoeba requires attachment to the glass slide and generally occurs within approximately one hour following procurement of the sample. The locomotive form is 20 × 25 μm and contains a characteristic cytoplasmic parasome adjacent to the nucleus.

Prevention of the disease is by avoidance and the reduction or elimination of potential risk factors that have been associated with the disease, as previously mentioned. Treatment of marine salmonids with the common chemical parasiticides is less effective than freshwater or brackish water baths (Munday et al. 1993), although the latter is not practical in netpen operations. In contrast, bath treatments of affected freshwater salmonids with the common chemical parasiticides, including formalin, are generally effective.

Diseases Due to Myxosporean Parasites

The myxozoans have historically been included within the phylum Protozoa but are a distinct and separate phylum (phylum Myxozoa) that includes various obligate parasites of lower vertebrates, especially fishes (class Myxosporea). The myxosporean life cycle is a complicated and indirect cycle with several sexual and asexual stages of development, although the complete life cycle has been determined for only a few myxosporeans (Noga 1995). Briefly, development within fish proceeds through several vegetative stages or trophozoites that eventually form multinucleated spores that are released into the environment. Status of the spores following release into the environment is not known for most myxosporean genera, although an aquatic oligochaete is required for completion of the life cycle and development of the infective stage in the *Myxobolus* spp. (El-Matbouli and Hoffmann 1990; Kent, Whitaker, and Margolis 1993; Ruidisch, El-Matbouli, and Hoffman 1991), including the causative agent of whirling disease—*M. cerebralis* (Wolf, Markiw, and Hiltunen 1986).

The myxosporeans can generally be classified as histozoic and coelozoic species. The histozoic species develop within tissues and are generally manifested as discrete, pseudocystic lesions that can result in significant disease, depending on the severity of the infection and the target tissue. Spores of the histozoic species are released into the environment following necrosis or rupture of the spore-containing plasmodium or following the death of the host. In contrast, coelozoic species develop within organ cavities, such as the coelomic cavity, swim bladder, gall bladder, and urinary bladder. Spores of the coelozoic species are released into the environment via the urine or feces. Classification of the myxosporeans is based on not only spore morphology and size but also the number and position of the polar capsules. Parasiticides for treatment of myxosporean infections are not commercially available, although treatment of various myxosporean infections with fumagillin and its derivatives has been successful in experimental trials, as previously summarized (Kent 1998b).

The most common myxosporean diseases in freshwater salmonids include whirling disease due to *M. cerebralis,* proliferative kidney disease (PKD) apparently due to a *Sphaerospora* sp. (Kent et al. 1998), and ceratomyxosis due to *Ceratomyxa shasta.* These diseases generally do not occur in commercial freshwater salmonid operations or do not have a significant economic impact in these operations. However, significant morbidity and mortality can occur secondary to these diseases in wild populations or in cultured populations raised in facilities that use surface water containing the parasite. Current reviews of whirling disease (Hedrick et al. 1998), proliferative kidney disease, and ceratomyxosis (Noga 1995) should be consulted for further information.

Diseases due to myxosporean parasites in marine salmonids have previously been summarized (Kent 1998b). The most important disease in marine salmonids is due to *Kudoa thyrsites,* which is a cosmopolitan parasite that can also infect various nonsalmonid marine fish (Moran and Kent 1999). Infection generally occurs in postsmolts after five to six months in seawater and is also more prevalent in grilses or reconditioned fish than in sexually immature fish (Moran and Kent 1999; St-Hilaire et al. 1998). Development of spores within the skeletal muscle is dependent on temperature but generally occurs five to six months following infection (Moran and Kent 1999). Progressive enlargement and subsequent rupture of the spore-containing pseudocysts result in severe inflammation. Severe infections and the development of soft flesh has a significant economic impact, due to downgrading or rejection of the carcass. Soft flesh occurs following storage of the fish on ice for three to six days or after the fillets have been smoked (St-Hilaire et al. 1997). Mortality in smolts due to *K. thyrsites* infection has also been reported (Harrell and Scott 1985).

A presumptive diagnosis is based on the occurrence of pseudocysts within the skeletal muscle and soft flesh, a diagnosis that is confirmed by wet-mount examination of muscle tissue and fluid. Wet-mount examination reveals stellate-shaped, 13 to 15 μm spores that contain four unequal polar capsules which converge at one end of the spore (Kent 1998b). Tissue imprints or histological sections stained with Giemsa can also be used for detection of the spores. A PCR assay has been developed (Hervio et al. 1997), although this is not a routine diagnostic procedure. Management of the disease is generally restricted to removal of grilses from the population prior to harvest, since these fish are more susceptible to infection (Kent 1998b). There are currently no commercially available parasiticides for treatment of *K. thyrsites* infections, nor is there a vaccine for prevention of the disease.

Diseases Due to Crustacean Parasites

Crustaceans that belong to the families Pennellidae, Ergasilidae, and Caligidae can cause disease in salmonids, although the caligid copepods, often referred to as sea lice, are the most common and have the greatest economic impact (Costello 1993). More specifically, the caligids *Lepeophtheirus salmonis* (sea louse) and *Caligus elongatus* are primarily responsible for severe disease in the Northern Hemisphere, whereas other *Caligus* sp. are important pathogens in the Southern Hemisphere (Johnson 1998).

Epizootiology

With the exception of *L. salmonis,* the parasitic copepods are generally not specific pathogens of salmonids (Johnson 1998). Production and development of the eggs is dependent on the species of copepod and host, age and condition of the host, and environmental factors including temperature and salinity (Tully 1989; Johnson and Albright 1991a; Costello 1993; Johnson 1998). Completion of the copepod life cycle is also influenced by salinity and temperature (Wootten, Smith, and Needham 1982; Tully 1989; Johnson and Albright 1991a; Johnson 1993, 1998). Generation times for *C. elongatus* and *L. salmonis* are approximately 50 days at 16°C and 48 days at 15.5°C, respectively (Tully 1989).

All species of copepods that are pathogenic to salmonids, except *C. elongatus,* have ten developmental stages (Piasecki 1996) that include two free-swimming, nonfeeding nauplius stages; one free-swimming, nonfeeding copepodid stage, which is the infective stage; four attached chalimus stages; two motile preadult stages; and one adult stage (Johnson and Albright 1991b; Schram 1993; Johnson 1998). Environmental distribution of the free-swimming nauplius and copepodid stages are influenced by environmental factors such as light, chemicals, pressure, and water flow (Bron, Sommerville, and Rae 1993). For example, the copepodid stage of *L. salmonis* may become concentrated at the surface during the day (Heuch, Parsons, and Boxaspen 1995), whereas decreased water movement will concentrate the nauplius and copepodid stages, which are weak swimmers (Costello, Costello, and Roche 1996). Although the chalimus stage is infectious, direct transmission of the preadult and adult stages may also occur, especially in high-density operations such as netpens (Ritchie 1997; Johnson 1998). Atlantic salmon and rainbow trout are most susceptible to infection with *L. salmonis,* whereas coho salmon are most resistant to infection (Johnson and Albright 1992a; Johnson 1998). Susceptibility is apparently related to the host response to infection (Jones, Sommerville, and Bron 1990; Johnson and Albright 1992a; MacKinnon 1993). Susceptibility increases with stress, concurrent disease conditions, and sexual maturation (Johnson 1998).

Clinical Disease

The copepodid and chalimus larvae have a diffuse external distribution and can also occur in the oropharyngeal cavity and gills but are most common on the fins. In contrast, the perianal area and dorsal integument posterior to the dorsal fin are the most common sites of the preadult and adult stages, especially with mild infections (Bron et al. 1991; Johnson and Albright 1992a, 1992b; Tully et al. 1993; Johnson et al. 1996; Johnson 1998). Disease occurs due to feeding of the attached stages on mucus, integument, and blood. However, clinical disease due to the copepodid and chalimus stages is generally less severe than disease due to the preadult and adult stages of development (Wootten, Smith, and Needham 1982; Bron et al. 1991; Johnson and Albright 1992a, 1992b). Severe infections result in focally extensive areas of hemorrhagic cutaneous erosion and ulceration, especially the dorsoanterior integument and perianal area (Wootten, Smith, and Needham 1982; Jónsdóttir et al. 1992; Johnson et al. 1996). Lesions can result in an increased susceptibility to other infectious diseases, compromised osmoregulation, downgrading or rejection of the carcass, and mortality (Wootten, Smith, and Needham 1982; Nylund, Wallace, and Hovland 1993; Tully et al. 1993). Infection with the chalimus larvae of *L. salmonis* can also result in severe erosion of the fins in Atlantic and chinook salmon.

Diagnosis

Diagnosis is based on detection of the copepods by direct examination, although magnification is required for identification of the copepodid and chalimus stages, due to their small (less then 4 mm) size (Johnson 1998). Guides to identification of the various copepod parasites and their developmental stages should be consulted for an accurate diagnosis (Kabata 1972;

Johnson and Albright 1991b; Schram 1993; Johnson and Margolis 1994; Piasecki 1996).

Prevention and Treatment

Prevention of disease due to copepod infections is the preferred management strategy (Johnson 1998). Prevention can be partially achieved by the location of netpens in areas of adequate water flow to prevent the accumulation of infectious copepodids. Likewise, adequate water flow through netpens should be maintained and not obstructed by fouling of the nets and/or a small size of the mesh. The segregation of year classes and fallowing of farm sites may also be an effective disease prevention strategy for *L. salmonis,* but not for *C. elongatus* (Bron et al. 1993). Exposure to wild salmonids and nonsalmonids that may transmit the infection should be avoided if possible, whereas stress and poor water quality should be eliminated.

Various chemical treatments used for copepod infections have previously been reviewed and summarized (Costello 1993; Roth, Richards, and Sommerville 1993; Johnson 1998). Briefly, the organophosphate azamethiphos has replaced the organophosphates dichlorvos and trichlorfon as the preferred treatment in Canada and Europe. However, successive bath treatments every two to four weeks is generally required for maximum efficacy, since the attached chalimus stages are not affected by organophosphate treatments (Roth et al. 1996). Furthermore, the use of organophosphates remains controversial due to the potential exposure of wild nontarget species. The synthetic pyrethroid cypermethrin is used in Norway as an alternative treatment for control of *L. salmonis* (Roth, Richards, and Sommerville 1993). Hydrogen peroxide and ivermectin have also been used for treatment of copepod infections, as previously reviewed (Johnson 1998), but are not recommended. Insect growth regulators that inhibit chitin synthesis, such as diflubenzuron and teflubenzuron, are currently being investigated for use in Norway (Johnson 1998). The advantage of this group of compounds is their effectiveness against the copepodid, chalimus, and preadult stages of development, but not against the adult stage. However, their use is also controversial due to the potential exposure and consequent adverse impact on nontarget species (Roth, Richards, and Sommerville 1993). Biological control using wrasses such as *Ctenolabrus exoletus* and *C. rupestris* is commonly included as a component of disease prevention and management programs but is expensive and generally not effective when used as the sole control strategy. Vaccines are not currently available but may be an effective management tool in the future, since experimental studies in Atlantic salmon injected with crude extracts of *L. salmonis* resulted in decreased egg production by the parasite (Grayson et al. 1995).

Diseases Due to Microsporidian Parasites

The condition referred to as *plasmacytoid leukemia* or *marine anemia* is an endemic disease of marine chinook salmon operations in British Columbia but has also been reported in coho and Atlantic salmon facilities in Chile and in freshwater chinook salmon facilities in California, Washington, and British Columbia (Kent 1998a, 1998b). The disease generally occurs in chinook salmon that have been in seawater for one year and may be associated with other disease conditions such as BKD (Kent 1998a). The causative agent has not been conclusively determined, although the intranuclear microsporidian *Nucleospora salmonis* and a retrovirus have been associated with the disease. Experimental transmission studies suggest that the disease can be transmitted by ingestion and cohabitation in freshwater (Baxa-Antonio, Groff, and Hedrick 1992; Kent 1998a, 1998b) but is not readily transmitted by cohabitation in seawater (Kent 1998b). Field observations also suggest a possible vertical transmission of the disease (Kent et al. 1993).

The disease generally results in a moderate, chronic mortality (Kent 1998a, 1998b). Affected fish often swim near the surface and may present with dark coloration, lethargy, pallor of the gills due to anemia, severe bilateral exophthalmos, and coelomic distension due to ascites. Internal lesions include enlargement of the spleen, kidney, and intestine, with petechial hemorrhage of the internal viscera. Hematological evaluation reveals variable anemia. Cytological examination of tissue imprints from the kidney, liver, or retrobulbar tissue stained reveals an abundance of immature plasma cells that are approximately 10 to 20 μm, with an increased nuclear to cytoplasmic ratio and occasional mitotic figures (Kent 1998a, 1998b). The spores of *N. salmonis* appear as clear, spherical, intranuclear vacuoles in Giemsa-stained imprints, whereas Gram-stained imprints reveal reniform, Gram-positive spores that are approximately 1 × 2 μm. The characteristic histological feature of the disease is a moderate to severe infiltration/proliferation of immature plasma cells in all tissues, but especially in the well-vascularized organs such as the spleen, kidney, and retrobulbar tissue (Kent et al. 1990). The spores of *N. salmonis* in histological sections stained with hematoxylin and eosin are eosinophilic, spherical structures that are approximately 2 to 4 μm and delineated by a rim of basophilic chromatin. Combined staining with Warthin-Starry and hematoxylin and eosin enhances detection of the parasite in histological sections (Kent et al. 1995).

A preliminary diagnosis of marine anemia is based on the characteristic gross and cytological findings, whereas a definitive diagnosis is based on the presence of typical histological lesions, according to the scheme proposed by Stephen and Ribble (1996). Specifically, the characteristic histological findings should be present in the kidney and at least one other nonhematopoietic organ, such as the liver, among other criteria. Prevention is difficult due to the uncertain etiology of the disease, although fish from infected sites should not be transferred to noninfected sites, and broodstock with a history of possible exposure should be avoided (Kent 1998a).

Finally, the microsporidian *Loma salmonae* is a common parasite of freshwater salmonids that generally does not result in significant disease conditions, although severe disease has been reported in marine salmonids (Kent et al. 1989b; Speare, Brackett, and Ferguson 1989). The primary site of infection is the gills, although infection may also involve the internal viscera, such as the heart, spleen, and kidney. Various aspects of the parasite and the disease have been summarized (Kent 1998b).

CONCLUSION

Similar to the production of any intensively managed species, the probability of disease in aquaculture species, including salmonids, has a direct correlation to the intensity of the production system. Regardless of etiology, disease conditions can threaten the economic viability of private-sector producers and associated industries, not only due to mortality but also due to morbidity that disrupts production schedules, decreases feed conversion and growth rates, increases the costs associated with diagnostic and treatment efforts, and may result in downgrading or rejection of the product. Therefore, prevention of disease is the preferred management strategy in aquaculture operations, since treatment of disease incurs an additional cost and is generally less efficient than disease prevention. Furthermore, the availability of antimicrobials and chemical therapeutants for treatment of infectious disease conditions is generally limited due to the lack of proper testing and registration of available therapeutants. For example, only a few antibiotics are presently approved for use in the United States, and these approved antibiotics are often not effective, depending on the specific disease condition. Although the continual development and registration of antibiotics is necessary, concerns such as antibiotic residues and the development of antibiotic-resistant bacteria are issues that cannot be ignored, due to the potential impact on human health.

The primary goal for disease prevention and control is the development of strategies that incorporate safe, effective, reliable, and economical methods. Development of these strategies is based on a thorough understanding of the epizootiology of any particular disease. The most common strategies that have been reiterated in this discussion include the reduction or elimination of stress, maintenance of proper water quality and nutrition, avoidance or elimination of potential carrier fish or other vectors of transmission, avoidance of water that may be contaminated with the pathogen, rejection and elimination of infected broodstock, segregation of year and brood classes, fallowing and/or disinfection of sites prior to reuse, proper disease diagnosis, and vaccination. However, the destruction of infected and potentially infected fish and eggs may have a greater economic impact on the aquaculture operation than the adverse effects of disease. Likewise, the practice of treating fertilized eggs with iodophor and raising the newly emergent embryos in pathogen-free water has been employed to prevent infection and the spread of certain viral diseases, but this is often not practical due to the inaccessibility of pathogen-free water in most aquaculture operations. Failure to properly diagnose a disease condition can result in additional economic losses, not only due to morbidity and mortality but also due to the possibility of disease recurrence in the population or facility.

Finally, vaccination programs are generally considered appropriate methods to achieve effective disease prevention and control but are only a component of an effective disease prevention program. However, vaccination also incurs an additional expense and needs to be considered in a cost-benefit context prior to incorporation of a vaccination protocol into the disease management program. Furthermore, few vaccines that provide protection of long duration are presently approved for use in the United States. Further development and registration of effective vaccines is therefore considered essential for the continual economic growth and viability of the commercial aquaculture industry.

REFERENCES

Almendras, F.E., I.C. Fuentealba, S.R.M. Jones, F. Markham, and E. Spangler. 1997. Experimental infection and horizontal transmission of *Piscirickettsia salmonis* in freshwater-raised Atlantic salmon, *Salmo salar* L. *Journal of Fish Diseases* 20:409-418.

Amend, D.F. 1970. Control of infectious hematopoietic necrosis virus by elevating water temperatures. *Journal of the Fisheries Research Board of Canada* 27:265-270.

Amend, D.F. 1976. Prevention and control of viral diseases of salmonids. *Journal of the Fisheries Research Board of Canada* 33:1059-1066.

Amend, D.F. and L. Smith. 1974. Pathophysiology of infectious hematopoietic necrosis virus disease in rainbow trout *(Salmo gairdneri):* Early changes in blood and aspects of the immune response after injection of IHN virus. *Journal of the Fisheries Research Board of Canada* 31:1371-1378.

Anderson, E.D., D.V. Mourich, S.E. Fahrenkrug, S.E. LaPatra, S. Shepard, and J.C. Leong. 1996. Genetic immunization of rainbow trout *(Oncorhynchus mykiss)* against hematopoietic necrosis virus. *Molecular Marine Biology and Biotechnology* 5:114-122.

Aoki, T., T. Kitao, N. Iemura, Y. Mitoma, and T. Nomura. 1983. The susceptibility of *Aeromonas salmonicida* strains isolated in cultured and wild salmonids to various chemotherapeutants. *Bulletin of the Japanese Society of Scientific Fisheries* 49:17-22.

Arakawa, C.K., R.E. Deering, K.H. Higman, K.H. Oshima, P.J. O'Hara, and J.R. Winton. 1990. Polymerase chain reaction (PCR) amplification of a nucleoprotein gene sequence of infectious hematopoietic necrosis virus. *Diseases of Aquatic Organisms* 8:165-170.

Armstrong, R.D., T.P.T. Evelyn, S.W. Martin, W. Dorward, and H.W. Ferguson. 1989. Erythromycin levels within eggs and alevins derived from spawning broodstock chinook salmon *(Oncorhynchus tshawytscha)* injected with the drug. *Diseases of Aquatic Organisms* 6:33-36.

Austin, B. and D.A. Austin. 1993. *Bacterial Fish Pathogens: Disease in Farmed and Wild Fish,* Second Edition. New York: Ellis Horwood.

Barnes, A.C., S.G.B. Amyes, T.S. Hastings, and C.S. Lewin. 1991. Fluoroquinolones display rapid bactericidal activity and low mutation frequencies against *Aeromonas salmonicida. Journal of Fish Diseases* 14:661-667.

Barnes, A.C., C.S. Lewin, T.S. Hastings, and S.G.B. Amyes. 1990a. Cross resistance between oxytetracycline and oxolinic acid in *Aeromonas salmonicida* associated with alterations in outer membrane proteins. *FEMS Microbiology Letters* 72:337-339.

Barnes, A.C., C.S. Lewin, T.S. Hastings, and S.G.B. Amyes. 1990b. *In vitro* activities of 4-quinolones against the fish pathogen *Aeromonas salmonicida. Antimicrobial Agents and Chemotherapy* 34:1819-1820.

Baxa-Antonio, D., J.M. Groff, and R.P. Hedrick. 1992. Experimental horizontal transmission of *Enterocytozoon salmonis* to chinook salmon, *Oncorhynchus tshawytscha. Journal of Protozoology* 39:699-702.

Bell, J.G., A.H. McVicar, and C.B. Cowey. 1987. Pyruvate kinase isozymes in farmed Atlantic salmon *(Salmo salar):* Pyruvate kinase and antioxidant parameters in pancreas disease. *Aquaculture* 66:33-41.

Bernardet, J.-F., and P.A.D. Grimont. 1989. Deoxyribonucleic acid relatedness and phenotypic characterization of *Flexibacter columnaris* sp. nov., nom. rev., *Flexibacter psychrophilus* sp. nov., nom. rev., and *Flexibacter maritimus* Wakabayashi, Hikida and Masumura 1986. *International Journal of Systematic Bacteriology* 39:346-354.

Bernardet, J.-F. and B. Kerouault. 1989. Phenotypic and genomic studies of *Cytophaga psychrophila* isolated from diseased rainbow trout *(Oncorhynchus mykiss)* in France. *Applied and Environmental Microbiology* 55:1796-1800.

Bernardet, J.-F., P. Segers, M. Vancanneyt, F. Berthe, K. Kersters, and P. Vandamme. 1996. Cutting a Gordian Knot: Emended classification and description of the Genus *Flavobacterium,* emended description of the family *Flavobacteriaceae,* and proposal of *Flavobacterium hydatis* nom. nov. (Basonym, *Cytophaga aquatilis* Strohl and Tait 1978). *International Journal of Systematic Bacteriology* 46:128-148.

Bernoth, E.-M., A.E. Ellis, P.J. Midtlyng, G. Olivier, and P.R. Smith. 1997. *Furunculosis—Multidisciplinary Fish Disease Research.* San Diego, CA: Academic Press.

Billi, J.L. and K. Wolf. 1969. Quantitative comparison of peritoneal washes and feces for detecting infectious pancreatic necrosis (IPN) virus in carrier brook trout. *Journal of the Fisheries Research Board of Canada* 26:1459-1465.

Branson, E.J. and D. Nieto Diaz-Munoz. 1991. Description of a new disease condition occurring in farmed coho salmon, *Oncorhynchus kisutch* (Walbaum), in South America. *Journal of Fish Diseases* 14:147-156.

Brocklebank, J.R., T.P.T. Evelyn, D.J. Speare, and R.D. Armstrong. 1993. Rickettsial septicemia in farmed Atlantic and chinook salmon in British Columbia: Clinical presentation and experimental transmission. *Canadian Veterinary Journal* 34:745-748.

Bron, J.E., C. Sommerville, M. Jones, and G.H. Rae. 1991. The settlement and attachment of early stages of the salmon louse, *Lepeophtheirus salmonis* (Copepoda: Caligidae) on the salmon host *Salmo salar. Journal of Zoology* 224:201-212.

Bron, J.E., C. Sommerville, and G.H. Rae. 1993. Aspects of the behavior of copepodid larvae of the salmon louse *Lepeophtheirus salmonis* (Kroyer, 1837). In G.A. Boxshall and D. Defaye (Eds.), *Pathogens of Wild and Farmed Fish: Sea Lice* (pp. 125-142). Chichester, England: Ellis Horwood.

Bron, J.E., C. Sommerville, R. Wootten, and G.H. Rae. 1993. Fallowing of marine Atlantic salmon, *Salmo salar* L., farms as a method for the control of sea lice, *Lepeophtheirus salmonis* (Kroyer, 1837). *Journal of Fish Diseases* 16:487-493.

Brown, L.L., T.P.T. Evelyn, G.K. Iwama, W.S. Nelson, and R.P. Levine. 1995. Bacterial species other than *Renibacterium salmoninarum* cross-react with antisera against *R. salmoninarum* but are negative for the p57 gene of *R. salmoninarum* as detected by the polymerase chain reaction (PCR). *Diseases of Aquatic Organisms* 21:227-231.

Brown, L.L., G.K. Iwama, T.P.T. Evelyn, W.S. Nelson, and R.P. Levine. 1994. Use of the polymerase chain reaction (PCR) to detect DNA from *Renibacterium salmoninarum* within individual salmonid eggs. *Diseases of Aquatic Organisms* 18:165-171.

Bruno, D.W. 1986. Histopathology of bacterial kidney disease in laboratory infected rainbow trout, *Salmo gairdneri* Richardson and Atlantic salmon, *Salmo salar* L., with reference to naturally infected fish. *Journal of Fish Diseases* 9:523-538.

Bruno, D.W., T.S. Hastings, and A.E. Ellis. 1986. Histology, bacteriology, and experimental transmission of cold-water vibriosis in Atlantic salmon *Salmo salar. Diseases of Aquatic Organisms* 1:163-168.

Bruno, D.W. and A.L.S. Munro. 1989. Immunity in Atlantic salmon, *Salmo salar* L., fry following vaccination against *Yersinia ruckeri,* and the influence of body weight and infectious pancreatic necrosis virus (IPNV) on the detection of carriers. *Aquaculture* 81:205-212.

Bullock, G.L., D.A. Conroy, and S.F. Snieszko. 1971. Bacterial diseases of fishes. In S.F. Snieszko and H.R. Axelrod (Eds.), *Diseases of Fishes,* book 2A (pp. 1-151). Neptune City, NJ: TFH Publications.

Bullock, G.L., B.R. Griffin, and H.M. Stuckey. 1980. Detection of *Corynebacterium salmoninus* by direct fluorescent antibody test. *Canadian Journal of Fisheries and Aquatic Sciences* 37:719-721.

Bullock, G.L., R.R. Rucker, D. Amend, K. Wolf, and H.M. Stuckey. 1976. Infectious pancreatic necrosis: Transmission with iodine-treated and nontreated eggs of brook trout *(Salvelinus fontinalis). Journal of the Fisheries Research Board of Canada* 33:1197-1198.

Bullock, G.L. and H.M. Stuckey. 1975a. *Aeromonas salmonicida:* Detection of asymptomatically infected trout. *Progressive Fish-Culturist* 37:237-239.

Bullock, G.L. and H.M. Stuckey. 1975b. Fluorescent antibody identification and detection of the *Corynebacterium* causing kidney disease of salmonids. *Journal of the Fisheries Research Board of Canada* 32:2224-2227.

Busch, R.A. and A.J. Lingg. 1975. Establishment of an asymptomatic carrier state infection of enteric redmouth disease in rainbow trout *(Salmo gairdneri). Journal of the Fisheries Research Board of Canada* 32:2429-2432.

Byrne, P.J., D.D. MacPhee, V.E. Ostland, G. Johnson, and H.W. Ferguson. 1998. Hemorrhagic kidney syndrome of Atlantic salmon, *Salmo salar* L. *Journal of Fish Diseases* 21:81-91.

Caswell-Reno, P., V. Lipipun, P.W. Reno, and B.L. Nicolson. 1989. Use of a group reactive and other monoclonal antibodies in an enzyme immunodot assay for identification and presumptive serotyping of aquatic birnaviruses. *Journal of Clinical Microbiology* 27:1924-1929.

Chowdhury, M.B.R. and H. Wakabayashi. 1988. Effects of sodium, potassium, calcium and magnesium ions on the survival of *Flexibacter columnaris* in water. *Fish Pathology* 23:231-235.

Costello, M.J. 1993. Review of methods to control sea lice (Caligidae: Crustacea) infestations of salmon *(Salmo salar)* farms. In G.A. Boxshall and D. Defaye (Eds.), *Pathogens of Wild and Farmed Fish: Sea Lice* (pp. 219-252). Chichester, England: Ellis Horwood.

Costello, M., J. Costello, and N. Roche. 1996. Planktonic dispersion of larval salmon-lice, *Lepeophtheirus salmonis,* associated with cultured salmon, *Salmo salar,* in western Ireland. *Journal of the Marine Biological Association of the United Kingdom* 76:141-149.

Cvitanich, J.D., O.N. Garate, and C.E. Smith. 1991. The isolation of a rickettsia-like organism causing disease and mortality in Chilean salmonids and its confirmation by Koch's postulate. *Journal of Fish Diseases* 14:121-145.

Dalsgaard, I. 1993. Virulence mechanisms in *Cytophaga psychrophila* and other *Cytophaga*-like bacteria pathogenic for fish. *Annual Review of Fish Diseases* 3:127-144.

Dannevig, B.H., K. Falk, and E. Namork. 1995. Isolation of the causal virus of infectious salmon anaemia (ISA) in a long-term cell line from Atlantic salmon head kidney. *Journal of General Virology* 76:1353-1359.

Daoust, P.Y. and H.W. Ferguson. 1985. Nodular gill disease: A unique form of proliferative gill disease in rainbow trout *Salmo gairdneri* Richardson. *Journal of Fish Diseases* 8:511-522.

Drolet, B.S., J.S. Rohovec, and J.C. Leong. 1994. The route of entry and progression of infectious hematopoietic necrosis virus in *Oncorhynchus mykiss* (Walbaum): A sequential immunohistochemical study. *Journal of Fish Diseases* 17:337-347.

Dubois-Darnaudpeys, A. 1977. Epidémiologie de la furonculose des salmonides. III. Ecologie de *Aeromonas salmonicida* proposition d'un modèle epidémiologique. *Bulletin Français de Pisciculture* 50:21-32.

Egidius, E., K. Andersen, E. Clausen, and J. Raa. 1981. Cold-water vibriosis or "Hitra disease" in Norwegian salmonid farming. *Journal of Fish Diseases* 4:353-354.

Egidius, E., R. Wiik, K. Andersen, K.A. Hoff, and B. Hjeltnes. 1986. *Vibrio salmonicida,* new species, a new fish pathogen. *International Journal of Systematic Bacteriology* 36:518-520.

Elliott, D.G. and T.Y. Barila. 1987. Membrane filtration-fluorescent antibody staining procedure for detecting and quantifying *Renibacterium salmoninarum* in coelomic fluid of chinook salmon *(Oncorhynchus tshawytscha). Canadian Journal of Fisheries and Aquatic Sciences* 44:206-210.

Elliott, D.G., R.J. Pascho, and G.L. Bullock. 1989. Developments in the control of bacterial kidney disease of salmonid fishes. *Diseases of Aquatic Organisms* 6:201-215.

Ellis, A.E. 1991. An appraisal of the extracellular toxins of *Aeromonas salmonicida* ssp. *salmonicida. Journal of Fish Diseases* 14:265-278.

El-Matbouli, M. and R.W. Hoffmann. 1990. Experimental transmission of two *Myxobolus* spp. developing bisporogeny via tubificid worms. *Parasitology Research* 75:461-464.

Enger, O., B. Husevag, and J. Goksoyr. 1989. Presence of the fish pathogen *Vibrio salmonicida* in fish farm sediments. *Applied and Environmental Microbiology* 55:2815-2818.

Evelyn, T.P.T. 1993. Bacterial kidney disease—BKD. In V. Inglis, R.J. Roberts, and N.R. Bromage (Eds.), *Bacterial Diseases of Fish* (pp. 177-195). Oxford, England: Blackwell Scientific Publications.

Evelyn, T.P.T., M.L. Kent, T.T. Poppe, and P. Bustos. 1998. Bacterial diseases. In M.L. Kent and T.T. Poppe (Eds.), *Diseases of Seawater Netpen-Reared Salmonid Fishes* (pp. 17-35). Nanaimo, British Columbia, Canada: Pacific Biological Station, Department of Fisheries and Oceans.

Evelyn, T.P.T., J.E. Ketcheson, and L. Prosperi-Porta. 1984. Further evidence for the presence of *Renibacterium salmoninarum* in salmonid eggs and for the failure of povidine-iodine to reduce the intraovum infection rate in water-hardened eggs. *Journal of Fish Diseases* 7:173-182.

Evelyn, T.P.T., J.E. Ketcheson, and L. Prosperi-Porta. 1986. Use of erythromycin as a means of preventing vertical transmission of *Renibacterium salmoninarum. Diseases of Aquatic Organisms* 2:7-12.

Evelyn, T.P.T., L. Prosperi-Porta, and J.E. Ketcheson. 1986. Experimental intra-ovum infection of salmonid eggs with *Renibacterium salmoninarum* and vertical transmission of the pathogen with such eggs despite treatment with erythromycin. *Diseases of Aquatic Organisms* 1:197-202.

Evensen, O., S. Espelid, and T. Hastein. 1991. Immunohistochemical identification of *Vibrio salmonicida* in stored tissues of Atlantic salmon *Salmo salar* from the first known outbreak of coldwater vibriosis (Hitra disease). *Diseases of Aquatic Organisms* 10:185-189.

Evensen, O., K.E. Thorud, and Y.A. Olsen. 1991. A morphological study of the gross and light microscopic lesions of infectious salmon anaemia in Atlantic salmon (*Salmo salar* L.). *Research in Veterinary Science* 51:215-222.

Ewing, W.H., A.J. Ross, D.J. Brenner, and G.R. Flanning. 1978. *Yersinia ruckeri* sp. nov., the redmouth (RM) bacterium. *International Journal of Systematic Bacteriology* 28:37-44.

Falk, K. and B. Dannevig. 1995. Demonstrations of infectious salmon anaemia (ISA) viral antigens in cell cultures and tissue sections. *Veterinary Research* 26:499-504.

Ferguson, H.W. and D.H. McCarthy. 1978. Histopathology of furunculosis in brown trout *Salmo trutta* L. *Journal of Fish Diseases* 1:165-174.

Ferguson, H.W., V.E. Ostland, P. Byrne, and J.S. Lumsden. 1991. Experimental production of bacterial gill disease in trout by horizontal transmission and by bath challenge. *Journal of Aquatic Animal Health* 3:118-123.

Ferguson, H.W., R.J. Roberts, R.H. Richards, R.O. Collins, and D.A. Rice. 1986. Severe degenerative cardiomyopathy associated with pancreas disease in Atlantic salmon, *Salmo salar* L. *Journal of Fish Diseases* 20:95-98.

Fijan, N.N. 1968. The survival of *Chondrococcus columnaris* in waters of different quality. *Bulletin de l'Office International des Epizootics* 69:1158-1166.

Fijan, N.N. and G. Giorgetti. 1978. Infectious pancreatic necrosis: Isolation of virus from eyed eggs of rainbow trout *Salmo gairdneri* Richardson. *Journal of Fish Diseases* 1:269-270.

Fjolstad, M. and A.L. Heyeraas. 1985. Muscular and myocardial degeneration in cultured Atlantic salmon, *Salmo salar,* suffering from "Hitra disease." *Journal of Fish Diseases* 8:367-372.

Frelier, P.F., R.A. Elston, J.K. Loy, and C. Mincher. 1994. Macroscopic and microscopic features of ulcerative stomatitis in farmed Atlantic salmon *Salmo salar. Diseases of Aquatic Organisms* 18:227-231.

Frerichs, G.N., S.D. Millar, and C. McManus. 1992. Atypical *Aeromonas salmonicida* isolated from healthy wrasse *(Ctenolabrus rupestris). Bulletin of the European Association of Fish Pathologists* 12:48-49.

Frerichs, G.N. and R.J. Roberts. 1989. The bacteriology of teleosts. In R.J. Roberts (Ed.), *Fish pathology* (pp. 289-319). London: Baillière Tindall.

Fryer, J.L., C.N. Lannan, L.H. Garcés, J.J. Larenas, and P.A. Smith. 1990. Isolation of a rickettsiales-like organism from diseased coho salmon *(Oncorhynchus kisutch)* in Chile. *Fish Pathology* 25:107-114.

Fryer, J.L., C.N. Lannan, S.J. Giovannoni, and N.D. Wood. 1992. *Piscirickettsia salmonis* gen. nov., sp. nov., the causative agent of an epizootic disease in salmonid fishes. *International Journal of Systematic Bacteriology* 42:120-126.

Fryer, J. and J. Sanders. 1981. Bacterial kidney disease of salmonid fish. *Annual Review of Microbiology* 35:273-298.

Grayson, T.H., R.J. John, S. Wadsworth, K. Greaves, D. Cox, J. Roper, A.B. Wrathmell, M.L. Gilpin, and J.E. Harris. 1995. Immunization of Atlantic salmon against the salmon louse: Identification of antigens and effects on louse fecundity. *Journal of Fish Biology* 47(Suppl. A):85-94.

Grisez, L., R. Ceusters, and F. Ollevier. 1991. The use of API 20E for the identification of *Vibrio anguillarum* and *V. ordalii. Journal of Fish Diseases* 14:351-365.

Groberg, W.J., R.H. McCoy, K.S. Pilcher, and J.L. Fryer. 1978. Relation of water temperature to infections of coho salmon *(Oncorhynchus kisutch),* chinook salmon *(O. tshawytscha)*, and steelhead trout *(Salmo gairdneri)* with *Aeromonas salmonicida* and *A. hydrophila*. *Journal of the Fisheries Research Board of Canada* 35:1-7.

Gustafson, C.E., C.J. Thomas, and T.J. Trust. 1992. Detection of *Aeromonas salmonicida* from fish by using polymerase chain reaction amplification of the virulence surface array protease gene. *Applied and Environmental Microbiology* 58:3816-3825.

Handlinger, J., M. Soltani, and S. Percival. 1997. The pathology of *Flexibacter maritimus* in aquaculture species in Tasmania, Australia. *Journal of Fish Diseases* 20:159-168.

Harrell, L.W. and T.M. Scott. 1985. *Kudoa thyrsitis* (Myxosporea: Multivalvulida) in Atlantic salmon *Salmo salar*. *Journal of Fish Diseases* 8:329-332.

Hastings, T.S. and A. McKay. 1987. Resistance of *Aeromonas salmonicida* to oxolinic acid. *Aquaculture* 61:165-171.

Hazen, T.C., G.W. Esch, R.V. Dimock, and A. Mansfield. 1982. Chemotaxis of *Aeromonas hydrophila* to the surface mucus of fish. *Current Microbiology* 7:371-375.

Hazen, T.C., R.P. Fliermans, R.P. Hirsch, and G.W. Esch. 1978. Prevalence and distribution of *Aeromonas hydrophila* in the United States. *Applied and Environmental Microbiology* 36:731-738.

Hedrick, R.P., M. El-Matbouli, M.A. Adkison, and E. MacConnell. 1998. Whirling disease: Re-emergence among wild trout. *Immunological Reviews* 166:365-376.

Heo, G.-J., K. Kasai, and H. Wakabayashi. 1990. Occurrence of *Flavobacterium branchiophila* associated with bacterial gill disease at a trout hatchery. *Fish Pathology* 25:99-105.

Hervio, D.M.L., M.L. Kent, J. Khattra, J. Sakanari, H. Yokoyama, and R.H. Devlin. 1997. Taxonomy of *Kudoa* species (Myxosporea), using a small-subunit ribosomal DNA sequence. *Canadian Journal of Zoology* 75:2112-2119.

Hetrick, F.M., J.L. Fryer, and M.D. Knittel. 1979. Effect of water temperature on the infection of rainbow trout *Salmo gairdneri* Richardson with infectious hematopoietic necrosis virus. *Journal of Fish Diseases* 2:253-257.

Heuch, P.A., A. Parsons, and K. Boxaspen. 1995. Diel vertical migration: A possible host-finding mechanism in salmon louse *(Lepeophtheirus salmonis)* copepodids. *Canadian Journal of Fisheries and Aquatic Sciences* 52:681-689.

Hill, B.J. and K. Way. 1995. Serological classification of infectious pancreatic necrosis (IPN) virus and other aquatic birnaviruses. *Annual Review of Fish Diseases* 5:55-77.

Hiney, M., M.T. Dawson, D.M. Heery, P.R. Smith, F. Gannon, and R. Powell. 1992. DNA probe for *Aeromonas salmonicida*. *Applied and Environmental Microbiology* 58:1039-1042.

Hiney, M.P., J.J. Kilmartin, and P.R. Smith. 1994. Detection of *Aeromonas salmonicida* in Atlantic salmon with asymptomatic furunculosis infections. *Diseases of Aquatic Organisms* 19:161-167.

Hjeltnes, B. and R.J. Roberts. 1993. Vibriosis. In V. Inglis, R.J. Roberts, and N.R. Bromage (Eds.), *Bacterial Diseases of Fish* (pp. 109-121). Oxford, England: Blackwell Scientific Publications.

Hodgkinson, J.J., D. Bucke, and B. Austin. 1987. Uptake of the fish pathogen, *Aeromonas salmonicida,* by rainbow trout (*Salmo gairdneri* L.). *FEMS Microbiology Letters* 40:207-210.

Hoie, S., H. Heum, and O.F. Thoresen. 1997. Evaluation of a polymerase chain reaction-based assay for the detection of *Aeromonas salmonicida* ssp. *salmonicida* in Atlantic salmon *Salmo salar. Diseases of Aquatic Organisms* 30:27-35.

Holm, K.O., E. Strom, K. Stensvag, J. Raa, and T. Jorgensen. 1985. Characteristics of a *Vibrio* sp. associated with the Hitra disease of Atlantic salmon in Norwegian fish farms. *Fish Pathology* 20:125-130.

Holt, R.A., J.S. Rohovec, and J.L. Fryer. 1993. Bacterial cold-water disease. In V. Inglis, R.J. Roberts, and N.R. Bromage (Eds.), *Bacterial Diseases of Fish* (pp. 3-22). Oxford, England: Blackwell Scientific Publications.

Huh, G.-J. and H. Wakabayashi. 1987. Detection of *Flavobacterium* sp., a pathogen of bacterial gill disease, using indirect fluorescent antibody technique. *Fish Pathology* 22:215-220.

Hunter, V.A., M.D. Knittel, and J.L. Fryer. 1980. Stress-induced transmission of *Yersinia ruckeri* infection from carriers to recipient steelhead trout, *Salmo gairdneri* Richardson. *Journal of Fish Diseases* 3:467-472.

Inglis, V. and R.H. Richards. 1991. The *in vitro* susceptibility of *Aeromonas salmonicida* and other fish-pathogenic bacteria to 29 antimicrobial agents. *Journal of Fish Diseases* 14:641-650.

Inglis V., R.J. Roberts, and N.R. Bromage. 1993. *Bacterial Diseases of Fish.* Oxford, England: Blackwell Scientific Publications.

Jarp, J., A.G. Gjevre, A.B. Olsen, and T. Bruheim. 1995. Risk factors for furunculosis, infectious pancreatic necrosis and mortality in post-smolt of Atlantic salmon, *Salmo salar* L. *Journal of Fish Diseases* 18:67-78.

Johnson, S.C. 1993. A comparison of development and growth rates of *Lepeophtheirus salmonis* (Copepoda: Caligidae) on naive Atlantic *(Salmo salar)* and chinook *(Oncorhynchus tshawytscha)* salmon. In G.A. Boxshall and D. Defaye (Eds.), *Pathogens of Wild and Farmed Fish: Sea Lice* (pp. 68-80). Chichester, England: Ellis Horwood.

Johnson, S.C. 1998. Crustacean parasites. In M.L. Kent and T.T. Poppe (Eds.), *Diseases of Seawater Netpen-Reared Salmonid Fishes* (pp. 80-90). British Columbia, Canada: Pacific Biological Station, Department of Fisheries and Oceans, Nanaimo.

Johnson, S.C. and L.J. Albright. 1991a. Development, growth, and survival of *Lepeophtheirus salmonis* (Copepoda: Caligidae) under laboratory conditions. *Journal of the Marine Biological Association of the United Kingdom* 71:425-436.

Johnson, S.C. and L.J. Albright. 1991b. The developmental stages of *Lepeophtheirus salmonis* (Kroyer, 1837) (Copepoda: Caligidae). *Canadian Journal of Zoology* 69:929-950.

Johnson, S.C. and L.J. Albright. 1992a. Comparative susceptibility and histopathology of the response of naive Atlantic, chinook, and coho salmon to experimental infection with *Lepeophtheirus salmonis* (Copepoda: Caligidae). *Diseases of Aquatic Organisms* 14:179-193.

Johnson, S.C. and L.J. Albright. 1992b. Effects of cortisol implants on the susceptibility and the histopathology of the responses of naive coho salmon *Oncorhynchus kisutch* to experimental infection with *Lepeophtheirus salmonis* (Copepoda: Caligidae). *Diseases of Aquatic Organisms* 14:195-205.

Johnson, S.C., R.B. Blaylock, J. Elphick, and K.D. Hyatt. 1996. Disease induced by the sea louse *(Lepeophtheirus salmonis)* (Copepoda: Caligidae) in wild sockeye salmon (*Oncorhynchus nerka)* stocks of Alberni Inlet, British Columbia. *Canadian Journal of Fisheries and Aquatic Sciences* 53:2888-2897.

Johnson, S.C. and L. Margolis. 1994. Sea lice. In J.C. Thoesen, (Ed.), *Suggested Procedures for the Detection and Identification of Certain Finfish and Shellfish Pathogens,* Fourth edition. Bethesda, MD: Fish Health Section, American Fisheries Society.

Jones, M.W., C. Sommerville, and J. Bron. 1990. The histopathology associated with the juvenile stages of *Lepeophtheirus salmonis* on the Atlantic salmon, *Salmo salar* L. *Journal of Fish Diseases* 13:303-310.

Jónsdóttir, H., J.E. Bron, R. Wootten, and J.F. Turnbull. 1992. The histopathology associated with the preadult and adult stages of *Lepeophtheirus salmonis* on the Atlantic salmon, *Salmo salar* L. *Journal of Fish Diseases* 15:521-527.

Jorgensen, T., K. Midling, S. Espelid, R. Nilsen, and K. Stensvag. 1989. *Vibrio salmonicida,* previously registered as a pathogen in salmonids, does also cause mortality in netpen captured cod *(Gadus morhua). Bulletin of the European Association of Fish Pathologists* 9:42-43.

Kabata, Z. 1972. Developmental stages of *Caligus clemensi* (Copepoda: Caligidae). *Journal of the Fisheries Research Board of Canada* 29:1571-1593.

Kanno, T., T. Nakai, and K. Muroga. 1989. Mode of transmission of vibriosis among ayu, *Plecoglossus altivelis*. *Journal of Aquatic Animal Health* 1:2-6.

Kent, M.L. 1998a. Neoplastic diseases and related disorders. In M.L. Kent and T.T. Poppe (Eds.), *Diseases of Seawater Netpen-Reared Salmonid Fishes* (pp. 106-113). British Columbia, Canada: Pacific Biological Station, Department of Fisheries and Oceans, Nanaimo.

Kent, M.L. 1998b. Protozoa and myxozoa. In M.L. Kent and T.T. Poppe (Eds.), *Diseases of Seawater Netpen-Reared Salmonid Fishes* (pp. 49-67). British Columbia, Canada: Pacific Biological Station, Department of Fisheries and Oceans, Nanaimo.

Kent, M.L., C.F. Dungan, R.A. Elston, and R. A. Holt. 1988. *Cytophaga* sp. (Cytophagales) infection in seawater pen-reared Atlantic salmon *Salmo salar. Diseases of Aquatic Organisms* 4:173-180.

Kent, M.L., D.G. Elliott, J.M. Groff, and R.P. Hedrick. 1989. *Loma salmonae* (Protozoa: Microspora) infections in seawater-reared coho salmon *(Oncorhynchus kisutch). Aquaculture* 80:211-222.

Kent, M.L., J.M. Groff, J.K. Morrison, W.T. Yasutake, and R.A. Holt. 1989. Spiral swimming behavior due to cranial and vertebral lesions associated with *Cytophaga psychrophila* infections in salmonid fishes. *Diseases of Aquatic Organisms* 6:11-16.

Kent, M.L., J.M. Groff, G.S. Traxler, J.G. Zinkl, and J.W. Bagshaw. 1990. Plasmacytoid leukemia in seawater-reared chinook salmon *Oncorhynchus tshawytscha. Diseases of Aquatic Organisms* 8:199-209.

Kent, M.L., J. Khattra, D.M.L. Hervio, and R.H. Devlin. 1998. Ribosomal DNA sequence analysis of isolates of the PKX myxosporean and their relationship to members of the genus *Sphaerospora. Journal of Aquatic Animal Health* 10:12-21.

Kent, M.L., G.C. Newbound, S.C. Dawe, C. Stephen, W.D. Eaton, G.S. Traxler, D. Kieser, and R.F. Markham. 1993. Observations on the transmission and range of plasmacytoid leukemia of chinook salmon. *Fish Health Section Newsletter/American Fisheries Society* 21(2):1-3.

Kent, M.L. and T.T. Poppe. 1998. *Diseases of Seawater Netpen-Reared Salmonid Fishes.* British Columbia, Canada: Pacific Biological Station, Department of Fisheries and Oceans, Nanaimo.

Kent, M.L., V. Rantis, J.W. Bagshaw, and S.C. Dawe. 1995. Enhanced detection of *Enterocytozoon salmonis* (Microspora), an intranuclear microsporean of salmonid fishes, with the Warthin-Starry stain combined with hematoxylin and eosin. *Diseases of Aquatic Organisms* 23:235-237.

Kent, M.L., T.K. Sawyer, and R.P. Hedrick. 1988. *Paramoeba pemaquidensis* (Sarcomastigophora: Paramoebidae) infestation of the gills of coho salmon *Oncorhynchus kisutch* reared in seawater. *Diseases of Aquatic Organisms* 5:163-170.

Kent, M.L., D.J. Whitaker, and L. Margolis. 1993. Transmission of *Myxobolus arcticus* (Pugachev and Khokhlov, 1979), a myxosporean of Pacific salmon, via a triactinomyxon from the aquatic oligochaete *Stylodrilus heringianus* (Lumbriculidae). *Canadian Journal of Zoology* 71:1207-1211.

King, C.H. and E.B. Shotts. 1988. Enhancement of *Edwardsiella tarda* and *Aeromonas salmonicida* through ingestion by the ciliated protozoan *Tetrahymena pyriformis. FEMS Microbiology Letters* 51:95-100.

Kitao, T., T. Aoki, and K. Muroga. 1984. Three new O-serotypes of *Vibrio anguillarum. Bulletin of the Japanese Society of Scientific Fisheries* 50:1955-1956.

Lannan, C.N., S.A. Ewing, and J.L. Fryer. 1991. A fluorescent antibody test for detection of the rickettsia causing disease in Chilean salmonids. *Journal of Aquatic Animal Health* 3:229-234.

Lannan, C.N. and J.L. Fryer. 1994. Extracellular survival of *Piscirickettsia salmonis. Journal of Fish Diseases* 17:545-548.

LaPatra, S.E. 1996. The use of serological techniques for virus surveillance and certification of fin fish. *Annual Review of Fish Diseases* 6:15-28.

LaPatra, S.E. 1998. Factors affecting pathogenicity of infectious hematopoietic necrosis virus (IHNV) for salmonid fish. *Journal of Aquatic Animal Health* 10:121-131.

LaPatra, S.E., K.A. Lauda, G.R. Jones, S.C. Walker, B.S. Shewmaker, and A.W. Morton. 1995. Characterization of IHNV isolates associated with neurotropism. *Veterinary Research* 26:433-437.

LeBlanc, D., K.R. Mittal, G. Olivier, and R. Lallier. 1981. Serogrouping of motile aeromonas species isolated from healthy and moribund fish. *Applied and Environmental Microbiology* 42:56-60.

Lee, E.G.-H. and T.P.T. Evelyn. 1994. Prevention of vertical transmission of the bacterial kidney disease agent *Renibacterium salmoninarum* by broodstock injection with erythromycin. *Diseases of Aquatic Organisms* 18:1-4.

Leong, J.C., Y.L. Hsu, H.M. Engelking, and D. Mulcahy. 1981. Strains of infectious hematopoietic necrosis (IHN) virus may be identified by structural protein differences. *Development of Biological Standards* 49:43-55.

MacDonnell, M.T. and R.R. Colwell. 1985. Phylogeny of the Vibrionaceae and recommendations for two new genera, *Listonella* and *Shewanella*. *Systematic and Applied Microbiology* 6:171-182.

MacKinnon, B.M. 1993. Host response of Atlantic salmon *(Salmo salar)* to infection by sea lice *(Caligus elongatus)*. *Canadian Journal of Fisheries and Aquatic Sciences* 50:789-792.

Martinsen, B., E. Myhr, E. Reed, and T. Håstein. 1991. *In vitro* antimicrobial susceptibility of sarafloxacin against clinical isolates of bacteria pathogenic to fish. *Journal of Aquatic Animal Health* 3:235-241.

Mauel, M.J., S.J. Giovannoni, and J.L. Fryer. 1996. Development of polymerase chain reaction assays for detection, identification, and differentiation of *Piscirickettsia salmonis*. *Diseases of Aquatic Organisms* 26:189-195.

McCarthy, D.H. 1975a. Detection of *Aeromonas salmonicida* antigen in diseased fish tissue. *Journal of General Microbiology* 88:384-386.

McCarthy, D.H. 1975b. Fish furunculosis. *Journal of the Institute of Fisheries Management* 6:13-18.

McCarthy, D.H. 1980. Some ecological aspects of the bacterial fish pathogen—*Aeromonas salmonicida*. *Aquatic Microbiology: Symposium of the Society of Applied Bacteriology* 6:299-324.

McCarthy, D.H. and R.J. Roberts. 1980. Furunculosis of fish—the present state of our knowledge. In M.A. Droop and H.W. Jannasch (Eds.), *Advances in Aquatic Microbiology* (pp. 292-341). London, England: Academic Press.

McKnight, I.J. and R.J. Roberts. 1976. The pathology of infectious pancreatic necrosis virus. I. The sequential histopathology of the naturally occurring condition. *British Veterinary Journal* 132:76-85.

McLoughlin, M.F., R.T. Nelson, H.M. Rowley, D.J. Cox, and A.N. Grant. 1996. Experimental pancreas disease in Atlantic salmon *Salmo salar* post-smolts induced by salmon pancreas disease virus (SPDV). *Diseases of Aquatic Organisms* 26:117-124.

McLoughlin, M.F., H.M. Rowley, and C.E. Doherty. 1998. A serological survey of salmon pancreas disease virus (SPDV) antibodies in farmed Atlantic salmon *Salmo salar* L. *Journal of Fish Diseases* 21:305-307.

McVicar, A.H. 1987. Pancreas disease of farmed Atlantic salmon, *Salmo salar*, in Scotland: Epidemiology and early pathology. *Aquaculture* 67:71-78.

Meyers, T.R., J.B. Thomas, J.E. Follett, and R.R. Saft. 1990. Infectious hematopoietic necrosis virus: Trends in prevalence and the risk management approach in Alaskan sockeye salmon cultures. *Journal of Aquatic Animal Health* 2:85-98.

Miyata, M., V. Inglis, and T. Aoki. 1996. Rapid identification of *Aeromonas salmonicida* subspecies *salmonicida* by the polymerase chain reaction. *Aquaculture* 141:13-24.

Mjaaland, S., E. Rimstad, K. Falk, and B.H. Dannevig. 1997. Genomic characterization of the virus causing infectious salmon anaemia in Atlantic salmon (*Salmo salar* L.): An orthomyxo-like virus in a teleost. *Journal of Virology* 71:7681-7686.

Moffitt, C.M. 1991. Oral and injectable applications of erythromycin in salmonid fish culture. *Veterinary and Human Toxicology* 33(Suppl. 1):49-53.

Moran, J.D.W. and M.L. Kent. 1999. *Kudoa thyrsites* (Myxozoa: Myxosporea) infections in pen-reared Atlantic salmon in the Northeast Pacific Ocean, with a survey of potential nonsalmonid reservoir hosts. *Journal of Aquatic Animal Health* 11:101-109.

Morrison, C., J. Cornick, G. Shun, and B. Zwicker. 1981. Microbiology and histopathology of "saddleback" disease of underyearling Atlantic salmon, *Salmo salar* L. *Journal of Fish Diseases* 4:243-258.

Mulcahy, D., D. Klaybor, and W.N. Batts. 1990. Isolation of infectious hematopoietic necrosis virus from a leech *(Piscicola salmositica)* and a copepod (*Salmincola* sp.), ectoparasites of sockeye salmon *Oncorhynchus nerka. Diseases of Aquatic Organisms* 8:29-34.

Mullins, J.E., D. Groman, and D. Waldowska. 1998. Infectious salmon anaemia in salt water Atlantic salmon (*Salmo salar* L.) in New Brunswick, Canada. *Bulletin of the European Association of Fish Pathologists* 18:110-114.

Munday, B.L., C.K. Foster, F.R. Roubal, and R.G.J. Lester. 1990. Paramoebic gill infection and associated pathology of Atlantic salmon, *Salmo salar,* and rainbow trout, *Salmo gairdneri,* in Tasmania. In F.O. Perkins and T.C. Cheng (Eds.), *Pathology in Marine Science* (pp. 215-222). San Diego, CA: Academic Press.

Munday, B.L., K. Lange, C. Foster, R.G.J. Lester, and J. Handlinger. 1993. Amoebic gill disease of sea-caged salmonids in Tasmania. *Tasmanian Fish Research* 28:14-19.

Munro, A.L.S. and T.S. Hastings. 1993. Furunculosis. In V. Inglis, R.J. Roberts, and N.R. Bromage (Eds.), *Bacterial Diseases of Fish* (pp. 122-142). Oxford, England: Blackwell Scientific Publications.

Nelson, R.T., M.F. McLoughlin, H.M. Rowley, M.A. Platten, and J.J. McCormick. 1995. Isolation of a toga-like virus from farmed Atlantic salmon *Salmo salar* with pancreas disease. *Diseases of Aquatic Organisms* 22:25-32.

Nese, L. and R. Enger. 1993. Isolation of *Aeromonas salmonicida* from salmon lice *Lepeophtheirus salmonis* and marine plankton. *Diseases of Aquatic Organisms* 16:79-81.

Nieto, T.P., M.J.R. Corcobado, A.E. Toranzo, and J.L. Barja. 1985. Relation of water temperature to infection of *Salmo gairdneri* with motile *Aeromonas. Fish Pathology* 20:99-106.

Nieto, T.P., Y. Santos, L.A. Rodriguez, and A.E. Ellis. 1991. An extracellular acetylcholinesterase produced by *Aeromonas hydrophila* is a major lethal toxin for fish. *Microbial Pathogenesis* 11:101-110.

Noga, E.J. 1995. *Fish Disease: Diagnosis and Treatment.* St. Louis, MO: Mosby.

Nylund, A., T. Hovland, K. Hodneland, F. Nilsen, and P. Lervik. 1994. Mechanisms for transmission of infectious salmon anaemia (ISA). *Diseases of Aquatic Organisms* 19:95-100.

Nylund, A. and P. Jakobsen. 1995. Sea trout as a carrier of infectious salmon anemia virus. *Journal of Fish Biology* 47:174-176.

Nylund, A., C. Wallace, and T. Hovland. 1993. The possible role of *Lepeophtheirus salmonis* (Kroyer) in the transmission of infectious salmon anemia. In G.A. Boxshall and D. Defaye (Eds.), *Pathogens of Wild and Farmed Fish: Sea Lice* (pp. 367-373). Chichester, England: Ellis Horwood.

O'Halloran, J. and R. Henry. 1993. *Vibrio salmonicida* (Hitra disease) in New Brunswick. *Bulletin of the Aquaculture Association of Canada* 93/94:96-98.

Ostland, V.E., D.G. McGrogan, and H.W. Ferguson. 1997. Cephalic osteochondritis and necrotic scleritis in intensively reared salmonids associated with *Flexibacter psychrophilus*. *Journal of Fish Diseases* 20:443-451.

Pascho, R.J., D. Chase, and C.L. McKibben. 1998. Comparisons of the membrane filtration-fluorescent antibody test, the enzyme-linked immunosorbent assay, and the polymerase chain reaction to detect *Renibacterium salmoninarum* in salmonid ovarian fluid. *Journal of Veterinary Diagnostic Investigation* 10:60-66.

Pascho, R.J. and D. Mulcahy. 1987. Enzyme-linked immunosorbent assay for a soluble antigen of *Renibacterium salmoninarum,* the causative agent of salmonid bacterial kidney disease. *Canadian Journal of Fisheries and Aquatic Sciences* 44:183-191.

Payne, M.A., R.E. Baynes, S.F. Sundlof, A. Craigmill, A.I. Webb, and J.E. Riviere. 1999. Drugs prohibited from extralabel use in food animals. *Journal of the Veterinary Medical Association* 215:28-32.

Piasecki, W. 1996. The developmental stages of *Caligus elongatus* von Nordmann, 1832 (Copepoda: Caligidae). *Canadian Journal of Zoology* 74:1459-1478.

Plumb, J.A. 1994. *Health Maintenance of Cultured Fishes: Principal Microbial Diseases.* Boca Raton, FL: CRC Press.

Rabb, L., J.W. Cornick, and L.A. McDermott. 1964. A macroscopic slide agglutination test for the presumptive diagnosis of furunculosis in fish. *Progressive Fish Culturist* 26:118-120.

Ransom, D.P., C.N. Lannan, J.S. Rohovec, and J.L. Fryer. 1984. Comparison of histopathology caused by *Vibrio anguillarum* and *V. ordalii* in three species of Pacific salmon. *Journal of Fish Disease* 7:107-115.

Ristow, S.S. and J.M. Arnzen. 1989. Development of monoclonal antibodies that recognize a type-2 specific and a common epitope on the nucleoprotein of infectious hematopoietic necrosis virus. *Journal of Aquatic Animal Health* 1:119-125.

Ritchie, G. 1997. The host transfer ability of *Lepeophtheirus salmonis* (Copepoda: Caligidae) from farmed Atlantic salmon, *Salmo salar. Journal of Fish Diseases* 20:153-157.

Roberts, R.J. 1989. Pathophysiology and systemic pathology of teleosts. In R.J. Roberts (Ed.), *Fish Pathology* (pp. 56-134). London: Baillière Tindall.

Roberts, R.J. 1993. Motile aeromonad septicaemia. In V. Inglis, R.J. Roberts, and N.R. Bromage (Eds.), *Bacterial Diseases of Fish* (pp. 143-155). Oxford, England: Blackwell Scientific Publications.

Rodgers, C.J. 1992. Development of a selective-differential medium for the isolation of *Yersinia ruckeri* and its application in epidemiological studies. *Journal of Fish Diseases* 15:243-254.

Rodgers, C.J. and B. Austin. 1982. Oxolinic acid for control of enteric redmouth disease in rainbow trout. *Veterinary Record* 112:83.

Rose, A.S., A.E. Ellis, and A.J.S. Munro. 1989. The infectivity by different routes of exposure and shedding rates of *Aeromonas salmonicida* ssp. *salmonicida* in Atlantic salmon, *Salmo salar* L., held in sea water. *Journal of Fish Diseases* 12:573-578.

Rose A.S., A.E. Ellis, and A.J.S. Munro. 1990. The survival of *Aeromonas salmonicida* ssp. *salmonicida* in sea water. *Journal of Fish Diseases* 13:205-214.

Ross, A.J., R.R. Rucker, and W.H. Ewing. 1966. Description of a bacterium associated with redmouth disease of rainbow trout *(Salmo gairdneri). Canadian Journal of Microbiology* 12:763-770.

Roth, M., R.H. Richards, D.P. Dobson, and G.H. Rae. 1996. Field trials on the efficacy of the organophosphorous compound azamethiphos for the control of sea lice (Copepoda:Caligidae) infestations of farmed Atlantic salmon *(Salmo salar). Aquaculture* 140:217-239.

Roth, M., R.H. Richards, and C. Sommerville. 1993. Current practices in the chemotherapeutic control of sea lice infestation in aquaculture: A review. *Journal of Fish Diseases* 16:1-26.

Roubal, F.R., R.J.G. Lester, and C.K. Foster. 1989. Studies on cultured and gill-attached *Paramoeba* sp. (Gymnamoebae: Paramoebidae) and the cytopathology of paramoebic gill disease in Atlantic salmon, *Salmo salar* L., from Tasmania. *Journal of Fish Diseases* 12:481-492.

Ruidisch, S., M. El-Matbouli, and R.W. Hoffmann. 1991. The role of tubificid worms as an intermediate host in the life cycle of *Myxobolus pavlovskii* (Akhmerov, 1954). *Parasitology Research* 77:663-667.

Sakai, M., S. Atsuta, and M. Kobayashi. 1986. Comparative sensitivities of several diagnostic methods to detect fish furunculosis. *Kitasato Archives of Experimental Medicine* 59:1-6.

Salte, R., K.-A. Rorvik, E. Reed, and K. Norberg. 1994. Winter ulcers of the skin in Atlantic salmon, *Salmo salar* L.: Pathogenesis and possible aetiology. *Journal of Fish Diseases* 17:661-665.

Schäperclaus, W. 1991. *Fish Diseases.* Volume 1, Fifth Edition. New Delhi, India: Amerind Publishing.

Schiewe, M.H., T.J. Trust, and J.H. Crosa. 1981. *Vibrio ordalii* sp. nov.: A causative agent of vibriosis in fish. *Current Microbiology* 6:343-348.

Schram, T.A. 1993. Supplementary descriptions of the developmental stages of *Lepeophtheirus salmonis* (Kroyer, 1837) (Copepoda:Caligidae). In G.A. Boxshall and D. Defaye (Eds.), *Pathogens of Wild and Farmed Fish: Sea Lice* (pp. 30-47). Chichester, England: Ellis Horwood.

Shotts, E.B. and J.D. Teska. 1989. Bacterial pathogens of aquatic vertebrates. In B. Austin and D.A. Austin (Eds.), *Methods for the Microbiological Examination of Fish and Shellfish* (pp. 164-186). New York: Wiley and Sons.

Smail, D.A., D.W. Bruno, G. Dear, L.A. McFarlane, and K. Ross. 1992. Infectious pancreatic necrosis (IPN) virus Sp serotype in farmed Atlantic salmon, *Salmo salar* L., post-smolts associated with mortality and clinical disease. *Journal of Fish Diseases* 15:77-83.

Smith, P.R., G.M. Brazil, E.M. Drinan, J. O'Kelly, R. Palmer, and A. Scallan. 1982. Lateral transmission of furunculosis in sea water. *Bulletin of the European Association of Fish Pathologists* 3:41-42.

Sorensen, U.B.S. and J.L. Larsen. 1986. Serotyping of *Vibrio anguillarum*. *Applied and Environmental Microbiology* 51:593-597.

Speare, D.J., J. Brackett, and H.W. Ferguson. 1989. Sequential pathology of the gills of coho salmon with a combined diatom and microsporidian gill infection. *Canadian Veterinary Journal* 30:571-575.

Speare, D.J., H.W. Ferguson, F.W.M. Beamish, J.A. Yager, and S. Yamashiro. 1991a. Pathology of bacterial gill disease: Sequential development of lesions during natural outbreaks of disease. *Journal of Fish Diseases* 14:21-32.

Speare, D.J., H.W. Ferguson, F.W.M. Beamish, J.A. Yager, and S. Yamashiro. 1991b. Pathology of bacterial gill disease: Ultrastructure of branchial lesions. *Journal of Fish Diseases* 14:1-20.

Speare, D.J., V.E. Ostland, and H.W. Ferguson. 1993. Pathology associated with meningoencephalitis during bacterial kidney disease of salmonids. *Research in Veterinary Science* 54:25-31.

Stephen, C. and C.S. Ribble. 1996. Marine anemia in farmed chinook salmon *(Oncorhynchus tshawytscha)*: Development of a working case definition. *Preventative Veterinary Medicine* 25:259-269.

Stevenson, R., D. Flett, and B.T. Raymond. 1993. Enteric redmouth (ERM) and other enterobacterial infections of fish. In V. Inglis, R.J. Roberts, and N.R. Bromage (Eds.), *Bacterial Diseases of Fish* (pp. 80-106). Oxford, England: Blackwell Scientific Publications.

St-Hilaire, S., M. Hill, M.L. Kent, D.J. Whitaker, and C.S. Ribble. 1997. A comparative study of muscle texture and intensity of *Kudoa thyrsites* infection in farm-reared Atlantic salmon *Salmo salar* on the Pacific coast of Canada. *Diseases of Aquatic Organisms* 31:221-225.

St-Hilaire, S., C. Ribble, D.J. Whitaker, and M.L. Kent. 1998. Prevalence of *Kudoa thyrsites* in sexually mature and immature pen-reared Atlantic salmon *(Salmo salar)* in British Columbia, Canada. *Aquaculture* 162:69-77

Stoskopf, M.K. 1993. *Fish Medicine*. Philadelphia, PA: Saunders.

Tajima, K., Y. Ezura, and T. Kimura. 1985. Studies on the taxonomy and serology of causative organisms of fish vibriosis. *Fish Pathology* 20:131-142.

Teska, J.D., E.B. Shotts, and T.C. Hsu. 1989. Automated biochemical identification of bacterial fish pathogens using the Abbott Quantum II. *Journal of Wildlife Diseases* 25:103-107.

Thoesen, J.C. 1994. *Suggested Procedures for the Detection and Identification of Certain Fish and Shellfish Pathogens,* Fourth Edition, Version 1. Bethesda, MD: Fish Health Section, American Fisheries Society.

Thorud, K. and H.O. Djupvik. 1988. Infectious anaemia in Atlantic salmon (*Salmo salar* L.). *Bulletin of the European Association of Fish Pathologists* 8:109-111.

Thune, R.L., L.A. Stanley, and R.K. Cooper. 1993. Pathogenesis of Gram-negative bacterial infections in warmwater fish. *Annual Review of Fish Diseases* 3:37-68.

Toranzo, A.E., J.L. Barja, M.L. Lemos, and F.M. Hetrick. 1983. Stability of infectious pancreatic necrosis virus (IPNV) in untreated, filtered, and autoclaved estuarine water. *Bulletin of the European Association of Fish Pathologists* 3:51-53.

Toranzo, A.E., Y. Santos, T.P. Nieto, and J.L. Barja. 1986. Evaluation of different assay systems for identification of environmental *Aeromonas* strains. *Applied and Environmental Microbiology* 51:652-656.

Totland, G.K., B.K. Hjeltnes, and P.R. Flood. 1996. Transmission of infectious salmon anaemia (ISA) through natural secretions and excretions from infected smolts of Atlantic salmon *Salmo salar* during their presymptomatic phase. *Diseases of Aquatic Organisms* 26:25-31.

Traxler, G.S., M.L. Kent, and T.T. Poppe. 1998. Viral diseases. In M.L. Kent and T.T. Poppe (Eds.), *Diseases of Seawater Netpen-Reared Salmonid Fishes* (pp. 36-45). British Columbia, Canada: Pacific Biological Station, Department of Fisheries and Oceans, Nanaimo.

Traxler, G.S., J.R. Roome, and M.L. Kent. 1993. Transmission of infectious hematopoietic necrosis virus in seawater. *Diseases of Aquatic Organisms* 16:111-114.

Traxler, G.S., J.R. Roome, K.A. Lauda, and S. LaPatra. 1996. Appearance of infectious hematopoietic necrosis virus (IHNV) and neutralizing antibodies in sockeye salmon *Oncorhynchus nerka* during their migration and maturation period. *Diseases of Aquatic Organisms* 28:31-38.

Treasurer, J. and D. Cox. 1991. The occurrence of *Aeromonas salmonicida* in wrasse (Labridae) and implications for Atlantic salmon farming. *Bulletin of the European Association of Fish Pathologists* 11:208-210.

Trust, T.J., I.D. Courtice, A.G. Khouri, J.H. Crosa, and M.H. Schiewe. 1981. Serum resistance and hemagglutination ability of marine vibrios pathogenic for fish. *Infection and Immunity* 34:702-707.

Trust, T.J. and R.A.H. Sparrow. 1974. The bacterial flora in the alimentary tract of freshwater salmonid fishes. *Canadian Journal of Microbiology* 20:1219-1228.

Tsoumas, A., D.J. Alderman, and C.J. Rodgers. 1989. *Aeromonas salmonicida*: Development of resistance to 4-quinolone antimicrobials. *Journal of Fish Diseases* 12:493-508.

Tully, O. 1989. The succession of generations and growth of the caligid copepods *Caligus elongatus* and *Lepeophtheirus salmonis* parasitizing farmed Atlantic salmon smolts (*Salmo salar* L.). *Journal of the Marine Biological Association of the United Kingdom* 69:279-288.

Tully, O., W.R. Poole, K.F. Whelan, and S. Merigoux. 1993. Parameters and possible causes of epizootics of *Lepeophtheirus salmonis* (Kroyer) infesting sea trout

(*Salmo trutta* L.) off the west coast of Ireland. In G.A. Boxshall and D. Defaye (Eds.), *Pathogens of Wild and Farmed Fish: Sea Lice* (pp. 202-213). Chichester, England: Ellis Horwood.

Turnbull, J.F. 1993. Bacterial gill disease and fin rot. In V. Inglis, R.J. Roberts, and N.R. Bromage (Eds.), *Bacterial Diseases of Fish* (pp. 40-58). Oxford, England: Blackwell Scientific Publications.

Ugajin M. 1979. Studies on the taxonomy of major microflora on the intestinal contents of salmonids. *Bulletin of the Japanese Society of Scientific Fisheries* 45:721-731.

Vallejo, A.N., Jr., and A.E. Ellis. 1989. Ultrastructural study of the response of eosinophil granule cells to *Aeromonas salmonicida* extracellular products and histamine liberators in rainbow trout *Salmo gairdneri* Richardson. *Developmental and Comparative Immunology* 13:133-148.

Wakabayashi, H. 1993. Columnaris disease. In V. Inglis, R.J. Roberts, and N.R. Bromage (Eds.), *Bacterial Diseases of Fish* (pp. 23-39). Oxford, England: Blackwell Scientific Publications.

Wakabayashi, H., K. Kanai, T-C. Hsu, and S. Egusa. 1981. Pathogenic activities of *Aeromonas hydrophila* biovar *hydrophila* (Chester) Popoff and Veron, 1976 to fishes. *Fish Pathology* 15:319-325.

Waltman, W.D. and E.B. Shotts. 1984. A medium for the isolation and differentiation of *Yersinia ruckeri*. *Canadian Journal of Fisheries and Aquatic Sciences* 41:804-806.

Wards, B.J., H.H. Patel, C.D. Anderson, and G.W. de Lisle. 1991. Characterization by restriction endonuclease analysis and plasmid profiling of *Vibrio ordalii* strains from salmon (*Oncorhynchus tshawytscha* and *Oncorhynchus nerka*) with vibriosis in New Zealand. *New Zealand Journal of Marine and Freshwater Research* 25:345-350.

Winton, J.R. 1991. Recent advances in detection and control of infectious hematopoietic necrosis virus in aquaculture. *Annual Review of Fish Diseases* 1:83-93.

Winton, J.R., C.K. Arakawa, C.N. Lannan, and J.L. Fryer. 1988. Neutralizing monoclonal antibodies recognize antigenic variants among isolates of infectious hematopoietic necrosis virus. *Diseases of Aquatic Organisms* 4:199-204.

Wolf, K. 1988. *Fish Viruses and Fish Viral Diseases*. Ithaca, NY: Cornell University Press.

Wolf, K., M.E. Markiw, and J.K. Hiltunen. 1986. Salmonid whirling disease: *Tubifex tubifex* (Muller) identified as the essential oligochaete in the protozoan life cycle. *Journal of Fish Diseases* 9:83-85.

Wolf, K., M.C. Quimby, and A.D. Bradford. 1963. Egg-associated transmission of IPN virus of trouts. *Virology* 21:317-321.

Wolke, R.E. 1975. Pathology of bacterial and fungal diseases affecting fish. In W.E. Ribelin and G. Migaki (Eds.), *The Pathology of Fishes* (pp. 33-116). Madison, WI: University of Wisconsin Press.

Wootten, R., J.W. Smith, and E.A. Needham. 1982. Aspects of the biology of the parasitic copepods *Lepeophtheirus salmonis* and *Caligus elongatus* on farmed salmonids, and their treatment. *Proceedings of the Royal Society of Edinburgh,* Section B (Biology) 81:185-197.

Yasutake, W.T. 1970. Comparative histopathology of epizootic salmonid virus disease. In S.F. Snieszko (Ed.), *A Symposium on Diseases of Fishes and Shellfishes* (pp. 341-350). Washington, DC: Special Publication No. 5, American Fisheries Society.

Yasutake, W.T. 1975. Fish viral diseases: Clinical, histopathological, and comparative aspects. In W.E. Ribelin and G. Migaki (Eds.), *The Pathology of Fishes* (pp. 247-271). Madison, WI: University of Wisconsin Press.

Yasutake, W.T. 1978. Histopathology of yearling sockeye salmon *(Oncorhynchus nerka)* infected with infectious hematopoietic necrosis (IHN). *Fish Pathology* 14:59-64.

Yasutake, W.T., T.J. Parisot, and G.W. Klontz. 1965. Virus diseases of the Salmonidae in western United States. II. Aspects of pathogenesis. *Annals of the New York Academy of Science* 126:520-530.

Yasutake, W.T. and C.J. Rasmussen. 1968. Histopathogenesis of experimentally induced viral hemorrhagic septicemia in fingerling rainbow trout *(Salmo gairdneri). Bulletin de l'Office International des Epizootics* 69:977-984.

Yoshimizu, M., S. Direkbusarakom, Y. Ezura, and T. Kimura. 1993. Monoclonal antibodies against *Aeromonas salmonicida* for serological diagnosis of furunculosis. *Bulletin of the Japanese Society of Scientific Fisheries* 59:333-338.

Yu, K.K., R.D. MacDonald, and A.R. Moore. 1982. Replication of infectious pancreatic necrosis virus in trout leucocytes and detection of the carrier state. *Journal of Fish Diseases* 5:401-410.

Chapter 3

The Penaeid Shrimp Viruses TSV, IHHNV, WSSV, and YHV: Current Status in the Americas, Available Diagnostic Methods, and Management Strategies

Donald V. Lightner

INTRODUCTION

Twenty-five years have elapsed since John Couch described *Baculovirus penaei,* the first recognized virus of penaeid shrimp, in *Penaeus duorarum* (northern pink shrimp) from the Gulf of Mexico in Florida (Couch 1974). Since then the list of viruses infecting this group of marine invertebrate animals has grown to include nearly 20 viruses (see Table 3.1), and viruses have emerged as important pathogens of penaeid shrimp virtually everywhere in the world where penaeid shrimp are cultured (Lightner 1996a). Currently, nine viruses (or groups of closely related viruses) are known to be enzootic in Western Hemisphere penaeids, and five of these pathogens have emerged as serious pathogens in one or more species of cultured shrimp (see Table 3.2). Virus diseases have also severely impacted the shrimp-farming industries of the Eastern Hemisphere. In the shrimp-growing regions of the Indo-Pacific and East Asia at least 12 viruses (or groups of closely related viruses) are recognized. Of the 12 virus groups, five have been documented

Funding for this research was provided by the Gulf Coast Research Laboratory Consortium Marine Shrimp Farming Program, Cooperative State Research, Education, and Extension Service (CSREES), USDA under Grant No. 88-38808-3320, the National Sea Grant Program, USDC under Grant No. NA56RG0617, and a special grant from the National Fishery Institute. Also thankfully acknowledged are my colleagues and students who have contributed to this review in immeasurable ways. They are: J. R. Bonami, R. M. Redman, B. T. Poulos, K. F. J Tang, L. M. Nunan, J. L. Mari, C. R. Pantoja, Q. Wang, J. L. Zhou, S. Durand, K. W. Hasson, L. L. Mohney, and B. L. White.

TABLE 3.1. Viruses of Penaeid Shrimp (As of January 1999)

DNA VIRUSES		
PARVOVIRUSES:		
IHHNV	=	infectious hypodermal and hematopoietic necrosis virus
HPV	=	hepatopancreatic parvovirus
SMV	=	spawner-isolated mortality virus
LPV	=	lymphoidal parvo-like virus
BACULOVIRUSES and BACULO-LIKE VIRUSES:		
BP-type	=	*Baculovirus penaei*-type viruses (PvSNPV type sp.): BP strains from the Gulf of Mexico, Hawaii, and Eastern Pacific
MBV-type	=	*Penaeus monodon*-type baculoviruses (PmSNPV type sp.): MBV strains from East and Southeast Asia, Australia, the Indo-Pacific, and India
BMN-type	=	baculoviral midgut gland necrosis-type viruses:
BMN	=	from *P. japonicus* in Japan
TCBV	=	type C baculovirus of *P. monodon*
PHRV	=	hemocyte-infecting nonoccluded baculo-like virus
WHITE SPOT SYNDROME BACULO-LIKE VIRUSES (PmNOBII-type):		
SEMBV	=	systemic ectodermal and mesodermal baculo-like virus
RV-PJ	=	rod-shaped virus of *P. japonicus*
PAV	=	penaeid acute viremia virus
HHNBV	=	hypodermal and hematopoietic necrosis baculo-like virus; agent of "SEED" (shrimp explosive epidermic disease)
WSBV	=	white spot baculo-like virus
WSSV/WSV	=	white spot syndrome virus
IRIDOVIRUS:		
IRDO	=	shrimp iridovirus
RNA VIRUSES		
PICORNAVIRUS:		
TSV	=	Taura syndrome virus
REOVIRUSES:		
REO-III and IV	=	reo-like virus type III and IV
TOGA-LIKE VIRUS:		
LOVV	=	lymphoid organ vacuolization virus
RHABDOVIRUS:		
RPS	=	rhabdovirus of penaeid shrimp
YELLOW-HEAD VIRUS GROUP:		
YHV/"YBV"	=	yellow-head ("yellow-head baculovirus") virus of *P. monodon*
GAV	=	gill associated virus of *P. monodon*
LOV	=	lymphoid organ virus of *P. monodon*

Sources: Modified from Lightner 1996a; Lightner et al. 1998.

to be responsible for serious disease epizootics regionally, with two of these viruses having caused panzootics throughout much of the industry in the Indo-Pacific and East Asia. Although there are nearly 20 different viruses recognized in penaeid shrimp, only four seem to stand out as being especially important from their historical, current, and potential future adverse effects on the international shrimp-farming industries. These four viruses are white spot syndrome virus (WSSV), yellow-head virus (YHV), Taura syndrome virus (TSV), and infectious hypodermal and hematopoietic necrosis virus (IHHNV). TSV and IHHNV have caused serious disease problems in many of the shrimp-growing countries in the Americas. Widespread epizootics due to YHV and WSSV emerged early in this decade in the Asian shrimp-farming industry, and together they have caused catastrophic disease losses (with a total negative economic impact averaging from U.S.$1 to 3 billion per year since 1994) to major shrimp-growing countries such as China, Thailand, India, Indonesia, Bangladesh, Malaysia, Taiwan, Vietnam,

TABLE 3.2. Viruses of Penaeid Shrimp Reported from the Eastern (Asia, Australia, Europe, and Africa) and Western Hemispheres (the Americas and Hawaii)

Virus or Virus Group	Eastern Hemisphere	Western Hemisphere
Baculo and baculo-like viruses	MBV-group BMN-group WSSV PHRV	BP WSSV
Parvo and parvo-like viruses	IHHNV HPV SMV LPV	IHHNV HPV
Picornavirus	none	TSV
Rod-shaped ssRNA viruses	YHV LOV/GAV	none
Reo-like viruses	REO-III REO-IV	REO-III
Toga-like viruses	none	LOVV
Rhabdovirus	none	RPS
Iridovirus	none	IRDO

Sources: Modified from Lightner 1996a; Lightner et al. 1998.

and Japan (ADB/NACA in press). This chapter will discuss the biology, hosts, available detection and diagnostic methods, and what are perceived as the best management practices for the shrimp culture industry of the Western Hemisphere to most effectively deal with these viruses. The taxonomic (Latin binomial and common) names used here are according to Holthuis (1980).

THE VIRUSES OF CONCERN: BIOLOGY, RANGE, AND HOSTS

Infectious Hypodermal and Hematopoietic Necrosis Virus (IHHNV)

The virion of IHHN is among the smallest animal viruses, and it is the smallest of the known penaeid shrimp viruses. The IHHN virion is a nonenveloped icosahedron of 22 nm in diameter with a 4.1 Kb (kilobases) ssDNA genome and nuclear replication. Its characteristics place IHHNV within the family Parvoviridae (Bonami et al. 1990). IHHNV was first recognized in 1981 when it was shown to be the cause of acute, catastrophic epizootics with cumulative mortality rates of ~60 to 90 percent in semi-intensively or intensively cultured juvenile *P. stylirostris* (blue shrimp) stocks that originated from Mexico, Ecuador, or Panama (Lightner et al. 1983; Lightner, Redman, and Bell 1983; Bell and Lightner 1987; Lightner 1996a, b). In *P. vannamei* (white leg shrimp) IHHNV was recognized soon after it was discovered to infect and cause disease, but not significant mortalities (Bell and Lightner 1984). Despite their relative resistance to IHHN disease, cultured *P. vannamei* can be chronically infected with IHHNV and suffer runt deformity syndrome (RDS) as a consequence. RDS was linked by epizootiological data to infection by IHHNV. Shrimp with RDS often show greatly reduced growth rates and a variety of cuticular deformities affecting the rostrum ("bent rostrum"), antennae, and other cephalothoracic and abdominal areas of the exoskeleton (Kalagayan et al. 1991; Browdy et al. 1993). RDS is an economically significant disease of cultured *P. vannamei,* that has been observed in virtually every country in the Americas where the species is cultured (Brock and Lightner 1990; Brock and Main 1994; Lightner 1996a). Cultured populations affected by RDS may contain up to 30 percent runts and consequently a wide distribution of size ("count" or the number of shrimp per pound) classes. Because runted shrimp have a lower market value than unaffected shrimp, RDS significantly reduces the market value of affected *P. vannamei* crops, resulting in revenue losses that can range from 10 to 50 percent of the value of similar IHHNV-free (and RDS-free) crops (Wyban et al. 1992).

Natural infections by IHHNV have been observed in *P. stylirostris, P. vannamei, P. occidentalis* (western white shrimp), *P. californiensis* (yellow-leg brown shrimp), *P. monodon* (giant tiger prawn), *P. semisulcatus* (green tiger prawn), and *P. japonicus* (Kuruma or Japanese tiger prawn). Because other penaeid species including *P. setiferus* (northern white shrimp), *P. duorarum,* and *P. aztecus* (northern brown shrimp) have been experimentally infected, natural infections by the virus probably occur in a number of other penaeid species. However, species such as *P. indicus* (Indian white prawn) and *P. merguiensis* (banana prawn) seem to be refractory to IHHNV (Lightner 1996a). While all life stages of susceptible host species may be infected by IHHNV, the juvenile stages are the most severely affected (see Table 3.3).

IHHNV has been documented to occur in East and Southeast Asia (Japan, Singapore, Malaysia, Indonesia, Thailand, and the Philippines) in shrimp culture facilities using only captive wild *P. japonicus* and *P. monodon* broodstock, and where American penaeids had not been introduced. Except for a single report from the Philippines that implicated

TABLE 3.3. Susceptibility and Severity of Disease of Important American Penaeids to the Viruses IHHNV, TSV, BP, WSSV, and YHV As Determined from Natural and Experimental Infections

Species	**IHHNV**				**TSV**				**WSSV**				**YHV**			
	L	**PL**	**J**	**A**	**L**	**PL**	**J**	**A**	**L**	**PL**	**J**	**A**	**L**	**PL**	**J**	**A**
P. vannamei	-	+	+	+	-	++	++	++	?	++	++	?	-	-	++	?
P. stylirostris	-	+	++	+	-	-	+	-	?	?	++	?	?	?	++	?
P. schmitti	-	?	+	?	-	-	+	-	?	?	?	?	?	?	?	?
P. setiferus	-	-	+	?	-	++	+	?	?	++	++	?	?	-	++	?
P. aztecus	-	-	+	?	-	+	+	?	?	++	+/-	?	?	-	++	?
P. duorarum	-	-	+	?	-	-	-	?	?	++	-	?	?	-	++	?
P. californiensis	-	-	+	+	-	-	-	?	?	?	+/-	?	?	?	?	?

Source: Modified from Lightner, Redman, Poulos, Nunan, Mari, and Hasson 1997; Lightner and Redman 1998b.

Key to symbols for each pathogen and life stages:

L = larvae
PL= postlarvae
J = juvenile
A = adult
- = infection occurs but without disease expression in this life stage
? = no data available
+ = infection accompanied by expression of moderate disease
++= infection accompanied by expression of significant disease

IHHNV as the cause of a serious epizootic in *P. monodon* (Rosenberry 1992), the virus is increasingly viewed as a generally insignificant pathogen in Asia (Baticados et al. 1990; Lightner et al. 1992; Flegel, Fegan, and Sriurairatana 1995; Lightner 1996a, 1996b; Flegel 1997). The occurrence of IHHNV in captive wild broodstocks and in their cultured progeny suggests that East and Southeast Asia is within the virus's natural geographic range and that *P. monodon* and *P. japonicus* may be among its natural host species. Based on the apparently stable host-pathogen relationship of IHHNV to its shrimp hosts in Asia versus the more serious history of IHHNV in the Americas, it has been suggested that IHHNV was introduced into the Americas from Asia during the early 1970s (Lightner 1996b). The hypothesis that IHHNV was introduced into the Americas with Asian *P. monodon* in the mid-1970s goes a long way toward explaining why the developing shrimp culture industry in Latin America and in the United States shifted from culturing *P. stylirostris* to *P. vannamei* during that period.

Taura Syndrome Virus (TSV)

TSV has been classified as a picornavirus based on its virion structure (i.e., 30 nm diameter, nonenveloped, icosahedron with cytoplasmic replication, and an ssRNA genome) (Bonami et al. 1997; Mari, Bonami, and Lightner 1998). Taura syndrome (TS) was first recognized in shrimp farms located near the mouth of the Taura River in the Gulf of Guayaquil, Ecuador, in mid-1992 (Jimenez 1992; Lightner et al. 1995), where the disease caused catastrophic disease losses with cumulative mortality rates of 60 to >90 percent of affected pond-cultured juvenile *P. vannamei.* Following its recognition as a distinct disease of cultured *P. vannamei* in Ecuador in 1992, TSV spread rapidly to virtually all of the shrimp-growing regions of Latin America and to parts of the United States (Brock et al. 1995, 1997; Lightner 1996a, 1996b; Lightner et al. 1998; Hasson et al. 1995; Hasson, Lightner, et al. 1997; Hasson, Hasson, et al. 1997; Hassan, Lightner, Mari, et al. 1999). The epidemiological and laboratory studies that followed its discovery and spread from Ecuador showed that TSV had a viral etiology and that, of the American (*P. vannamei, P. stylirostris, P. schmitti* [the southern white shrimp], *P. setiferus,* and *P. aztecus*) and Asian (*P. monodon, P. japonicus,* and *P. chinensis* [fleshy prawn or Chinese white shrimp]) penaeid species naturally or experimentally infected by the virus, *P. vannamei* was by far the most severely affected (Brock et al. 1995, 1997; Hasson et al. 1995; Hasson, Lightner, et al. 1997; Hasson, Lightner, Mohney, et al. 1999; Lightner 1996a; Lightner, Redman, Poulos, Nunan, Mari, Hassan, and Bonami 1997; Overstreet et al. 1997).

Because shrimp culture in the Americas was, and still is, a virtual monoculture based on *P. vannamei* (Rosenberry 1993, 1998), TS has seriously impacted the shrimp culture industry. In Ecuador, the disease resulted in production losses that reached between 15 and 30 percent of the country's production in 1993 and 1994 (Rosenberry 1994a, 1994b; Wigglesworth 1994). At shrimp wholesale prices from 1992 to 1994 (~$13.00/kg for 31/35 count tails), a 30 percent reduction in production relative to Ecuador's 1991 production of 100,000 t translates to nearly a $400 million loss in revenue per year. TS has had a similarly devastating impact on the farms of other countries as it spread from Ecuador (Lightner 1996b; Brock et al. 1997).

Since shortly after the discovery of TSV, a controversy over the etiology of TS began and it has persisted to the present. Although both toxic and infectious etiologies have been proposed for TS, the disease is caused by a virus (Bonami et al. 1997; Hasson et al. 1995); this has been demonstrated and confirmed by several independent laboratories (Brock et al. 1995, 1997; Hasson, Lightner, Mari, et al. 1999). The hypothesis that TS has a toxic etiology has not held up to scientific scrutiny, and the hypothesis has endured only in the interest of ongoing litigation. Unfortunately, the controversy on the etiology of TS has also contributed to the geographic spread of the disease (Lightner 1996b; Hasson, Lightner, Mari, et al. 1999). The transfer of TSV from Ecuador into other countries of the region with TSV-infected postlarval *P. vannamei* may not have happened so rapidly, if at all, had there not been so much reluctance by the Ecuadorian shrimp-farming industry to accept the notion that TS had a viral etiology. Other important factors in the biology of TSV that have contributed to its spread include its being carried in the gut contents and feces (at least within and among adjacent farms) by aquatic insects such as the water boatman and by gulls (Lightner 1996b; Garza et al. 1997). That survivors of TSV infections (in either susceptible *P. vannamei* or in resistant *P. stylirostris*) remain persistently infected by the virus, perhaps for life, provides the virus with the opportunity for both horizontal and vertical transmission (Hasson, Lightner, Mohney, et al. 1999b).

White Spot Syndrome Baculo-Like Virus (WSSV) Complex

At least five viruses in the white spot syndrome (WSS) complex have been named in the literature (for a review see Lightner 1996a; Lightner et al. 1998). They appear to be very similar viruses. The names of the viruses and the diseases they cause are summarized in Table 3.1. All are very similar in morphology and replicate in the nuclei of infected cells, which are typically in tissues of ectodermal and mesodermal origin. Infected nuclei in enteric tissues (i.e., midgut mucosa and hepatopancreatic tubule epithelium) are rarely, if ever, present (Lightner 1996a).

Isolated virions from this WSS complex, when contrasted by negative staining and viewed by TEM, are enveloped, elliptical rods, averaging approximately 130 nm in diameter by 350 nm in length with size variations ranging from 100 to 140 nm and 270 to 420 nm, respectively. Some virions possess a tail-like appendage at one extremity that is an extension of the envelope. Nucleocapsids are rod shaped with blunted ends, measure 90 nm by 360 nm (range of 70 to 95 nm by 300 to 420 nm, respectively), and display a superficially segmented appearance with an angle of 90° to the long axis of the particle. The nucleic acid of WSS viruses is a large single molecule of circular dsDNA that is between 100 and 200 Kb in length (Wongteerasupaya et al. 1995; Durand et al. 1996; Lo et al. 1996). The characteristics of the WSSV complex are most like members of the family Baculoviridae (Francki et al. 1991; Murphy et al. 1995). However, the nonoccluded viruses in this family were removed from the family and placed temporarily in a group of viruses of uncertain taxonomic position (Murphy et al. 1995).

The WSSV complex infects and causes serious disease in many species of penaeid shrimp and in a variety of other decapod crustaceans. Among the Asian penaeids reported to be infected by WSSV complex are *P. monodon, P. semisulcatus, P. japonicus, P. chinensis, P. penicillatus* (redtail prawn), *P. indicus, P. merguiensis, Trachypenaeus curvirostris* (southern rough shrimp), and *Metapenaeus ensis* (greasyback shrimp) (Chang, Chen, and Wang 1998; Lightner 1996a; Wang et al. 1998). Western Hemisphere penaeids are also susceptible to WSSV infection and disease (Lightner 1996a; Tapay et al. 1997; Lightner et al. 1998). WSSV infects and can cause serious disease in *Macrobrachium rosenbergii* (giant river prawn) and in the North American crayfish, *Procambarus clarkii* (red swamp crayfish). WSSV also can infect but does not seem to cause significant disease in a variety of marine crabs and spiny lobsters (Chang, Chen, and Wang 1998; Wang et al. 1998).

Following its appearance in 1992-1993 in Northeast Asia, the WSSV has spread very rapidly throughout most of the shrimp-growing regions of Asia and the Indo-Pacific. Documented reports of the WSSV epizootics in Asia now include Taiwan, China, Korea, Thailand, Indonesia, Vietnam, Malaysia, India, Sri Lanka, and Bangladesh (Inouye et al. 1994, 1996; Nakano et al. 1994; Takahashi et al. 1994; Chen 1995; Chou et al. 1995; Huang et al. 1995; Wang et al. 1995; Wongteerasupaya et al. 1995; Lo et al. 1996; Chang, Chen, and Wang 1998; Kasornchandra, Boonyaratpalin, and Itami 1998; Wang et al. 1998).

In the American penaeids, natural infections by WSSV have been documented every year since an initial outbreak in late 1995 involving *P. setiferus* cultured in Texas (Lightner 1996a; Lightner et al. 1998; Lightner and Redman 1998a). Since that initial outbreak, WSSV was detected on two separate occasions in 1996 and 1997 in captive wild crayfish being held at the National Zoo in Washington, DC, where the virus was associated with severe disease and mortalities in the crayfish, *Orochnectes punctimanus* and *Procambarus* sp. (Richman et al. 1997). In penaeids, WSSV has been found in two separate regions of the United States since 1995. The virus was detected in wild *P. setiferus* and *P. duorarum* collected off the coast of Texas in 1997 and 1998. In South Carolina, WSSV was found in 1997 and 1998 in wild *P. setiferus* and in a variety of other wild decapod crustaceans, in cultured stocks of *P. setiferus* in 1997, and in cultured stocks of *P. stylirostris* and *P. vannamei* in 1998. In the latter two cases, WSSV infections occurred at shrimp farms and were associated with >95 percent cumulative losses to the affected farms within days to weeks after the onset of disease.

Laboratory studies have provided some information on the susceptibility of American penaeids to WSSV. Postlarval (PL) and juvenile stages of *P. vannamei, P. stylirostris, P. setiferus, P. aztecus,* and *P. duorarum* were experimentally infected by a WSSV isolate that was originally derived from cultured *P. monodon* from Thailand. Challenge of PL stages with the virus resulted in severe infections in *P. setiferus* and *P. vannamei* and in less severe infections in PL *P. aztecus* and PL *P. duorarum.* The results from WSSV challenge studies with juveniles followed a similar trend to that observed with the PL stages. Challenge resulted in positive infections in juveniles of all four species, with severe infections and 100 percent cumulative mortalities resulting in juvenile *P. setiferus* and *P. vannamei,* while only moderate infection, disease, and mortalities occurred in challenged juvenile *P. aztecus,* and no signs of disease or mortality resulted in challenged juvenile *P. duorarum* (Lightner et al. 1998).

Yellow-Head Virus Group (YHV)

Yellow-head virus (YHV) from Southeast Asia (Boonyaratpalin et al. 1993; Chantanachookin et al. 1993; Wongteerasupaya et al. 1995) and the morphologically similar lymphoid organ virus (LOV) (Spann, Vickers, and Lester 1995) and gill-associated virus (GAV) from Australian, *P. monodon* (Spann et al. 1997) are rod-shaped, enveloped viruses that replicate in the cytoplasm of infected cells. For the purpose of this chapter these viruses will be called YHV. The YHV virion is enveloped and measures 44 by 173 nm in length (with a range of 38 to 50 nm by 160 to 186 nm, respectively), contains a cylindrical nucleocapsid of ~15 nm in diameter and a single piece of ssRNA as its genome. While not yet adequately characterized, YHV has been suggested to be a member of the families *Rhabdoviridae, Paramyxoviridae* (Boonyaratpalin et al. 1993; Chantanachookin et al. 1993; Flegel et al. 1995; Kasornchandra, Boonyaratpalin, and Supametaya 1995; Spann, Vickers, and Lester 1995; Lightner 1996a), and most recently, the *Coronaviridae* (P. J. Walker, personal communication, CSIRO, Brisbane, Australia).

Yellow-head virus causes a serious disease in *P. monodon* in intensive culture systems in Southeast Asia and India. YHV is widespread in cultured stocks of *P. monodon* in the southeast Asian and Indo-Pacific countries of Thailand, China, Malaysia, Indonesia, India (Flegel et al. 1995; Flegel 1997; Lightner 1996a). The virus was found in Taiwan in *P. japonicus* that were co-infected with WSSV (Wang et al. 1998). Likewise, the closely related virus GAV causes serious disease in *P. monodon* cultured in Australia (Spann et al. 1997). Disease due to YHV and GAV infection typically occurs in juveniles and subadults (Boonyaratpalin et al. 1993; Flegel et al. 1995; Lightner 1996a). The brackish water shrimps, *Palaemon styliferus* (grass shrimp) and *Euphausia superba* and *Acetes* sp. (planktonic shrimps), which are often resident in shrimp ponds in Thailand, were found in bioassays with healthy *P. monodon* to carry YHV (Flegel et al. 1995; Alday de Graindorge and Flegel 1999). Although resistant to YHV in ponds, *P. merguiensis* and *M. ensis* were found to be experimentally infected by YHV in laboratory challenge studies (Flegel et al. 1995).

American penaeids were found to be highly susceptible to experimental infection by YHV. Although the PL stages were refractory to infection, juveniles of *P. vannamei, P. stylirostris, P. setiferus, P. aztecus,* and *P. duorarum* were found to be susceptible to challenge by the virus and to suffer significant disease (Lu et al. 1995; Lightner 1996a; Lightner et al. 1998).

DIAGNOSTIC METHODS

Methods for the detection of penaeid shrimp viruses and diagnosis of the diseases that they cause can be as simple as the observation of specific gross signs that accompany infection by particular viruses (see Table 3.4). Of the four viruses being reviewed in the present chapter, all four can cause unique signs of infection that may aid in diagnosis. Specifically, chronic infections by IHHNV typically result in RDS, with its unique cuticular deformities and high numbers of undersized shrimp in affected populations, in *P. vannamei* and sometimes in *P. stylirostris.* TSV produces unique cuticular melanized spots in the transitional phase of its disease cycle in *P. vannamei.* As the name implies, some shrimp in the recovery phase of infection by WSSV may display prominent subcuticular white spots. Likewise, in populations of *P. monodon* with epizootic disease due to YHV, some affected shrimp display yellowish discoloration of the gills and hepatopancreatic region of the gnathothorax (Lightner 1996a). However, in American penaeids, infection by WSSV or YHV may result in acute disease and near 100 percent cumulative mortalities, and, hence, these diagnostic gross signs may seldom, if ever, be present.

Histology is among the most important and commonly used diagnostic method for penaeid diseases in general (Lightner 1996a; Bell and Lightner 1988), and each of the four viruses reviewed in the present chapter (IHHNV, TSV, WSSV, and YHV) produces unique lesions at some phase of its infection cycles to provide a definitive diagnosis (see Table 3.4). These specific lesions and methods used for histological diagnosis of infections by these viruses are readily available from a number of sources (Baticados et al. 1990; Brock and Lightner 1990; Lightner 1993, 1996a; Brock and Main 1994; Chanratchakool et al. 1994; Johnson 1995; OIE 1997).

Several serodiagnostic methods have been developed for use in shrimp disease diagnosis (see Table 3.4). However, despite the considerable research and development efforts (by research groups in Asia, Europe, and North America) that have attempted to develop monoclonal antibodies (MABs) to

TABLE 3.4. Summary of Diagnostic and Detection Methods for the Major Viruses of Concern to the Shrimp Culture Industries of the Americas

METHOD	IHHNV	TSV	YHV	WSSV
Direct BF/LM	-	++	++	++
Phase contrast	-	+	-	+
Dark-field LM	-	-	-	++
Histopathology	++	+++	+++	++
Enhancement/histology	++	+	+	+
Bioassay/histology	+	+++	+++	++
Transmission EM	+	+	+	+
Fluorescent antibody	r&d	r&d	-	r&d
ELISA with PABs/MABs	r&d	r&d	r&d	-
DNA Probes	+++/K	+++/K	+++/K	++/K
PCR/RT-PCR	+++	+++	+++	+++

Sources: Modified from Lightner 1996a; Lightner and Redman 1998b.

Definitions for virus acronyms and symbols:

- \- = no known or published application of technique
- \+ = application of technique is known or published
- ++ = application of technique is considered to provide sufficient diagnostic accuracy or pathogen detection sensitivity for most applications
- +++ = technique provides a high degree of sensitivity in pathogen detection
- K = diagnostic kit available from DiagXotics, Inc. (Wilton, Connecticut)
- r&d = techniques in research and development phase

Methods:

- before = bright field LM of tissue impression smears, wet-mounts, stained whole mounts
- LM = light microscopy
- EM = electron microscopy of sections or of purified or semi-purified virus
- ELISA = enzyme-linked immunosorbent assay
- PABs = polyclonal antibodies
- MABs = monoclonal antibodies
- PCR = DNA amplification by polymerase chain reaction
- RT-PCR = PCR after reverse transcription of viral RNA genome

the penaeid viruses IHHNV, YHV, TSV, and WSSV, only for TSV has a serodiagnostic method been developed and made available commercially to the industry (Poulos et al. in press). While serodiagnostic methods for IHHNV were developed (Poulos et al. 1994), the MABs developed were of the IGM class. These antibodies reacted specifically with purified IHHNV or its capsid proteins in Western blots. However, they also reacted nonspecifically with components in normal shrimp tissue, resulting in false positive reactions with uninfected shrimp tissue samples in ELISA-based assays (Poulos et al. 1994; Lightner and Redman 1998c). Although the development of serological tests for the more important shrimp pathogens has lagged behind the development of molecular detection and classical diagnostic methods, it is very likely that the use of tests based on polyclonal and monoclonal antibodies will become much more common in shrimp diagnostic laboratories in the next few years. Because of their speed, versatility, relatively low cost, simplicity, and reasonably good sensitivity, monoclonal antibody-based tests are potentially very useful as routine diagnostic tests, even in the most modestly equipped diagnostic laboratories (Reddington and Lightner 1994).

Molecular methods (gene probes and DNA amplification using the polymerase chain reaction [PCR]) are increasingly becoming the standard for the detection and diagnosis of shrimp viruses. Nearly a decade has passed since the first nonradioactively labeled gene probe was developed and applied to the diagnosis of the IHHNV (see Table 3.4) (Mari, Bonami, and Lightner 1993). Since then, molecular probe methods for all of the important penaeid shrimp viruses have been developed and made commercially available as diagnostic kits and as labeled probes marketed under the product name ShrimProbes (DiagXotics, Wilton, Connecticut*).

As genome sequence information became available for each of the viruses of concern, PCR and reverse transcriptase PCR (RT-PCR) methodologies were developed that have added even more potential sensitivity to the detection capabilities for shrimp viruses. In PCR, small, often undetectable amounts of DNA can be amplified to produce detectable quantities of the target DNA. This is accomplished by using unique and specific oligonucleotide primers designed to bind to the target sequences of the positive and negative sense strands of the DNA molecule. The primers, along with a buffered reaction mixture containing nucleotides and DNA polymerase, fill in the specific nucleotide sequence of the positive and negative DNA strands that lie between the primers. Using a programmable thermal cycler, the process of PCR is repeated from 20 to 40 times. In the case of the RNA viruses (YHV and TSV), a step is added to the assay in which the ssRNA strand is converted to cDNA by reverse transcription. After that step, the resulting cDNA is amplified by PCR, as would be the specific gene sequence from a DNA virus. The resultant PCR or RT-PCR product may then be compared to a known standard for the virus being assayed, using gel electrophoresis. Alternatively, the PCR product may be blotted onto a membrane (or transferred using the Southern transfer method) and tested with a specific genomic probe by reaction with a specific DNA probe. In some applications PCR products themselves may be labeled with DIG and used as specific DNA probes (Innis et al. 1990; Perkin Elmer 1992). Details of these methods are available from a number of recent publications: IHHNV (Lightner

*Use of trade or manufacturer's name does not imply endorsement.

1996a); TSV (Nunan, Poulos, and Lightner 1998b); YHV (Wongteerasupaya, Boonsaeng, et al. 1997; Tang and Lightner 1999); and WSSV (Kimura et al. 1996; Lo et al. 1996; Takahashi et al. 1996; Nunan and Lightner 1997).

THE THREAT OF WSSV AND YHV TO THE AMERICAS

As discussed earlier in this review, WSSV and YHV were thought to be limited to Asia, until WSSV appeared in November 1995 in cultured *P. setiferus* in Texas (Lightner, Redman, Poulos, Nunan, Mari, and Hasson 1997, 1998; Lotz 1997; Nunan, Poulos, and Lightner 1998a). Since that initial recognition of the virus in Texas, it has been found every year in wild or cultured penaeids in Texas or South Carolina, as well as in wild crabs and in captive wild freshwater crayfish (see the WSSV section). Following the WSSV outbreak in Texas, various sources for introduction of the pathogen were considered. The stocks being cultured at the affected farm had been produced from captive wild broodstock of *P. setiferus,* and it was at the end of the culture cycle (the affected shrimp were late juvenile to subadults). The occurrence of a WSSV-caused epizootic late in the culture cycle of a highly susceptible species indicated that the virus was introduced to the farm during culture, rather than being introduced with the PL stages when the farm was stocked months earlier. A similar scenario applies to the WSSV epizootics of 1997 and 1998 at the affected farms in South Carolina. The affected farms were culturing domesticated SPF lines (Wyban et al. 1992; Lotz et al. 1995) of *P. vannamei* or *P. stylirostris* with no prior history of WSSV, and since only SPF (i.e., free of TSV, IHHNV, WSSV, YHV, and other important pathogens) *P. vannamei* had been cultured in Texas and South Carolina for two to three years preceding the initial appearance in these states, it is extremely unlikely that these viruses were not introduced with the SPF shrimp stocks. Their source had to have been from some other activity.

That source could be imported commodity shrimp. Despite the presence of large monocultures of highly susceptible *P. vannamei* and *P. stylirostris* being grown in a number of major shrimp-growing countries in Latin America, nowhere but in the United States has WSSV been detected in wild or cultured shrimp stocks. The difference between these WSSV-negative countries and the United States is imports. The United States is a major market for shrimp, and each year it imports thousands of tons of cultured penaeid shrimp from Asian countries (Filose 1995), where WSSV and YHV have been enzootic and causing serious epizootics since 1992. In marked contrast, the other countries of the Americas where penaeid shrimp are farmed or fished do not import significant quantities of Asian shrimp. According to U.S. Department of Commerce (USDC) data, imports of penaeid shrimp for the U.S. market have increased significantly in recent years, and since 1995 over half of the imported shrimp marketed in the United States now comes from farm-cultured stocks (Filose 1995; New 1997; USDC 1997). More than half of those imports come from countries (Thailand, India, and China) that, since 1992, have been severely impacted by WSSV and YHV. Imported commodity shrimp are distributed throughout the United States, and some of the imported shrimp are reprocessed at shrimp-packing plants situated on coastal bays and estuaries where native penaeid nursery grounds also occur. Imported small-sized, heads-on shrimp are also packed specifically for sale as bait to U.S. sport fishermen (JSA 1997; USDC 1997; Lightner, Redman, Poulos, Nunan, Mari, and Hassan 1997).

Emergency harvests are commonly employed in Asia to salvage marketable shrimp crops with developing epizootics due to these viruses (Flegel, Fegan, and Sriurairatana 1995; Flegel et al. 1995; Jory 1996; Lightner, Redman, Poulos, Nunan, Mari, and Hasson 1997). Because of this practice, small count size (40 to 90 count) *P. monodon* displaying gross signs of WSSV infection (e.g., cuticular white spots and reddish pigmentation) began to appear in U.S. retail outlets around 1994 and are still frequently found in U.S. retail outlets (Nunan, Poulos, and Lightner 1998a; Lightner et al. in press). PCR assays of samples of these shrimp confirmed the presence of WSSV and YHV. Live shrimp bioassays with SPF *P. vannamei* or *P. stylirostris* as the indicator for infectious virus have given positive results for both WSSV and YHV (see Table 3.5).

These studies have confirmed that frozen imported commodity shrimp do in fact contain nonindigenous viruses that are infectious to Western Hemisphere penaeids and other decapods.

Genetic and virulence comparisons of WSSV isolates from Asia, Texas, South Carolina, the National Zoo, and imported commodity shrimp have shown little or no difference among the isolates (Lo et al. 1999; Wang et al. 1999). These results suggest that WSSV is pandemic and that the same or a few very closely related strains of the virus have spread from the initial epicenter of the disease in East Asia.

STRATEGIES FOR CONTROL OF VIRUS DISEASE

Numerous strategies have been attempted for the control of viral diseases in penaeid shrimp aquaculture. These strategies have ranged from the use of improved husbandry practices to stocking specific-pathogen-free (SPF) or specific-pathogen-resistant (SPR) species or stocks (Lotz 1997). In the Americas, many strategies have been employed in efforts to reduce production losses due to the enzootic viruses *Baculovirus penaei* (BP) (Couch 1974; Lightner 1996a), IHHNV, and TSV. Improved husbandry practices have been successfully used to control BP, and for nearly a decade, this virus has seldom been reported as an economic constraint to successful shrimp culture (OIE 1997; Lightner and Redman 1998b, c).

Until recently, the popularity and use of the relatively IHHNV-resistant species *P. vannamei,* in preference to the culture of the more IHHNV-

TABLE 3.5. Results of PCR Assays for WSSV and YHV Using Samples of Frozen Imported *Penaeus monodon* Tails from Thailand and Other Southeast Asian Sources*

UAZ ID #	Source/ Origin	Species/ Size	Gross Signs	PCR Primers WSSV	PCR Primers YHV	PCR Results	Bioassay UAZ ID #	Bioassay Indicator Species	Bioassay Results
95-204	Tucson Retail Thailand	*P. monodon* 51/60 count	reddish w/spots	F-6581 R-7632	N/A	++++WSSV	95-204	*P. stylirostris*	+++ YHV + WSSV
95-319	California Retail / Asia	*P. monodon* ~40 count	reddish w/spots	F-6581 R-7632	N/A	+++ WSSV	N/A	N/A	N/A
95-320	Washington Retail / Asia	*P. monodon* 26/30 count	none	F-6581 R-7632	N/A	Negative WSSV	N/A	N/A	N/A
96-87	Tucson Retail Thailand	*P. monodon* ~40 count	reddish w/spots	F-6581 R-7632	N/A	++++ WSSV	N/A	N/A	N/A
96-101	Tucson Retail Thailand	*P. monodon* ~40 count	none	F-6581 R-7632	N/A	Negative WSSV	N/A	N/A	N/A
96-104	Texas Retail Asia	*P. monodon* ~40 count	none	F-6581 R-7632	N/A	Negative WSSV	N/A	N/A	N/A
96-109	Tucson Retail Thailand	*Macro-brachium rosenbergii*	w/spots	F-6581 R-7632	N/A	+ WSSV	N/A	N/A	N/A
96-115	Tucson Retail Thailand	*P. monodon* ~40 count	reddish w/spots	F-6581 R-7632	N/A	+++ WSSV	N/A	N/A	N/A
98-78	Tucson Retail Asia	*P. monodon* 90 count	reddish w/spots	E-12 E-14	Thai	++++ WSSV + YHV	B98-210/2	*P. vannamei*	++++ WSSV (100% mortality) Negative YHV
98-111/A	Blue Sky Thailand	*P. monodon*	N/D	E-12 E-14	Thai	+++ WSSV Negative YHV	B98-146	*P. vannamei*	Negative WSSV Negative YHV

TABLE 3.5 *(continued)*

98-111/B	Suratthani Thailand	"White Tiger"**	N/D	E-12 E-14	Thai	+++ WSSV Negative YHV	B98-146	*P. vannamei*	Negative WSSV Negative YHV
98-111/C	Transamut Thailand	*P. monodon*	N/D	E-12 E-14	Thai	+ WSSV Negative YHV	N/A	N/A	N/A
98-111/D	Transamut Thailand	*P. monodon* & "White Tiger"	N/D	E-12 E-14	Thai	+ WSSV Negative YHV	N/A	N/A	N/A
98-111/E	Blue Sky Thailand	*P. monodon*	N/D	E-12 E-14	Thai	+ WSSV + YHV	B98-146	*P. vannamei*	Negative WSSV + YHV (PCR +; no mortality)
98-111/F	Suratthani Thailand	"White Tiger"	N/D	E-12 E-14	Thai	Negative WSSV Negative YHV	N/A	N/A	N/A
98-131	Tucson Retail Thailand	*P. monodon* 40 count	reddish w/spots	E-12 E-14	Thai	++++ WSSV Negative YHV	B98-210/3	*P. vannamei*	++++ WSSV (100% mortality) Negative YHV

* 98 to 111 samples were provided by the National Fisheries Institute; the remaining samples were purchased from retail outlets in Tucson, AZ, Washington State, California, and Texas.
** "White tiger" shrimp may be a marketing name for a pale variety of *P. monodon* or *P. semisulcatus.*
N/D = not described
N/A = not analyzed
+ = a weak positive test result; +++ = strong positive test result; ++++ = very strong positive test result
F-6581 and R-7632 from Nunan and Lightner (1997)
E-12 and E-14 = WSSV primers developed by S. Durand (unpublished, University of Arizona)
Thai = YHV primers developed by Wongteerasupaya et al. (1997)

susceptible *P. stylirostris,* was characteristic of the shrimp-farming industries of the Americas. The popularity of *P. vannamei* began to decline when TSV emerged as a very serious pathogen of this species in 1992 and then spread to virtually all of the shrimp-growing regions of the Americas during the ensuing four years. Because *P. stylirostris* was found to be innately TSV resistant, at least two domesticated, genetically selected SPR strains of this species, which are resistant to IHHN disease, are currently being developed and marketed in the Americas. In some regions, these SPR stocks of TSV and IHHNV-resistant *P. stylirostris* are replacing *P. vannamei* stocks in culture. Other shrimp-farming interests are using wild or domesticated stocks of *P. vannamei* that show improved resistance to TSV. Although resistance to TSV was used as a selection criterion for the domesticated stocks of *P. vannamei,* natural selection for TSV resistance appears to be occurring in wild stocks where TSV has been enzootic for several years. The same selective process for IHHNV resistance seems to be occurring in some wild stocks of *P. stylirostris.* Through selective breeding programs by broodstock producers, we will very likely see continued improvements in TSV and IHHNV resistance in domesticated lines of *P. vannamei* and *P. stylirostris.* However, because of natural selection where these viruses have become enzootic in wild populations, improved resistance in wild stocks of these species is very likely developing now and will improve with time (Lightner and Redman 1998b, c; Lightner et al. in press).

Control strategies for WSSV and YHV in the Western Hemisphere should be based on prevention by exclusion. Success with this depends in large part upon the availability and the use of sensitive diagnostic tools and on avoiding the introduction of these viruses through exclusion of live and frozen shrimp products from regions or facilities that have a history of infections by these viruses.

REFERENCES

ADB/NACA. In press. *Final Report on the Regional Study and Workshop on Aquaculture Sustainability and the Environment* (RETA 5534). Asian Development Bank and Network of Aquaculture Centres in Asia-Pacific. Bangkok, Thailand: NACA.

Alday de Graindorge, V. and T.W. Flegel. 1999. *Diagnosis of Shrimp Diseases.* Bangkok, Thailand: FAO and Mulitmedia Asia Co., Ltd.

Baticados, M.C.L., E.R. Cruz-Lacierda, M.C. de la Cruz, R.C. Duremdez- Fernandez, R.Q. Gacutan, C.R. Lavilla-Pitogo, and G.D. Lio-Po. 1990. Diseases of penaeid shrimps in the Philippines. *Aquaculture Extension Manual* No. 16. Iloilo, Philippines: Aquaculture Department, Southeast Asian Fisheries Development Center (SEAFDEC), Tigbauan.

Bell, T.A. and D.V. Lightner. 1984. IHHN virus: Infectivity and pathogenicity studies in *Penaeus stylirostris* and *Penaeus vannamei. Aquaculture* 38:185-194.

Bell, T.A. and D.V. Lightner. 1987. IHHN disease of *Penaeus stylirostris:* Effects of shrimp size on disease expression. *Journal of Fish Diseases* 10:165-170.

Bell, T.A. and D.V. Lightner. 1988. *A Handbook of Normal Shrimp Histology.* Special Publication No. 1. Baton Rouge, LA: World Aquaculture Society.

Bonami, J.R., M. Brehelin, M., J. Mari, B. Trumper, and D.V. Lightner. 1990. Purification and characterization of IHHN virus of penaeid shrimps. *Journal of General Virology* 71:2657-2664.

Bonami J.R., K.W. Hasson, J. Mari, B.T. Poulos, D.V. Lightner. 1997. Taura syndrome of marine penaeid shrimp: Characterization of the viral agent. *Journal of General Virology* 78:313-319.

Boonyaratpalin, S., K. Supamataya, J. Kasornchandra, S. Direkbusarakom, U. Ekpanithanpong, and C. Chantanachookin. 1993. Non-occluded baculo-like virus, the causative agent of yellow head disease in the black tiger shrimp *(Penaeus monodon). Fish Pathology* 28:103-109.

Brock, J.A., R.B. Gose, D.V. Lightner, and K.W. Hasson. 1995. An overview on Taura syndrome, an important disease of farmed *Penaeus vannamei.* In C.L. Browdy and J.S. Hopkins (Eds.), *Swimming Through Troubled Water. Proceedings of the Special Session on Shrimp Farming, Aquaculture '95* (pp. 84-94). Baton Rouge, LA: World Aquaculture Society.

Brock, J.A., R. B. Gose, D. V. Lightner, and K. Hasson. 1997. Recent developments and an overview of Taura syndrome of farmed shrimp in the Americas. In T.W. Flegel and I.H. MacRae (Eds.), *Diseases in Asian Aquaculture III.* Manila, Philippines: Fish Health Section, Asian Fisheries Society.

Brock, J.A. and D.V. Lightner. 1990. Diseases of crustacea. Diseases caused by microorganisms. In O. Kinne (Ed.), *Diseases of Marine Animals,* Volume III (pp. 245-349). Hamburg, Germany: Biologische Anstalt Helgoland.

Brock, J.A. and K. Main. 1994. *A Guide to the Common Problems and Diseases of Cultured* Penaeus vannamei. Honolulu, HI: The Oceanic Institute, Makapuu Point.

Browdy, C.L., J.D. Holloway, C.O. King, A.D. Stokes, J.S. Hopkins, and P.A. Sandifer. 1993. IHHN virus and intensive culture of *Penaeus vannamei:* Effects of stocking density and water exchange rates. *Journal of Crustacean Biology* 13:87-94.

Chang, P.S., H.C. Chen, and Y.C. Wang. 1998. Detection of white spot syndrome associated baculovirus in experimentally infected wild shrimp, crabs and lobsters by in situ hybridization. *Aquaculture* 164: 233-242.

Chanratchakool, P., J.F. Turnbull, S. J. Funge-Smith, I.H. MacRae, and C. Limsuwan. 1994. *Health Management in Shrimp Ponds.* Bangkok, Thailand: Aquatic Animal Health Research Institute, Department of Fisheries, Kasetsart Univeristy.

Chantanachookin, C., S. Boonyaratpalin, J. Kasornchandra, S. Direkbusarakom, U. Ekpanithanpong, K. Supamataya, S. Sriurairatana, and T.W. Flegel. 1993. History and ultrastructure reveal a new granulosis-like virus in *Penaeus monodon* affected by "yellow head" disease. *Diseases of Aquatic Organisms* 17:145-157.

Chen, S.N. 1995. Current status of shrimp aquaculture in Taiwan. In C.L. Browdy and J.S. Hopkins (Eds.), *Swimming Through Troubled Water. Proceedings of the*

Special Session on Shrimp Farming, Aquaculture '95 (pp. 29-34). Baton Rouge, LA: World Aquaculture Society.

Chou, H.Y., C.Y. Huang, C.H. Wang, H.C. Chiang, and C.F. Lo. 1995. Pathogenicity of a baculovirus infection causing white spot syndrome in cultured penaeid shrimp in Taiwan. *Diseases of Aquatic Organisms* 23:165-173.

Couch, J.A. 1974. Free and occluded virus similar to *Baculovirus* in hepatopancreas of pink shrimp. *Nature* 247:229-231.

Durand, S., D.V. Lightner, L.M. Nunan, R.M. Redman, J. Mari, and J.R. Bonami. 1996. Application of gene probes as diagnostic tools for the white spot baculovirus (WSBV) of penaeid shrimp. *Diseases of Aquatic Organisms* 27:59-66.

Filose, J. 1995. Factors affecting the processing and marketing of farmed raised shrimp. In C.L. Browdy and J.S. Hopkins (Eds.), *Swimming Through Troubled Water. Proceedings of the Special Session on Shrimp Farming, Aquaculture '95* (pp. 227-234). Baton Rouge, LA: World Aquaculture Society.

Flegel, T.W. 1997. Major viral diseases of the black tiger prawn *(Penaeus monodon)* in Thailand. In *Proceedings of the NRIA International Workshop, New Approaches to Viral Diseases of Aquatic Animals* (pp. 167-189). Mie, Japan: National Institute of Aquaculture, Nansei, Watarai.

Flegel, T.W., D.F. Fegan, and S. Sriurairatana. 1995. Environmental control of infectious diseases in Thailand. In M. Shariff, J.R. Arthur, and R.P. Subasinghe (Eds.), *Diseases in Asian Aquaculture II* (pp. 65-79). Manila, Philippines: Fish Health Section, Asian Fisheries Society.

Flegel, T.W., S. Sriurairatana, C. Wongteerasupaya, V. Boonsaeng, S. Panyim, and B. Withyachumnarnkul. 1995. Progress in characterization and control of yellow-head virus of *Penaeus monodon.* In C.L. Browdy and J.S. Hopkins (Eds.), *Swimming Through Troubled Water, Proceedings of the Special Session on Shrimp Farming, Aquaculture '95* (pp. 76-83). Baton Rouge, LA: World Aquaculture Society.

Francki, R.I.B., C.M. Fauquet, D.L. Knudson, and F. Brown (Eds.). 1991. *Classification and Nomenclature of Viruses. Archives of Virology.* Wien, NY: Springer-Verlag.

Garza, J.R., K.W. Hasson, B.T. Poulos, R.M. Redman, B.L. White and D.V. Lightner. 1997. Demonstration of infectious Taura syndrome virus in the feces of sea gulls collected during an epizootic in Texas. *Journal of Aquatic Animal Health* 9:156-159.

Hasson, K.W., J. Hasson, H. Aubert, R.M. Redman, and D.V. Lightner. 1997. A new RNA-friendly fixative for the preservation of penaeid shrimp samples for virological detection using cDNA genomic probes. *Journal of Virological Methods* 66:227-236.

Hasson, K.W., D.V. Lightner, J. Mari, J.R. Bonami, B.T. Poulos, L.L. Mohney, R.M. Redman, and J.A. Brock. 1997. The geographic distribution of Taura syndrome virus in the Americas: Determination by histopathology and in situ hybridization using TSV-specific cDNA probes. In *Proceedings of the IV Central American Symposium on Aquaculture, April 22-24, 1997* (pp. 154-155). Tegucigalpa, Honduras: Asociacion Nacional de Acicultores de Honduras.

Hasson, K.W., D.V. Lightner, J. Mari, J.R. Bonami, B.T. Poulos, L.L. Mohney, R.M. Redman, and J.A. Brock. 1999. The geographic distribution of Taura syndrome virus (TSV) in the Americas: Determination by histopathology and in situ hybridization using TSV-specific cDNA probes. *Aquaculture* 171:13-26.

Hasson, K.W., D.V. Lightner, L.L. Mohney, R.M. Redman, B.T. Poulos, and B.L. White. 1999. The Taura syndrome virus lesion development and disease cycle in the Pacific white shrimp, *Penaeus vannamei. Diseases of Aquatic Organisms* 36:81-93.

Hasson, K.W., D.V. Lightner, B.T. Poulos, R.M. Redman, B.L. White, J.A. Brock, and J.R. Bonami. 1995. Taura syndrome in *Penaeus vannamei:* Demonstration of a viral etiology. *Diseases of Aquatic Organisms* 23:115-126.

Holthuis, L.B. 1980. *FAO species catalog.* FAO Fisheries Synopsis No. 125, Volume 1. Rome, Italy: Food and Agriculture Organization of the United Nations.

Huang, J., X.L. Song, J. Yu, and C.H. Yang. 1995. Baculoviral hypodermal and hematopoietic necrosis—study on the pathogen and pathology of the shrimp explosive epidemic disease of shrimp. *Marine Fisheries Research* 16:1-10.

Innis, M.A., D.H. Gelfand, J.J. Sninsky, and T.J. White (Eds.). 1990. *PCR Protocols: A Guide to Methods and Applications.* Berkeley, CA: Academic Press, Inc.

Inouye, K., S. Miwa, N. Oseko, H. Nakano, T. Kimura, K. Momoyama, and M. Hiraoka. 1994. Mass mortalities of cultured Kuruma shrimp *Penaeus japonicus* in Japan in 1993: Electron microscopic evidence of the causative virus. *Fish Pathology* 29:149-158.

Inouye, K., K. Yamano, N. Ikeda, T. Kimura, H. Nakano, K. Momoyama, J. Kobayashi, and S. Miyajima. 1996. The penaeid rod-shaped DNA virus (PRDV), which causes penaeid acute viremia (PAV). *Fish Pathology* 31:39-45.

Jimenez, R. 1992. Sindrome de Taura (Resumen). In *Acuacultura del Ecuador* (pp. 1-16). Guayaquil, Ecuador: Camara Nacional de Acuacultura.

Johnson, S.K. 1995. *Handbook of Shrimp Diseases.* TAMU-SG-90-601(r). College Station, TX: Texas A&M Sea Grant College Program, Texas A&M University.

Jory, D.E. 1996. Marine shrimp farming in the Kingdom of Thailand: Part II. *Aquaculture Magazine* 22:71-78.

JSA. 1997. *An Evaluation of Potential Shrimp Virus Impacts on Wild Shrimp Populations in the Gulf of Mexico and Southeastern U.S. Atlantic coastal waters.* Washington, DC: Joint Subcommittee on Aquaculture.

Kalagayan, G., D. Godin, R. Kanna, G. Hagino, J. Sweeney, J. Wyban, and J. Brock. 1991. IHHN virus as an etiological factor in runt-deformity syndrome of juvenile *Penaeus vannamei* cultured in Hawaii. *Journal of the World Aquaculture Society* 22:235-243.

Kasornchandra, J., S. Boonyaratpalin, and T. Itami. 1998. Detection of white-spot syndrome in cultured penaeid shrimp in Asia: Microscopic observation and polymerase chain reaction. *Aquaculture* 164:243-251.

Kasornchandra, J., S. Boonyaratpalin, and K. Supamataya. 1995. Electron microscopic observations on the replication of yellow-head baculovirus in the lymphoid organ of *Penaeus monodon.* In M. Shariff, J.R. Arthur, and R.P. Subasinghe (Eds.), *Diseases in Asian Aquaculture* (pp. 99-106). Manila, Philippines: The Fish Health Section, Asian Fisheries Society.

Kimura, T., K. Yamano, H. Nakano, K. Momoyama, M. Hiraoka, and K. Inouye. 1996. Detection of penaeid rod-shaped DNA virus (PRDV) by PCR. *Fish Pathology* 31:93-98.

Lightner, D.V. 1993. Diseases of penaeid shrimp. In J.P. McVey (Ed.), *CRC Handbook of Mariculture: Crustacean Aquaculture.* Second edition (pp. 393-486). Boca Raton, FL: CRC Press.

Lightner, D.V. 1996a. *A Handbook of Shrimp Pathology and Diagnostic Procedures for Diseases of Cultured Penaeid Shrimp.* Baton Rouge, LA: World Aquaculture Society.

Lightner, D.V. 1996b. The penaeid shrimp viruses IHHNV and TSV: Epizootiology, production impacts, and role of international trade in their distribution in the Americas. *Revue Scientifique et Technique, Office International des Épizooties* 15:579-601.

Lightner D.V., T.A. Bell, R.M. Redman, L.L. Mohney, J.M. Natividad, A. Rukyani, and A. Poernomo. 1992. A review of some major diseases of economic significance in penaeid prawns/shrimps of the Americas and IndoPacific. In M. Shariff, R. Subasinghe and J.R. Arthur (Eds.), *Diseases in Asian Aquaculture I* (pp. 57-80). Manila, Philippines: Fish Health Section, Asian Fisheries Society.

Lightner, D.V., K.W. Hasson, R.M. Redman, and B.L. White. 1998. Experimental infection of Western Hemisphere penaeid shrimp (Crustacea: Decapoda) with Asian isolates of white spot and yellow head syndrome viruses. *Journal of Aquatic Animal Health* 10:271-281.

Lightner, D.V. and R.M. Redman. 1998a. Emerging crustacean diseases. In A.S. Kane and S.L. Poynton (Eds.), *Proceedings Third International Symposium on Aquatic Animal Health, August 30th- September 3rd, 1998* (pp. 68-71). Baltimore, MD: APC Press.

Lightner, D.V. and R.M. Redman. 1998b. Shrimp diseases and current diagnostic methods. *Aquaculture* 164:201-220.

Lightner, D.V. and R.M. Redman. 1998c. Strategies for the control of viral diseases of shrimp in the Americas. *Fish Pathology* 33:165-180.

Lightner, D.V., R.M. Redman, and T.A. Bell. 1983. Observations on the geographic distribution, pathogenesis, and morphology of the baculovirus from *Penaeus monodon* Fabricius. *Aquaculture* 32:209-233.

Lightner, D.V., R.M. Redman, T.A. Bell, and J.A. Brock. 1983. Detection of IHHN virus in *Penaeus stylirostris* and *P. vannamei* imported into Hawaii. *Journal of the World Mariculture Society* 14:212-225.

Lightner D.V., R.M. Redman, K.W. Hasson, and C.R. Pantoja. 1995. Taura syndrome in *Penaeus vannamei:* Histopathology and ultrastructure. *Diseases of Aquatic Organisms* 21:53-59

Lightner, D.V., R.M. Redman, B.T. Poulos, L.M. Nunan, J.L. Mari, and K.W. Hasson. 1997. Risk of spread of penaeid shrimp viruses in the Americas by the international movement of live and frozen shrimp. *Revue Scientifique et Technique, Office International des Épizooties* 16:146-160.

Lightner, D.V., R.M. Redman, B.T. Poulos, L.M. Nunan, J.L. Mari, K.W. Hasson, and J.R. Bonami. 1997. Taura syndrome: Etiology, pathology, hosts and geographic distribution, and detection methods. In *Proceedings of the NRIA Interna-*

tional Workshop, New Approaches to Viral Diseases of Aquatic Animals (pp. 190-202). Mie, Japan: National Institute of Aquaculture, Nansei, Watarai.

Lightner, D.V., R.M. Redman, B.T. Poulos, L.M. Nunan, L.L Mohney, J.L. Mari, K.W. Hasson, C. R. Pantoja, K.T. Nelson, J.L. Zhou, Q. Wang, J. Garza, and B.L. White. In press. Viral diseases of shrimp in the Americas: Diagnosis, distribution, and control strategies. In *Memory Book for the Congreso Latinoamericano de Camaricultura,* Panama.

Lo, C.F., H.C. Hsu, M.F. Tsai, C.H. Ho, S.E. Peng, G.H. Kou, and D.V. Lightner. 1999. Specific genomic DNA fragment analysis of different geographical clinical samples of shrimp white spot syndrome virus. *Diseases of Aquatic Organisms* 35:1175-185.

Lo, C.F., J.H. Leu, C.H. Chen, S.E. Peng, Y.T. Chen, C.M. Chou, P.Y. Yeh, C.J. Huang, H.Y. Chou, C.H. Wang, and G.H. Kou. 1996. Detection of baculovirus associated with white spot syndrome (WSBV) in penaeid shrimps using polymerase chain reaction. *Diseases of Aquatic Organisms* 25:133-141.

Lotz, J.M. 1997. Special topic review: Viruses, biosecurity, and specific pathogen-free stocks in shrimp aquaculture. *World Journal of Microbiology and Biotechnology* 13:405-413.

Lotz J.M., C.L. Browdy, W.H. Carr, P.F. Frelier, and D.V. Lightner. 1995. USMSFP suggested procedures and guidelines for assuring the specific pathogen status of shrimp broodstock and seed. In C.L. Browdy and J.S. Hopkins (Eds.), *Swimming Through Troubled Water. Proceedings of the Special Session on Shrimp Farming, Aquaculture '95* (pp. 66-75). Baton Rouge, LA: World Aquaculture Society.

Lu, Y., L.M. Tapay, P.C. Loh, and J.A. Brock. 1995. Infection of the yellow head baculo-like virus in two species of penaeid shrimp, *P. stylirostris* and *P. vannamei. Journal of Fish Diseases* 17:649-656.

Mari, J., J.R. Bonami, and D.V. Lightner. 1993. Partial cloning of the genome of infectious hypodermal and hematopoietic necrosis virus, an unusual parvovirus pathogenic for penaeid shrimps; diagnosis of the disease using a specific probe. *Journal of General Virology* 74:2637-2643.

Mari, J., J.R. Bonami, and D.V. Lightner. 1998. Taura syndrome of penaeid shrimp: Cloning of viral genome fragments and development of specific gene probes. *Diseases of Aquatic Organisms* 33:11-17.

Murphy, F.A., C.M. Fauquet, M.A. Mayo, A.W. Jarvis, S.A. Ghabrial, M.D. Summers, G.P. Martelli, and D.H.L. Bishop (Eds.). 1995. *The Classification and Nomenclature of Viruses. Archives of Virology.* Wein, NY: Springer-Verlag.

Nakano, H., H. Koube, S. Umezawa, K. Momoyama, M. Hiraoka, K. Inouye, and N. Oseko. 1994. Mass mortalities of cultured Kuruma shrimp, *P. japonicus,* in Japan in 1993: Epizootiological survey and infection trials. *Fish Pathology* 29:135-139.

New, M.B. 1997. Aquaculture and the capture fisheries—Balancing the scales. *World Aquaculture* 28:11-30.

Nunan, L.M. and D.V. Lightner. 1997. Development of a non-radioactive gene probe by PCR for detection of white spot syndrome virus (WSSV). *Journal of Virological Methods* 63:193-201.

Nunan, L.M., B.T. Poulos, and D.V. Lightner. 1998a. The detection of white spot syndrome virus (WSSV) and yellow head virus (YHV) in imported commodity shrimp. *Aquaculture* 160:19-30.

Nunan, L.M., B.T. Poulos, and D.V. Lightner. 1998b. Reverse transcription polymerase chain reaction (RT-PCR) used for the detection of Taura syndrome virus (TSV) in experimentally infected shrimp. *Diseases of Aquatic Organisms* 34:87-91.

OIE. 1997. *Diagnostic Manual for Aquatic Diseases.* Paris: Office International des Epizooties.

Overstreet, R.M., D.V. Lightner, K.W. Hasson, S. McIlwain, and J. Lotz. 1997. Susceptibility to TSV of some penaeid shrimp native to the Gulf of Mexico and Southeast Atlantic Ocean. *Journal of Invertebrate Pathology* 69:165-176.

Perkin, Elmer. 1992. *DNA Thermal Cycler 480 User's Manual.* Norwalk, CT: The Perkin Elmer Corporation.

Poulos, B.T., R. Kibler, D. Bradley-Dunlop, L.L. Mohney, and D.V. Lightner. in press. Production and use of antibodies for the detection of the Taura syndrome virus in penaeid shrimp. *Diseases of Aquatic Organisms* 37:99-106.

Poulos, B.T., D.V. Lightner, B. Trumper, and J.R. Bonami. 1994. Monoclonal antibodies to the penaeid shrimp parvovirus, infectious hypodermal and hematopoietic necrosis virus (IHHNV). *Journal of Aquatic Animal Health* 6:149-154.

Reddington, J. and D. Lightner. 1994. Diagnostics and their application to aquaculture. *World Aquaculture* 25:1-48.

Richman, L.K., R.J. Montali, D.K. Nichols, and D.V. Lightner. 1997. A newly recognized fatal baculovirus infection in freshwater crayfish. In C.K. Baer (Ed.) *Proceedings of the American Association of Zoo Veterinarians* (pp. 262-264).

Rosenberry, B. 1992. IHHN virus hits Philippines. *World Shrimp Farming* 17(3):6-7.

Rosenberry, B. 1993. Taura syndrome hits farms in Ecuador—Again. *Shrimp News International* 18(3):6.

Rosenberry, B. 1994a. Update on Taura syndrome in Ecuador. *Shrimp News International* 19(3):2-4.

Rosenberry, B. 1994b. *World Shrimp Farming 1994.* San Diego, CA: Shrimp News International.

Rosenberry, B. (Ed.). 1998. *World Shrimp Farming 1998.* San Diego, CA: Shrimp News International.

Spann, K.M., J.A. Cowley, P.J. Walker, and R.J.G. Lester. 1997. A yellow-head- like virus from *Penaeus monodon* cultured in Australia. *Diseases of Aquatic Organisms* 31:169-179.

Spann, K.M., J.E. Vickers, and R.J.G. Lester. 1995. Lymphoid organ virus of *Penaeus monodon. Diseases of Aquatic Organisms* 26:127-134.

Takahashi, Y., T. Itami, M. Kondo, M. Maeda, R. Fujii, S. Tomonaga, K. Supamataya, and S. Boonyaratpalin. 1994. Electron microscopic evidence of bacilliform virus infection in Kuruma shrimp *(Penaeus japonicus). Fish Pathology* 29:121-125.

Takahashi, Y., T. Itami, M. Maeda, N. Suzuki, J. Kasornchandra, K. Supamataya, R. Khongpradit, S. Boonyaratpalin, M. Kondo, K. Kawai, I. Hirono, T. Aoki. 1996. Polymerase chain reaction (PCR) amplification of baciliform virus (PV-PJ)

DNA in *Penaeus japonicus* Bate and systemic ectodermal and mesodermal baculovirus (SEMBV) DNA in *Penaeus monodon* Fabricius. *Journal of Fish Disease* 19:399-403.

Tang, K.F.J. and D.V. Lightner. 1999. A yellow-head virus gene probe: Application to in situ hybridization and determination of its nucleotide sequence. *Diseases of Aquatic Organisms* 35:165-173.

Tapay, L.M., Y. Lu, R.B. Gose, E.C.B. Nadala Jr., J.A. Brock, and P.C. Loh. 1997. Infection of white-spot baculovirus-like virus (WSBV) in two species of penaeid shrimp *Penaeus stylirostris* (Stimpson) and *P. vannamei* (Boone). In T.W. Flegel and I.H. MacRae (Eds.), *Diseases in Asian Aquaculture III* (pp. 297-303) Manila, Philippines: Fish Health Section, Asian Fisheries Society.

USDC. 1997. *Shrimp Imports*. Washington, DC: U.S. Department of Commerce, Bureau of the Census, Foreign Trade Division.

Wang, C.H., C.F. Lo, J.H. Leu, C.M. Chou, P.Y. Yeh, H.Y. Chou, M.C. Tung, C.F. Chang, M.S. Su, and G.H. Kou. 1995. Purification and genomic analysis of baculovirus associated with white spot syndrome (WSBV) of *Penaeus monodon. Diseases of Aquatic Organisms* 23:239-242.

Wang, Q., B.L. White, R.M. Redman, and D.V. Lightner. 1999. Per os challenge of *Litopenaeus vannamei* postlarvae and *Farafantepenaeus duorarum* juveniles with six geographic isolates of white spot syndrome virus (WSSV). *Aquaculture* 170:179-194.

Wang, Y.C., C.F. Lo, P.S. Chang, and G.H. Kou. 1998. White spot syndrome associated virus (WSSV) infection in cultured and wild decapods in Taiwan. *Aquaculture* 164:221-231.

Wigglesworth J. 1994. "Taura syndrome" hits Ecuador farms. *Fish Farmer* 17(3):30-31.

Wongteerasupaya, C., V. Boonsaeng, S. Panyim, A. Tassanakajon, B. Withyachumnarnkul, and T.W. Flegel. 1997. Detection of yellow-head virus (YHV) of *Penaeus monodon* by RT-PCR amplification. *Diseases of Aquatic Organisms* 31:181-186.

Wongteerasupaya, C., J.E. Vickers, S. Sriurairatana, G.L. Nash, A. Akarajamorn, V. Boonsaeng, S. Panyim, A. Tassanakajon, B. Withyachumnarnkul, and T.W. Flegel. 1995. A non-occluded, systemic baculovirus that occurs in cells of ectodermal and mesodermal origin and causes high mortality in the black tiger prawn, *Penaeus monodon. Diseases of Aquatic Organisms* 21:69-77.

Wyban, J.A., J.S. Swingle, J.N. Sweeney, and G.D. Pruder. 1992. Development and commercial performance of high health shrimp using specific pathogen free (SPF) broodstock *Penaeus vannamei*. In J. Wyban (Ed.), *Proceedings of the Special Session on Shrimp Farming* (pp. 254-260). Baton Rouge, LA: World Aquaculture Society.

Chapter 4

Overview of Nutritional Strategies Affecting the Health of Marine Fish

Wendy M. Sealey
Delbert M. Gatlin III

INTRODUCTION

An increased demand for fish by health-conscious consumers in recent years has created a need that can no longer be met by existing wild fish stocks. Commercial aquaculture production of a variety of species has subsequently increased to meet this growing demand (Stickney 1994). Total U.S. aquacultural production was valued at over $760 million in 1990 and continues to increase, primarily through intensifying culture practices. As a result, production losses due to disease also have increased and indicate the need for precisely formulated, economical diets that adequately meet the needs of fish grown in high-density production systems.

Infectious diseases are the major cause of economic loss in commercial aquaculture (Lovell 1996). Current methods for treating diseases include a limited number of government-approved antibiotics and chemotherapeutics that are marginally effective. Problems have arisen with the improper administration of antibiotics (such as feeding prior to proper disease diagnosis or feeding for an inappropriate length of time) that have limited their effectiveness through the creation of antibiotic-resistant strains of bacteria. In addition, sick fish generally do not feed vigorously and therefore may not consume enough medicated feed to ensure the proper dose. Further, administration of chemotherapeutics is complicated by the fact that pond treatment requires large amounts of chemicals that are expensive and accumulate in the environment (Anderson 1992).

For these reasons, the aquaculture industry has begun to focus on prevention of diseases rather than treatment (Baulny et al. 1996). Vaccines are being used with some fish species and show promise for the future prevention of many diseases of many other species. Currently, however, vaccines are costly and are not effective against many commercially important diseases (Raa et al. 1992). The industry, therefore, has begun exploring other methods of preventing disease, including immunostimulation through alteration of the diet and feeding practices.

DIETARY APPROACHES

A proper diet has long been recognized as a necessity in preserving an animal's health and maintaining its ability to resist diseases (Lall and Olivier 1993). Nutrients may alter immune responses by acting on immune cells directly or indirectly through metabolic, neurological, or endocrine pathways (Reddy and Frey 1992). Dietary modulation of the immune response has been accomplished in a variety of fish species using various approaches, including manipulating levels of various nutrients, adding certain nonnutrient immunostimulatory compounds, and altering feeding strategies.

Effect of Nutrient Levels

Nutrient deficiencies or excesses may have profound effects on fish survival and disease development. Although requirements generally do not vary greatly among fish species, differences in nutrient needs are evident between warm-water and cold-water, freshwater and marine fishes (National Research Council 1993). Therefore, dietary requirements of a variety of marine fish for energy, protein, lipids, vitamins, and minerals have been or are currently being established.

Experiments conducted to define nutrient requirements typically estimate the minimum amount of the nutrient necessary to maintain normal growth and survival or some specific metabolic function. Although these minimum dietary requirements are generally adequate to prevent severe deficiency and death, they may not prevent marginal deficiencies that could predispose fish to infection. However, information on the effects of marginal nutrient deficiencies on fish is lacking because these conditions are difficult to characterize and often overlooked.

Overfortification

In contrast to marginal deficiency, overfortification of diets with large amounts or megadoses of specific micronutrients has received considerable attention in fish. Micronutrients, most frequently the antioxidants vitamin C and vitamin E, have been included at 10 to 100 times the established dietary requirement and examined for their ability to upregulate the immune response. One hypothesis relating to the enhancement of immune responses suggests that the presence of these nutrients above maintenance levels provides a readily available reservoir for use by the host defense system upon activation. Such nutrients are most likely involved in various metabolic activities associated with immune responses, including protection of tissues from damage by free radicals resulting from phagocytic activities of cellular defense mechanisms. However, the influences of many nutrients on the immune responses of fish have not yet been fully characterized.

In freshwater fish such as the channel catfish, *Ictalurus punctatus,* megadose levels of vitamin C added to the diet has been shown to improve antibody response, complement activity, and survival following infection with *Edwardsiella tarda* and *E. ictaluri* (Li and Lovell 1985; Liu et al. 1989). Similar improvements in immune response of rainbow trout, *Oncorhynchus mykiss,* following dietary supplementation with vitamin C (Blazer 1982; Navarre and Halver 1989; Verlhac et al. 1998) or vitamin E (Blazer and Wolke 1984a, 1984b; Verlhac et al. 1993) also have been observed. However, other studies have failed to show positive responses due to overfortification of vitamin C (Li, Johnson, and Robinson 1993; Li, Wise, and Robinson 1998).

Dietary supplementation of megadose levels of vitamin C and vitamin E have been investigated in a few marine species with varied results (see Table 4.1). In general, marine fish appear to be somewhat less responsive to dietary supplementation of megadoses of vitamins C and E, although some researchers have observed improvements in immune response. Hardie, Fletcher, and Secombes (1991) observed an increase in complement activity of Atlantic salmon, *Salmo salar,* following dietary supplementation with 2,750 mg vitamin C/kg. Verlhac and Gabaudan (1994) observed increased hydrogen peroxide production at a dietary supplementation level of 1,000 mg vitamin C/kg in Atlantic salmon. More promising are the results of Waagbo et al. (1993), who observed an increased resistance of Atlantic salmon to furunculosis following supplementation of 4,000 mg vitamin C/kg, although these results contrast with those of Lall (1989). The conflicting data on the effects of vitamin C and vitamin E on immune response and disease resistance indicate the need for additional research using standardized experiment protocols to resolve these issues in marine fish.

Supplementation of Nonnutritive Compounds

Immunostimulants and probiotics are two promising nonnutritive dietary supplements to potentially aid in disease control of marine fish. Immunostimulants increase disease resistance by causing upregulation of host defense mechanisms against pathogenic microorganisms, while probiotics decrease the frequency and abundance of pathogenic or opportunistically pathogenic organisms in the environment.

Immunostimulants

Immunostimulatory compounds are often grouped by either function or origin and consist of a heterogeneous group (Anderson 1992). Nonnutritive

TABLE 4.1. Summary of the Effects of Megadoses of Micronutrients on Immune Response and Disease Resistance of Marine Fish

Nutrient	Level in the Diet	Immune Response and Effect	Reference
Atlantic salmon, *Salmo salar*			
Vitamin C	50-2,000 mg/kg	No effect on immune response.	Lall (1989)
	5,000 mg/kg	No effect on antibody production.	Sandnes et al. (1990)
	2,980 mg/kg 4,770 mg/kg C2*	Increased antibody production to *Yersinia ruckeri,* no effect on resistance to *Yersinia ruckeri* or *Vibrio salmonicida.*	Erdal et al. (1991)
	2,750 mg/kg	Increased complement activity, no effect on superoxide anion, phagocytosis, or antibody production.	Hardie, Fletcher, and Secombes (1991)
	Megadose	No effect on resistance to *Aeromonas salmonicida* or *Vibrio anguillarum.*	Lall and Olivier (1993)
	4,000 mg/kg	Increased lysozyme production, increased resistance to furunculosis.	Waagbo et al. (1993)
	1,000 mg/kg	Increased hydrogen peroxide production, increased lymphocyte proliferation.	Verlhac and Gabaudan (1994)
Vitamin E	> requirement	No effect on resistance to *Aeromonas salmonicida.*	Lall (1988)
	800 mg/kg	No effect on superoxide, lysozyme, or antibody production.	Hardie, Fletcher, and Secombes (1990)
Pyridoxine	5 mg/kg	No effect on immune response or disease resistance.	Lall and Weerakoon (1990)
Chinook salmon, *Oncorhynchus tshawytscha*			
Vitamin C	2,500 mg/kg	No effect on resistance to *Aeromonas salmonicida* or *Renibacterium salmoninarum.*	Leith and Kaattari (1989)
Vitamin E	> req	No effect on resistance to bacterial kidney disease.	Leith and Kaattari (1989)
	300 mg/kg	No effect on resistance to *Renibacterium salmoninarum.*	Thorarinsson et al. (1994)

Nutrient	Level in the Diet	Immune Response and Effect	Reference
Pyridoxine	> requirement	Increased resistance to *Vibrio anguillarum.*	Hardy, Halver, and Brannon (1979)
	> requirement	No effect on immune response or resistance to *Aeromonas salmonicida* and *Renibacterium salmoninarum.*	Leith and Kaattari (1989)
Riboflavin	> requirement	No effect on immune response or resistance to *Aeromonas salmonicida* and *Renibacterium salmoninarum.*	Leith and Kaattari (1989)
Pantothenic acid	> requirement	No effect on immune response or resistance to *Aeromonas salmonicida* and *Renibacterium salmoninarum.*	Leith and Kaattari (1989)
Folic acid	> requirement	No effect on immune response or resistance to *Aeromonas salmonicida* and *Renibacterium salmoninarum.*	Leith and Kaattari (1989)
Coho salmon, *Oncorhynchus kisutch*			
Vitamin C	400-1,000 mg/kg	Improved wound healing.	Halver (1972)
Sockeye salmon, *Oncorhynchus nerka*			
Vitamin C	> requirement	No effect on bacterial kidney disease.	Bell et al. (1984)
Turbot, *Scopthalmus maximus*			
Vitamin C	800-1,200 mg/kg	Increased phagocytosis and lysozyme production.	Roberts et al. (1995)
Red sea bream, *Pargus major*			
Vitamin C	10,000 mg/kg	Increased phagocytosis, no effect on complement activity.	Yano et al. (1990)

*C2 = Ascorbyl 2-sulfate.

compounds that have been examined most frequently for their ability to increase the nonspecific immune responses of fish include glucans and the synthetic drug levamisole (see Table 4.2). Additionally, animal-derived products such as chitin (Sakai et al. 1992; Siwicki, Anderson, and Rumsey

TABLE 4.2. Summary of the Effects of Immunostimulants on Immune Response and Disease Resistance of Marine Fish

Immuno-stimulant	Dose/Route[1]	Immune Response and Effect	Reference
Atlantic salmon, *Salmo salar*			
ß-Glucans	15 mg/kg; Inj. 150 mg/kg; Oral and Anal	Increased superoxide anion and lysosomal acid phosphatase.	Dalmo et al. (1996)
	1 ml/fish; Inj.	Increased lysozyme and complement activity.	Engstad et al. (1992)
		Increased specific antibody production.	Aaker et al. (1994)
	50-200 μg/ml; Inj.	Increased survival following challenge with *Vibrio anguillarum,* increased survival following challenge with *Vibrio salmonicida,* no effect on survival following challenge with *Yersinia ruckeri.*	Dalmo et al. (1996)
IFA[2]	0.1 ml/fish; Inj.	Increased survival following challenge with *Aeromonas salmonicida.*	Olivier et al. (1985)
Coho salmon, *Oncorhynchus kisutch*			
ß-Glucans	5 and 15 mg/kg; Inj.	No effect on antibody production.	Nikl et al. (1991)
Levamisole	5 mg/kg; Inj.	No effect on antibody production, no effect on survival following challenge with *Aeromonas salmonicida.*	Nikl et al. (1991)
	0.1 ml/fish; Inj.	Increased survival following challenge with *Aeromonas salmonicida.*	Olivier et al. (1985)
DID[3]	12.5 mg/kg; Inj.	No effect on antibody production, no effect on survival following challenge with *Aeromonas salmonicida.*	Nikl et al. (1991)
MCFA[4]	5 mg/kg; Inj.	No effect on antibody production, no effect on survival following challenge with *Aeromonas salmonicida.*	Nikl et al. (1991)
	5 mg/kg; Inj.	Increased survival following challenge with *Aeromonas salmonicida.*	Olivier et al. (1985)
WY-18, 251[5]	10 mg/kg; Inj.	No effect on antibody production, no effect on survival following challenge with *Aeromonas salmonicida.*	Nikl et al. (1991)

Immuno-stimulant	Dose/Route[1]	Immune Response and Effect	Reference
MDP[6]	50 µg/fish; Inj.	No effect on survival following challenge with *Aeromonas salmonicida.*	Nikl et al. (1991)
Chinook salmon, *Oncorhynchus tshawytscha*			
ß-Glucans	1.0 percent; Oral	Increased survival following challenge with *Aeromonas salmonicida.*	Rorstad et al. (1993)
Dentex, *Dentex dentex*			
ß-Glucans	0.5 percent; Oral	No effect on hematocrit, leukocyte number, lysozyme, NBT, or total serum protein, increased survival to concurrent protozoan disease.	Efthimiou (1996)
Gilthead sea bream, *Sparus aurata*			
Levamisole	125-500 µg/kg; Oral	Increased phagocytosis, respiratory burst, lymphokine production and complement activation.	Mulero et al. (1998)
ß-Glucans	500 mg/kg; Oral	Increased survival following challenge with *Vibrio anguillarum.*	
Sockeye salmon, *Oncorhynchus nerka*			
CFA	5 mg/kg; Inj.	Increased antibody production.	Cipriano and Pyle (1985)
Turbot, *Scopthalmus maximus*			
ß-Glucans	2 gm/kg; Oral	Increased total leukocyte numbers. No effect on phagocytosis, complement activation, lysozyme levels, or survival following challenge with *Vibrio anguillarum.*	Baulny et al. (1996)
	1 gm/100ml; Oral	Increased survival following challenge with *Vibrio anguillarum.*	
Yellowtail, *Seriola quinqueradiata*			
ß-Glucans	2-10 mg/kg; Inj	Increased phagocytosis, complement activation, and lysozyme level.	Matsuyama et al. (1992)
	5 and 10 mg/kg; Inj.	Increased survival following challenge with *Streptococcus* sp., no effect on survival following challenge with *Pasteurella* sp.	Matsuyama et al. (1992)
	640 µg/ml; Inj.	No effect on survival following challenge with *Pasteurella piscida.*	Kawakami et al. (1998)

TABLE 4.2 *(continued)*

Immuno-stimulant	Dose/Route[1]	Immune Response and Effect	Reference
Chitin	4 mg/ml; Inj.	No effect on survival following challenge with *Pasteurella piscida.*	Kawakami et al. (1998)

Notes:
[1] Inj. = Injection, Imm.=Immersion.
[2] IFA = Modified Incomplete Freund's Adjuvant.
[3] DID = Diethyldithiocarbamate.
[4] MCFA = Modified Complete Freund's Adjuvant.
[5] WY-18, 251 = Commercial peptide preparation.
[6] MDP = Muramyl dipeptide.

1994) and abalone extract (Sakai et al. 1991), bacterial-derived products such as muramyl dipeptide (MDP), and alginates such as k-carrageenan (Fujiki et al. 1997a, 1997b) and spirulina (Duncan and Klesius, 1996) also have been examined. Much of the research on immunostimulants to date has focused on routes of administration other than through the diet, but information is presented here to indicate the potential application of these products as dietary supplements and the need for further research in the area of their oral administration.

Glucans appear to show the most promise of all immunostimulants thus far examined in fish. ß-glucans are insoluble polysaccharides consisting of repeating glucose units that can be joined through ß1-3 and ß1-6 linkages when derived from yeast and mycelia fungi (Yano, Mangindaan, and Matsuyama 1989) or through ß1-3 and ß1-4 linkages when derived from barley (Wang and Wang 1996, 1997). The source and extraction process from which these glucans are obtained can greatly affect their immunostimulatory capacity. Robertsen et al. (1990) reported an increase in nonspecific disease response of Atlantic salmon following injection with a commercial preparation of ß1-3, 1-6 glucan derived from the cell wall of *Saccharomyces cerevisiae* that was considerably higher than that obtained upon injection with *S. cerevisiae* glucan prepared by a different procedure. Wang and Wang (1996, 1997) also observed large differences between glucans in their ability to enhance disease resistance in freshwater fish such as grass carp, *Ctenopharyngodon idellus;* bighead carp, *Aristichthys nobilis;* milkfish, *Chanos chanos;* and blue tilapia, *Oreochromis aureus.*

Engstad, Robertsen, and Frivold (1992) suggested that the mechanism through which these glucans induced protection in Atlantic salmon was an increase in lysozyme and complement activation. Similar results have been observed in other marine fish species including turbot, *Scophthalmus maximus* (Baulny et al. 1996, Santarem, Novoa, and Figueras 1997), and yellowtail, *Seriola quinqueradiata* (Matsuyama, Mangindaan, and Yano

1992). Increased oxidative capacity of phagocytic cells also has been suggested as a mechanism through which ß-glucans enhance nonspecific immune response. Dalmo et al. (1996) observed enhanced production of superoxide anion in leukocytes obtained from Atlantic salmon that had been administered ß1-3 glucans. Efthimiou (1996) observed no effect on oxidative capacity or lysozyme activity of blood leukocytes of juvenile dentex, *Dentex dentex,* fed ß1-3, 1-6 glucans. Dalmo and Seljelid (1995) examined the effects of the addition of laminaran, a ß1-6 branched ß1-3 D glucan, to isolated head kidney macrophages of Atlantic salmon and observed that macrophages underwent spreading and membrane ruffling indicative of activation. In addition, these cells demonstrated increased pinocytosis, increased superoxide anion production, and elevated acid phosphatase levels. Jorgensen and Robertsen (1995), using Atlantic salmon head kidney macrophages, and Figueras, Santarem, and Novoa (1997), with turbot head kidney cells, reported similar results when cells were cultured with glucan in vitro. However, these studies also indicated a suppression of response at the highest glucan levels examined.

A few studies have addressed the potential of other immunostimulatory compounds in fish (see Table 4.2). Levamisole, which is a synthetic phenylimidazothiazole, has been shown to have the ability to upregulate nonspecific immune responses in several freshwater fish (Anderson 1992) and the marine gilthead sea bream, *Sparus aurata* (Mulero et al. 1998). Mulero and colleagues (1998) observed increased respiratory burst, complement activation, phagocytosis, and increased lymphokine secretion when gilthead sea bream were fed levamisole. Certain animal products, such as abalone extract and chitin, also have shown nonspecific immunostimulatory potential in freshwater fish (Anderson 1992) and also may have potential in marine fish species, though few studies have evaluated these products as dietary supplements to date.

Products derived from various algal species, such as the blue-green algae *Spirulina,* and carrageenans, also have shown limited nonspecific immunostimulatory potential in freshwater (Duncan and Klesius 1996) as well as in marine fish (Skjermo et al. 1995). Alginates with high mannuronic acid content enhanced resistance of juvenile turbot to *Vibrio anguillarum* infection when fed via *Artemia* (Skjermo et al. 1995).

Probiotics

Probiotics are live microorganisms that are often introduced into the food chain to shift the microbial balance from disease-causing microorganisms to beneficial microorganisms. Lactic acid bacteria such as *Carnobacterium* spp., which produce bactericins, are often used as probiotics (Nousianinen and Setala 1993). Lactic acid bacteria have been isolated from the intestine of Atlantic salmon and can inhibit the growth of pathogenic bacteria such as *V. anguillarum,* and *Aeromonas salmonicida* (Gildberg, Johansen, and Bogwald 1995), and *A. hydrophila* (Lewus, Kaiser, and Montville 1991). Challenge experiments have shown that feeding larval turbot with rotifers enriched in lactic acid bacteria (Gatesoupe 1994) or feeding cod fry, *Gadus morhua,* dry feed containing lactic acid bacteria (Gildberg et al. 1997) improved the survival of fish following challenge with *Vibrio* sp. However, no effect on protection of salmon fry against *A. salmonicida* was noted when lactic acid bacteria was added to dry feed (Gildberg, Johansen, and Bogwald 1995). In addition, Gatesoupe (1994) observed a negative correlation between high concentrations of lactic acid bacteria and survival in turbot larvae fed 2×10^7 colony forming units/ml, indicating the need for careful determination of inclusion levels of lactic acid bacteria in order to obtain beneficial, not detrimental, effects of these probiotics.

Manipulation of Feeding Strategies

Another approach to increasing disease resistance through dietary manipulation is the use of feeding regimens that involve periods of feed deprivation. In the freshwater channel catfish, Kim and Lovell (1995) found that year two (>0.5 kg) fish had increased resistance to enteric septicemia of catfish (ESC) caused by *E. ictaluri,* following a three-month period of food deprivation. In a separate report, Okwoche and Lovell (1996) stated that humoral and cell-mediated immunity were not affected significantly by feed deprivation, although disease resistance was improved. These authors noted a significant reduction in serum iron in the feed-deprived fish.

Recent studies with marine fish have demonstrated a relationship between a low-iron diet and increased resistance to disease caused by *V. anguilarum* in Atlantic salmon (Lall et al. 1996) without effect on humoral immunity and the complement system, but correlating with a lower level of iron in the organs of fish fed the low-iron diet, following experimental challenge. Taken together these studies seem to indicate that elevated tissue levels of certain micronutrients may predispose fish to infection and that temporary feed restriction through lowering tissue stores may improve disease resistance.

The nutritional status of fish may affect disease resistance by altering immune function or by retarding growth of pathogenic organisms (Lall and Olivier 1993). Although the mechanism by which temporary feed restriction increases disease resistance has not been conclusively determined, this regimen may serve as a useful method to decrease losses from certain bacterial infections of marine fish, such as *V. anguillarum,* but must be investigated throughly.

CONCLUSIONS

Variations in experiment protocols currently limit the ability to precisely determine the effects of dietary modulation on immune response and disease resistance. Differences in the compositions of diets, feed intake, growth rate, species, genetic variation, and rearing conditions are known to affect responses of fish (Lall and Olivier 1993). In addition, the antigen used, route and dose of infection, and immune responses measured are often varied and inappropriate. The establishment of a disease model system for marine fish is necessary to reduce inconsistency of results.

Dietary immunomodulation has the potential to greatly aid aquaculture production through prevention and/or improvement of disease resistance. The continued determination of nutrient requirements for a variety of marine species is necessary to refine dietary formulations, maintain fish health, and make commercial production of these fish more economical. Changes in current dietary recommendations, such as overfortification of marine fish diets with antioxidant vitamins and dietary supplementation with non-nutritive immunostimulatory compounds to optimize immune function and disease resistance, hold some promise and may be cost-effective if reductions in mortality can be demonstrated. However, additional study using standardized protocols is needed at this time to address questions of fish species, pathogen species and strain variability, nutrient and nonnutrient immunostimulatory compound dosages necessary to elicit protection, and duration of the various treatments.

REFERENCES

Aaker, R., H.I. Wergeland, P.M. Aasjord, and C. Endresen. 1994. Enhanced antibody response in Atlantic salmon (*Salmo salar* L.) to *Aeromonas salmonicida* cell wall antigens using a bacterin containing ß-1,3-M-glucan as an adjuvant. *Fish and Shellfish Immunology* 4:47-61.

Anderson, D.P. 1992. Immunostimulants, adjuvants, and vaccine carriers in fish: Applications to aquaculture. *Annual Review of Fish Diseases* 2:281-307.

Baulny, M.O., C. Quentel, V. Fournier, F. Lamour, and R.L. Gouvello. 1996. Effect of long-term oral administration of ß-glucan as an immunostimulant or an adjuvant on some non-specific parameters of the immune response of turbot *Scophthalmus maximus. Diseases of Aquatic Organisms* 26:139-147.

Bell, G.R., D.A. Higgs, and G.S. Traxler. 1984. The effect of dietary ascorbate, zinc, and manganese on the development of experimentally induced bacterial kidney disease in sockeye salmon *(Onchorhynchus nerka). Aquaculture* 36:293-311.

Blazer, V.S. 1982. The effects of marginal deficiencies of ascorbic acid and alpha-tocopherol on the natural resistance and immune response of rainbow trout *(Salmo gairdneri).* Doctoral dissertation, University of Rhode Island, Kingston, Rhode Island.

Blazer, V.S. and R.E. Wolke. 1984a. Effect of diet on the immune response of rainbow trout *(Salmo gairdneri). Canadian Journal of Fisheries and Aquatic Science* 44:1244-1247.

Blazer, V.S. and R.E. Wolke. 1984b. The effects of alpha-tocopherol on the immune response and nonspecific resistance factors of rainbow trout (*Salmo gairdneri* Richardson). *Aquaculture* 37:1-9.

Cipriano, R.C. and S.W. Pyle. 1985. Adjuvant-dependent immunity and the agglutinin response of fishes against *Aeromonas salmonicida* cause of furunculosis. *Canadian Journal of Fisheries and Aquatic Sciences* 42:1290-1295.

Dalmo, R.A., J. Bogwald, K. Ingebrigtsen, and R. Seljelid. 1996. The immunomodulary effect of laminaran [ß(1,3)-D-glucan] on Atlantic salmon, *Salmo salar* L., anterior kidney leucocytes after intraperitoneal, peroral, and peranal administration. *Journal of Fish Diseases* 19:449-457.

Dalmo, R.A. and R. Seljelid. 1995. The immunomodulary effect of laminaran [ß(1,3)-D-glucan] on Atlantic salmon, *Salmo salar* L., macrophages in vitro. *Journal of Fish Diseases* 18:175-185.

Duncan, P.L. and P.H. Klesius. 1996. Effects of feeding spirulina on specific and nonspecific immune responses of channel catfish. *Journal of Aquatic Animal Health* 8:308-313.

Efthimiou, S. 1996. Dietary intake of ß-1,3/1,6 glucans in juvenile dentex *(Dentex dentex),* Sparidae: Effects on growth performance, mortalities, and non-specific defense mechanisms. *Journal of Applied Ichthyology* 12:1-7.

Engstad, R.E., B. Robertsen, and E. Frivold. 1992. Yeast glucan induces increase in lysozyme and complement-mediated haemolytic activity in Atlantic salmon blood. *Fish and Shellfish Immunology* 2:287-297.

Erdal, J.I., O. Evensen, O.K. Kaurstad, A. Lillehaug, R. Solbakken, and K. Thorud. 1991. Relationship between diet and immune response in Atlantic salmon (*Salmo salar* L.) after feeding various levels of ascorbic acid and omega-3 fatty acids. *Aquaculture* 98:363-379.

Figueras, A., M.M. Santarem, and B. Novoa. 1997. In vitro immunostimulation of turbot *(Scophthalmus maximus)* leucocytes with ß-glucan and/or *Photobacterium damsela* bacterin. *Fish Pathology* 32:153-157.

Fujiki, K., D.H. Shin, M. Nakao, and T. Yano. 1997a. Effects of K-carrageenan on the nonspecific defense system of carp *Cyprinus carpio. Fisheries Science* 63:934-938.

Fujiki, K., D.H. Shin, M. Nakao, and T. Yano. 1997b. Protective effect of K-carrageenan against bacterial infections in carp *Cyprinus carpio. Journal of the Faculty of Agriculture, Kyushu University* 42:113-119.

Gatesoupe, F. 1994. Lactic acid bacteria increase the resistance of turbot larvae *Scophthalmus maximus,* against pathogenic vibrio. *Aquatic Living Resources* 7:277-282.

Gildberg, A., A. Johansen, and J. Bogwald. 1995. Growth and survival of Atlantic salmon *(Salmo salar)* fry given diets supplemented with fish protein hydrolysate and alctic acid bacteria during a challenge trial with *Aeromonas salmonicida. Aquaculture* 138:23-34.

Gildberg, A., H. Mikkelsen, E. Sandaker, and E. Ringo. 1997. Probiotic effect of lactic acid bacteria in the feed on growth and survival of fry of Atlantic cod *(Gadus morhua). Hydrobiologia* 352:279-285.

Halver, J.E. 1972. The role of ascorbic acid in fish disease and tissue repair. *Bulletin of the Japanese Society of Scientific Fisheries* 38:79-92.

Hardie, L.J., T.C. Fletcher, and C.J. Secombes. 1990. The effect of vitamin E on the immune response of the Atlantic salmon (*Salmo salar* L.). *Aquaculture* 87:1-13.

Hardie, L.J., T.C. Fletcher, and C.J. Secombes. 1991. The effect of dietary vitamin C on the immune response of the Atlantic salmon (*Salmo salar* L.). *Aquaculture* 95:201-214.

Hardy, R.W., J.E. Halver, and E.L. Brannon. 1979. Effects of dietary pyridoxine levels on growth and disease resistance of chinook salmon. In J.E. Halver and K. Tiews (Eds.), *Finfish Nutrition and Fish Feed Technology* (pp. 253-260). Berlin: H. Heenemann GmbH and Co.

Jorgensen, J.B. and B. Robertsen. 1995. Yeast ß-glucan stimulates respiratory burst activity of Atlantic salmon (*Salmo salar* L.) macrophages. *Developmental and Comparative Immunology* 19:43-57.

Kawakami, H., N. Shinohara, and M. Sakai. 1998. The non-specific immunostimulation and adjuvant effects of *Vibrio anguillarum* bacterin, M-glucan, chitin, and Freund's complete adjuvant against *Pasteurella piscida* infection in yellowtail. *Fish Pathology* 33:287-292.

Kim, M.K. and R.T. Lovell. 1995. Effect of overwinter feeding regimes on body weight, body composition, and resistance to *Edwardsiella ictaluri* in channel catfish, *Ictalurus punctatus. Aquaculture* 134:237-246.

Lall, S.P. 1988. Disease control through nutrition. *Proceedings of Aquaculture International Congress and Exposition* (pp. 607-610). British Columbia, Canada: B.C. Pavillion Corparation, Vancouver.

Lall, S.P. 1989. The minerals. In J.E. Halver (Ed.), *Fish Nutrition* (pp. 216-255). New York: Academic Press.

Lall, S.P., N. Naser, G. Olivier and R. Keith. 1996. Influence of dietary iron on immunity and disease resistance in Atlantic salmon, *Salmo salar.* Seventh International Symposium on Nutrition and Feeding of Fish, Texas A&M University, College Station, Texas.

Lall, S.P. and G. Olivier. 1993. Role of micronutrients in immune response and disease resistance of fish. In S.J. Kaushik and P. Luquet (Eds.), *Fish Nutrition in Practice* (pp. 101-118). Biarritz, France: INRA Editions, Les Colloques.

Lall, S.P. and D.E.M. Weerakoon. 1990. Vitamin B6 requirement of Atlantic salmon *(Salmo salar). Federation of the American Society of Experimental Biology Journal* 4:9-12.

Leith, D. and S. Kaattari. 1989. *Effects of vitamin nutrition on the immune response of hatchery-reared salmonids.* Final report. Portland, OR: U.S. Department of Energy, Bonneville Power Administration, Division of Fisheries and Wildlife.

Lewus, C.B., A. Kaiser, and T.J. Montville. 1991. Inhibition of food-borne bacterial pathogens by bacteriocins from lactic acid bacteria isolated from meat. *Applied Environmental Microbiology* 57:1683-1688.

Li, M.H., M.R. Johnson, and E.H. Robinson. 1993. Elevated dietary vitamin C concentrations did not improve resistance of channel catfish, *Ictalurus punctatus,* against *Edwardsiella ictaluri* infection. *Aquaculture* 117:303-312.

Li, M.H., D.J. Wise, and E.H. Robinson. 1998. Effect of dietary vitamin C on weight gain, tissue ascorbate concentration, stress response, and disease resistance of channel catfish *Ictalurus punctatus. Journal of the World Aquaculture Society* 29:1-8.

Li, Y. and R.T. Lovell. 1985. Elevated levels of dietary ascorbic acid increase immune response in channel catfish. *Journal of Nutrition* 115:123-131.

Liu, P.R., J.A. Plumb, M. Guerin, and R.T. Lovell. 1989. Effects of megadose levels of dietary vitamin C on the immune response of channel catfish, *Ictalurus punctatus,* in ponds. *Diseases of Aquatic Organisms* 7:191-194.

Lovell, R.T. 1996. Feed deprivation increases resistance of channel catfish to bacterial infection. *Aquaculture Magazine* 6:65-67.

Matsuyama, H., R.E.P. Mangindaan, and T. Yano. 1992. Protective effect of schizophyllan and scleroglucan against *Streptococcus* sp. infection in yellowtail *(Seriola quinqueradiata). Aquaculture* 101:197-203.

Mulero, V., M.A. Esteban, J. Munoz, and J. Meseguer. 1998. Dietary intake of levamisole enhances the immune response and disease resistance of the marine teleost gilthead seabream (*Sparus aurata* L.). *Fish and Shellfish Immunology* 8:49-62.

National Research Council, 1993. *Nutrient Requirements of Fish.* Washington, DC: National Academy Press.

Navarre, O. and J.E. Halver. 1989. Disease resistance and humoral antibody production in rainbow trout fed high levels of vitamin C. *Aquaculture* 79:207-221.

Nikl, L., L.J. Albright, and T.P.T. Evelyn. 1991. Influence of immunostimulants on the immune response of coho salmon to *Aeromonas salmonicida. Diseases of Aquatic Organisms* 12:712-730.

Nousianinen, J. and J. Setala. 1993. Lactic acid bacteria as animal probiotics. In S. Salmine and A. Von Wright (Eds.), *Lactic Acid Bacteria* (pp. 315-356). New York: Marcel Dekker.

Okwoche, V.O. and R.T. Lovell. 1996. Effects of winter feeding regimen on resistance to *Edwardsiella ictaluri* challenge by channel catfish, *Ictalurus punctatus.* Seventh International Symposium on Nutrition and Feeding of Fish, Texas A&M University, College Station, Texas.

Olivier, G., T.P.T. Evelyn, and R. Lallier. 1985. Immunity to *Aeromonas salmonicida* in coho salmon *(Oncorhynchus kisutch)* induced by modified Freund's complete adjuvant: Its nonspecific nature and the probable role of macrophages in the phenomenon. *Developmental and Comparative Immunology* 9:419-432.

Raa, J., G. Roerstad, R. Engstad, and B. Robertsen. 1992. The use of immunostimulants to increase resistance of aquatic organisms to microbial infections. *Diseases of Asian Aquaculture* 1:39-50.

Reddy, P.G. and R.A. Frey. 1992. Nutritional modulation of immunity in domestic food animals. *Advances in Veterinary Science and Comparative Medicine* 35:255-281.

Roberts, M.L., S.J. Davies, and A.C. Pulsford. 1995. The influence of ascorbic acid (vitamin C) on nonspecific immunity in the turbot (*Scophthalmus maximus* L.). *Fish and Shellfish Immunology* 5:27-38.

Robertsen, B., G. Rorstad, R. Engstad, and J. Raa. 1990. Enhancement of non-specific disease resistance in Atlantic salmon, *Salmo salar* L., by a glucan from *Saccharomyces cerevisiae* cell walls. *Journal of Fish Diseases* 13:391-400.

Rorstad, G., P.M. Aasjord, and B. Robertsen. 1993. Adjuvant effect of a yeast glucan in vaccines against furunculosis in Atlantic salmon (*Salmo salar* L.). *Fish and Shellfish Immunology* 3:179-190.

Sakai, M., H. Kamiya, S. Ishii, S. Atsuta, and M. Kobayashi. 1991. The immunomodulary effects in rainbow trout, *Oncorhynchus mykiss,* injected with the extract of abalone, *Haliotis discus hannai. Journal of Applied Ichthyology* 7:54-59.

Sakai, M., H. Kamiya, S. Ishii, S. Atsuta, and M. Kobayashi. 1992. The immunostimulating effects of chitin in rainbow trout, *Oncorhynchus mykiss. Diseases in Asian Aquaculture* 1:413-417.

Sandnes, K., T. Hansen, J.E.A. Killie, and R.Waagbo. 1990. Ascorbate 2-sulfate as a dietary vitamin C source for Atlantic salmon *(Salmo salar).* I. Growth, bioactivity, haematology and humoral immune response. *Fish Physiology and Biochemistry* 8:419-427.

Santarem, M., B. Novoa, and A. Figueras. 1997. Effects of ß-glucans on the non-specific immune response of turbot (*Scophthalmus maximus* L.). *Fish and Shellfish Immunology* 7:429-437.

Siwicki, A.K, D.P. Anderson, and G.L. Rumsey. 1994. Dietary intake of immunostimulants by rainbow trout affects nonspecific immunity and protection against furunculosis. *Veterinary Immunology and Immunopathology* 41:125-139.

Skjermo, J., T. Defoor, M. Dehasque, T. Espevik, Y. Olsen, G. Skjak-Braek, P. Sorgeloos, and O. Vadstein. 1995. Immunostimulation of juvenile turbot (*Scopthalmus maximus* L.) using an alginate with high mannuronic acid content administered via the live food organism Artemia. *Fish and Shellfish Immunology* 5:531-534.

Stickney, R.R. 1994. *Principles of Aquaculture.* New York: John Wiley and Sons.

Thorarinsson, R., M.L. Landolt, D.G. Elliott, R.J. Pascho, and R.W. Hardy. 1994. Effect of dietary vitamin E and selenium on growth, survival and the prevalence of *Renibacterium salmoninarum* infection in chinook salmon *(Oncorhynchus tshawytscha). Aquaculture* 121:343-358.

Verlhac, V. and J. Gabaudan. 1994. Influence of vitamin C on the immune system of salmonids. *Aquaculture and Fisheries Management* 25:21-36.

Verlhac, V., A. N'doyle, J. Gabaudan, D. Troutaud, and P. Deschaux. 1993. Vitamin nutrition and fish immunity: Influence of antioxidant vitamins (C and E) on immune response of rainbow trout. In S.J. Kaushik and P. Luquet (Eds.), *Fish Nutrition in Practice* (pp. 167-177). Biarritz, France: INRA Editions, Les Colloques.

Verlhac, V., A. Obach, J. Gabaudan, W. Schuep and R. Hole. 1998. Immunomodulation by dietary vitamin C and glucan in rainbow trout *(Oncorhynchus mykiss). Fish and Shellfish Immunology* 8:409-424.

Waagbo, R., J. Glette, E. Raa-Nilsen, and K. Sandnes. 1993. Dietary vitamin C, immunity, and disease resistance in Atlantic salmon *(Salmo salar). Fish Physiology and Biochemistry* 12:61-73.

Wang, W.S. and D.H. Wang. 1996. Use of glycans to increase resistance of bighead carp, *Aristichthys nobilis,* and milkfish, *Chanos chanos,* to bacterial infections. *Taiwan Journal of Veterinary Medicine and Animal Husbandry* 66:83-91.

Wang, W.S. and D.H. Wang. 1997. Enhancement of the resistance of tilapia and grass carp to experimental *Aeromonas hydrophila* and *Edwardsiella tarda* infections by several polysaccharides. *Comparative Immunology and Microbiology and Infectious Diseases* 20:261-270.

Yano, T., M. Furuichi, M. Nakao, and S. Ito. 1990. Effects of L-ascorbyl-2-phosphate Mg on the growth and nonspecific immune system of red sea bream *Pagrus major.* Abstract T29.12. World Aquaculture Society Meeting, June 10-14, 1990. Halifax, Canada.

Yano, T., R.E.P. Mangindaan, and H. Matsuyama. 1989. Enhancement of the resistance of carp *Cyprinus carpio* to experimental *Edwardsiella tarda* infection by some ß-1,3-glucans. *Nippon Suisan Gakkaishi* 55:1815-1819.

Chapter 5

Nutritional Aspects of Health and Related Components of Baitfish Performance

Rebecca Lochmann
Harold Phillips

INTRODUCTION

The golden shiner, *Notemigonus crysoleucas;* goldfish, *Carassius auratus;* and fathead minnow, *Pimephales promelas,* are the three principal species of fish cultured for bait in the United States. Total pond bank value of bait species in Arkansas, the main source of cultured baitfish, was approximately $38 million in 1998 (Bo Collins, personal communication).

Natural food organisms are important nutrient sources for baitfish in ponds (Lochmann and Phillips 1996), but prepared diets based on formulas for channel catfish have been used to double or triple production (Stone et al. 1997). Poor nutrition is frequently implicated in baitfish morbidity and mortality on commercial farms, but cause and effect are difficult to establish when other stressful conditions exist. Stress is a well-known predisposing factor for fish disease (Pickering 1981). Nutritional stress in baitfish may result from the use of diets that do not meet their nutritional requirements, which are not well established. However, feeding too infrequently or too little at each feeding can result in malnourished baitfish, regardless of the nu-

---

The authors thank Drs. Andrew Goodwin, Ken Davis, and Bill Simco for their willingness to experiment with stress measurement techniques in small baitfish. Dr. Konrad Dabrowski and Regis Moreau provided expertise on ascorbic acid nutrition of fish. Mr. Neil Anderson and Dr. Eric Park provided fish and valuable information on raising baitfish species under commercial conditions. Numerous student workers have assisted in the successful completion of baitfish nutrition studies at the University of Arkansas at Pine Bluff. The research reviewed in this chapter was supported in part by the Southern Regional Aquaculture Center, United States Department of Agriculture.

The manuscript was submitted with the approval of the Director of the Arkansas Agriculture Experiment Station #99093.

tritional quality of the diet. Baitfish producers sometimes reduce the amount or frequency of feeding to "hold" the fish at desirable market size. Reducing or withholding diet reduces fish growth, but the effect of this practice on fish health is unknown. Baitfish are also maintained at high densities to curtail growth, which imposes additional stress on the fish.

Diseases associated with parasites cause more mortalities of baitfish than bacteria or viruses, unlike in food fish (Stone et al. 1997). Infested fish that are not killed by parasites are frequently unmarketable. Malnourished fish are more susceptible to parasitic infestation, and heavily infested fish are subject to further malnourishment due to the parasites. For example, fish heavily infested by large numbers of *Ichthyophthirius multifiliis* (Ich) can become anorexic to the point of starvation (Stone et al. 1997). Bacterial and fungal infections also afflict malnourished baitfish more often and more severely than well-nourished fish. Nutrition is viewed primarily as a preventive measure against disease in baitfish. There is very little information on the role of nutrition in recovery of affected baitfish.

Nutritionally complete diets may not be necessary when baitfish can derive a large proportion of their essential nutrients from pond biota (Lochmann and Phillips 1996). However, nutritionally incomplete, or "supplemental," diets tend to contain large amounts of fiber and other ingredients that may not be palatable or digested or absorbed as well as complete diets. Unutilized components can degrade water quality, which is another source of stress for fish.

As with fish cultured for food, baitfish are stressed by culture practices such as stocking, harvesting, grading, holding, and hauling. However, baitfish are graded, held, and hauled repeatedly as they move from the production facility to distributors, the retail market, and finally to the consumer. The fish must remain alive, healthy, and vigorous throughout these procedures to retain their market value. Health maintenance under a variety of stressful conditions is a primary goal of baitfish production and marketing.

The potential for nutritional modulation of health in baitfish is great, but this area has not been examined specifically in most baitfish nutrition research. Baitfish nutrition studies typically include only gross indicators of dietary effect on health such as survival, yield, and condition index. Growth reduction also can indicate nutritional deficiencies or imbalances. However, reduced growth in the absence of pathology may be viewed as a positive feature in baitfish, as mentioned previously. Physical anomalies such as scoliosis and lordosis, exophthalmia, and fin erosion sometimes appear in fish during feeding trials conducted either in tanks or ponds. It is difficult to determine whether the affected fish in a given dietary treatment are more sensitive to nutritional deficiencies or imbalances than others, or whether the abnormalities are unrelated to diet. Diagnosis of nutritional diseases in baitfish is hindered also by the inability to perform quantitative chemical tests on small quantities of tissue samples.

There is comparatively little information on the nutritional requirements of baitfish. Results of both basic and applied studies have been directed toward the development of optimal commercial feeding strategies. Small-scale indoor studies have yielded basic information on the nutritional requirements of baitfish species in the absence of natural foods and other variables that are difficult to control in pond studies. Results of pond studies are more applicable to industry, since the conditions of the studies are more similar to those of commercial facilities.

NUTRIENT REQUIREMENTS

Proteins and Amino Acids

The essential amino acid ratios of whole-body golden shiners, goldfish, and fathead minnows are similar to those of channel catfish, *Ictalurus punctatus,* and common carp, *Cyprinus carpio* (Gatlin 1987), indicating possible similarities in essential amino requirements among these species. Lochmann and Phillips (1994a) determined that growth, survival, and feed efficiency of golden shiners and goldfish fed semipurified diets with 29 percent protein in aquaria was similar to that of fish fed diets with higher protein levels when fed at 4 to 7 percent of body weight.

There is a need to develop a purified diet that meets the nutritional requirements of baitfish to facilitate basic laboratory research. Growth of golden shiners fed casein-based or fish meal-based diets thought to be nutritionally complete was equally good, but severe fin erosion occurred in fish fed the casein diet (see Figure 5.1), indicating a lack of one or more essential nutrients in the casein-based diet (Lochmann and Phillips, unpublished data).

Gannam and Phillips (unpublished data) found no differences in weight gain or yield of golden shiners in ponds that were fed practical diets with only vegetable proteins versus a diet with 5 or 10 percent fish meal. The primary vegetable protein source for all diets was soybean meal. Soybean meal also is utilized well by channel catfish and other warm-water omnivores (Lovell 1989). In addition, consumption of pond biota by baitfish probably reduces the need for high-quality protein sources in the feed.

Lipids and Fatty Acids

Optimal dietary lipid level for juvenile golden shiners and goldfish was determined in feeding trials using graded levels of a 1:1 mixture of cod liver and soybean oils. The lipid mixture contained 18-carbon as well as 20- and 22-carbon fatty acids of the n-3 and n-6 families, which encompasses the essential fatty acid requirements of most fish species (Watanabe 1982). Weight gain of golden shiners fed diets containing 34 percent protein and 7 to 12 percent lipid was higher than that of fish fed diets with lower or higher lipid levels (SRAC 1998). Survival of golden shiners fed diets with 3

FIGURE 5.1. Golden Shiners Fed a Fish Meal Diet (Upper) and a Casein Diet (Lower) That Meet the Nutritional Requirements of Channel Catfish

Note: Weight gain of fish fed the two diets was similar. However, fin erosion was pronounced in golden shiners fed the casein diet. Photo by Andrew Goodwin.

to 15 percent lipid was 92 percent or higher, and feed efficiency was similar among diets. Weight gain and feed efficiency of goldfish fed diets with 3 to 6 percent lipid was optimal (SRAC 1998). Survival varied directly with dietary lipid level but was 93 percent or higher in all treatments. Statistically significant diet effects on weight gain were more difficult to obtain in goldfish compared to golden shiners because growth rates of individual goldfish were higher and more variable than those of golden shiners fed identical diets to slight excess under similar experimental conditions (Lochmann and Phillips, unpublished data). These species differences have been noted in a variety of laboratory experiments using purified and practical diets. However, growth of both golden shiners and goldfish fed diets with 15 percent lipid was significantly reduced compared to fish fed diets with lower lipid levels. Currently, commercial feeds for baitfish contain only 3 to 4 percent lipid. There is potential to increase the lipid content in baitfish diets to spare protein for growth or, perhaps, to maintain fish in good condition at a small size. Golden shiners in aquaria fed isonitrogenous and isocaloric practical diets with either 4 or 13 percent poultry fat for 7.5 weeks had similar growth (Lochmann and Phillips, unpublished data). However, survival of the fish fed the diet with 13 percent lipid was significantly higher than that of fish fed the diet with 4 percent lipid. The reason for the difference in survival is unknown, and further studies are needed to determine if the effect is reproducible in ponds.

Following the feeding trial in aquaria, groups of twenty fish of similar size from each 110-liter tank in each treatment were moved to 3.8-liter tanks with aeration and allowed to acclimate for one hour. Following acclimation, airstones were removed and the dissolved oxygen was monitored in each tank over time. After 4.5 hours, the dissolved oxygen level averaged 2.5 mg/liter, and one fish had died. Since the stress was intended to be sublethal, this was considered the endpoint of the stress test. The remaining fish were returned to their original 110-liter tanks, and survival was monitored for two weeks while the fish continued to receive their experimental diets. No mortality occurred during this time. Golden shiners in ponds exposed to dissolved oxygen levels of 1.0 to 1.5 mg/liter for prolonged periods suffer increased mortality (Stone et al. 1997). The stress imposed on the fish in this laboratory test was less severe, which might have influenced the results. However, very low dissolved oxygen is not always essential to elicit diet effects in golden shiners.

Another series of experiments was conducted with golden shiners in aquaria to establish their qualitative essential fatty acid (EFA) requirements. Golden shiners were fed semipurified diets containing 10 percent lipid as poultry fat, cod liver, soybean, olive, canola, or rice bran oils, or a mixture (50 percent each) of cod liver and soybean oils. Cod liver oil was the only lipid that contained 1 percent or more of the highly unsaturated fatty acids eicosapentaenoic acid [20:5(n-3)], docosahexaenoic acid [22:6(n-3)], and arachidonic acid [20:4(n-6)], which are essential for some marine fish (Watanabe 1982). Canola oil contained the highest level (10 percent) of linolenic acid [18:3(n-3)], which is essential for rainbow trout and some other freshwater fish (Castell et al. 1972; Henderson and Tocher 1987). Poultry fat contained about 1 percent linolenic acid and 17 percent linoleic [18:2(n-6)] acid. The latter is essential for tilapia (Takeuchi, Satoh, and Watanabe 1983) and some other freshwater fish (Henderson and Tocher 1987). The rice bran oil contained 34 percent linoleic and 2 percent linolenic acid. The olive oil contained 10 percent linoleic and 1 percent linolenic acids. The soybean oil contained a high level (54 percent) of 18:2(n-6), but also 7 percent of 18:3(n-3). The combination of cod liver and soybean oils was expected to meet the qualitative EFA requirement of golden shiners because it contained fatty acids with between 18 and 22 carbons from the n-3 and n-6 families of fatty acids. Following an 11-week feeding trial, there were no significant differences in weight gain or survival of golden shiners fed diets with different lipids. Whole-body lipid was higher in fish fed diets with vegetable versus animal lipid sources. A subset of the golden shiners from each treatment was subjected to a low dissolved oxygen stress test, as described previously. After two hours the dissolved oxygen averaged 3.8 mg/liter, and two fish in different tanks died (endpoint). Fish were returned to their original tanks and continued to receive their experimental diets. During the first 24 hours after the stress test mortalities were recorded.

No mortality occurred for seven days following the 24-hour period, and cumulative mortality during the stress test and for the next 24 hours was compared among fish fed diets with different lipid sources. Cumulative mortality was significantly ($P < 0.05$) higher in fish fed the diet with olive oil than in fish fed other diets. Fish fed the diet with soybean oil incurred no mortality and also had the highest (although nonsignificant) weight gain and prestress test survival. The dissolved oxygen test was effective in detecting stress responses of golden shiners to dietary lipids, even though the stressor (low dissolved oxygen) appeared to be less severe than in the previous experiment.

Another feeding trial using the same diets was conducted for eight months. Again, weight gain and survival of golden shiners did not differ among diets. A crowding stress test was conducted at the end of this trial. Before the stress was imposed, blood was drawn from golden shiners and serum cortisol and electrolytes were measured (resting levels). The water levels in the tanks were then reduced suddenly, so that the fish were still immersed in water but had limited mobility. Serum cortisol and electrolytes were measured immediately after the stress was imposed. The water levels were restored to their original levels and final serum cortisol and electrolyte measurements were taken two hours after the stress was removed. The cortisol and electrolyte levels were highly variable and no diet effects on stress response were apparent. It is not known whether the variability in the data was typical of these fish, or if the results were an artifact of the minute amount of serum available from small-sized (3 g) fish. A pattern resembling the typical stress response in fish (Mazeaud, Mazeaud, and Donaldson 1977) was discernible in the cortisol data, but more studies are needed to characterize the stress response in golden shiners.

The appearance of fish at the end of this eight-month trial differed for some treatments, but the effects were not quantifiable. Most fish fed the diet with olive oil had severe fin and opercular erosion, the integrity of the epithelium on the general body surface was poor, and some fish had exophthalmia. By contrast, fish fed the diet with canola oil maintained fin and skin integrity and exhibited no external abnormality. Although differences in external appearance were not accompanied by differences in growth and survival of the two groups, fish fed the diet with canola oil were clearly superior baitfish than those fed the diet with olive oil. Pozernick and Wiegand (1997) found that growth and survival of larval goldfish fed diets with cod liver or canola oils was equally good.

The fatty acid profiles of whole-body golden shiners resembled those of the diets. Large amounts of oleic acid [18:1(n-9)] were present only in fish fed diets with high levels of this fatty acid (rice bran oil or olive oil diets). Eicosatrienoic acid (20:3n-9), an indicator of EFA-deficiency (Henderson and Tocher 1987), was not detected in any of the fish. There was insufficient tissue to perform fatty acid analysis of the liver, which might yield more specific information on EFA utilization in golden shiners.

In summary, golden shiners fed diets with different lipid sources did not consistently indicate a specific requirement for n-3 or n-6 fatty acids, or for

18-carbon or longer-chain fatty acids of either family. The ratio of n-3:n-6 fatty acids in the diets ranged from 0.06 (rice bran oil) to 9.0 (cod liver oil). Because no diet effects were apparent in early trials, the ingredients (casein, gelatin, cellulose, dextrin, and carboxymethylcellulose) in the purified diets used in subsequent trials were extracted with boiling ethanol prior to incorporation into diets to further reduce extraneous sources of fatty acids (Satoh, Poe, and Wilson 1989a, 1989b). For unknown reasons, the diets containing ethanol-extracted ingredients improved weight gain or survival of golden shiners significantly in different trials, relative to fish fed diets with unextracted ingredients and the same lipid sources. However, extraction did not yield additional information on the utilization of different lipid sources in golden shiners.

Although the length of these studies varied, golden shiners were fed diets differing in lipid source for as long as eight months without developing the definitive signs of EFA deficiency suggested by Watanabe (1982). The results are reminiscent of earlier studies with common carp (Watanabe et al. 1975; Watanabe, Takeuchi, and Ogino 1975) and channel catfish (Dupree 1969; Stickney and Andrews 1972). Consistent responses to dietary lipids were not obtained in carp or catfish without the use of purified esters of fatty acids (Takeuchi and Watanabe 1977; Satoh, Poe, and Wilson 1989a, b).

Results of our studies with golden shiners appear to indicate that they require both n-3 and n-6 fatty acids for optimal performance and appearance. However, the determination of specific fatty acid requirements of goldfish, golden shiners, and other baitfish will require further studies, possibly with individual fatty acids. Radunz-Neto et al. (1996) found that trace amounts (0.05 to 0.1 percent dry diet) of n-3 fatty acids satisfied the EFA requirements of larval carp. Further studies are needed to determine if the quantitative requirement for n-3 and other fatty acids is equally low for golden shiners and other cyprinid baitfish.

Phospholipid supplementation of diets for baitfish species may be beneficial. Practical diets supplemented with soybean lecithin enhanced growth but did not affect survival of juvenile goldfish (Lochmann and Brown 1997) relative to diets containing lipid as triglyceride from either soybean or fish oils. Phospholipid supplementation of semipurified diets improved both growth and survival of larval goldfish (Szlaminska, Escafre, and Bergot 1993) and carp (Geurden, Radunz-Neto, and Bergot 1995). Phospholipids may facilitate lipid digestion, absorption, and transport in baitfish, as in other fish (Hertrampf 1992).

Practical constraints on diet production may limit the use of information on lipid nutrition in baitfish. Changes in diet formulation that result in increased cost must be carefully justified. Major sources of oil rich in n-3 fatty acids (cod liver and canola oils) are not produced near major baitfish-producing regions. Furthermore, diet producers are reluctant to use highly unsaturated lipids in commercial feeds for baitfish due to concerns about oxidative rancidity. Poultry fat is the primary fat source used in commercial diets for pond-reared baitfish due to the proximity of the baitfish and poultry industries in Arkansas. Increasing the fat content of extruded feeds (to 13 percent) reduces the rate of feed production and increases feed cost. Feed grade lecithin is prohibitively expensive to use in baitfish feeds, despite demonstrated benefits in goldfish. However, dietary lipid is known to have significant impacts on fish health (Blazer 1992; Fracalossi and Lovell 1994; Fracalossi, Craig-Schmidt, and Lovell 1994; Kiron et al. 1995; Thompson, Tatner, and Henderson 1996), and further research on lipid nutrition in baitfish is warranted.

Carbohydrates

Weight gain and survival of golden shiners fed semipurified isocaloric and isonitrogenous diets with 15 percent, 30 percent, or 45 percent starch was similar, indicating that they perform well over a wide range of dietary carbohydrate:lipid ratios (1:1 to 27:1). Weight gain of golden shiners fed diets with 15 percent carbohydrate from different sources improved with increasing complexity of the carbohydrate: starch dextrin sucrose = glucose. Survival was not affected by carbohydrate source. Results are similar to those for other warm-water omnivores (NRC 1993).

Vitamins and Minerals

Weight gain and total net yield of golden shiners in ponds that are fed diets with or without a combination vitamin and mineral supplement for eight weeks did not differ (Lochmann and Phillips 1994b). However, diet affected the yield of fish in different size classes, as determined by standard baitfish grading methods (Stone et al. 1997). There was a significant increase in yield of golden shiners that were 23/64 to 27/64 inches (0.91 to 1.07 cm) in width (grader size 23-27) in groups of fish fed the vitamin- and mineral-supplemented diet, which could be advantageous for marketing. No specific indices of health were examined in that study.

A study was performed at the University of Arkansas at Pine Bluff in aquaria to determine whether or not golden shiners have a dietary requirement for ascorbic acid (AA). Four diets were formulated for the study. Two diets contained fish meal as the main protein source and the other contained casein as the main protein source. Each type of diet was supplemented with either 0 or 250 ppm AA. Stay-C 40 (Roche) was used in the supplemented casein diet, and a mixture of fat-coated ascorbic acid, ascorbyl-2-monophosphate (BASF), Stay-C 25 (Roche), and Stay-C 40 was added to the supplemented fish meal diet. Diets were fed to golden shiners for 12.5 weeks. Weight gain was higher in golden shiners fed the fish meal diet supplemented with 250 ppm AA than in those fed the AA-unsupplemented diet with fish meal. Survival of fish fed diets with fish meal was not affected by AA supplementation. No external signs typical of AA deficiency (e.g., scoliosis, severe fin erosion, and so on) were evident in golden shiners fed either diet with fish meal. Survival of fish fed the casein diet supplemented with 250 ppm AA was higher than that of fish fed the AA-unsupplemented diet with casein.

Weight gain of golden shiners fed diets with casein was not affected by AA supplementation. There were significant differences in total AA, reduced AA, and percent AA content of whole bodies of golden shiners fed AA-supplemented and AA-unsupplemented diets. No scoliosis or lordosis was observed in golden shiners fed any of the diets, in contrast to common carp fed diets without AA (Dabrowski et al. 1988). Pronounced fin erosion occurred in fish fed both diets with casein, indicating that the effect was independent of dietary AA content. A bacterial challenge was attempted with *Cytophaga columnaris* but results were inconclusive because the golden shiners were infected with *C. columnaris* prior to challenge (Andrew Goodwin, UAPB, personal communication). Ascorbic acid nutrition in golden shiners needs further study, but results of this study indicated that they do have a dietary requirement for this vitamin.

Additional information on dietary requirements of baitfish species for other vitamins or minerals probably exists, especially for ornamental species such as goldfish. However, limited information is available from published sources.

Feeding Regime

Baitfish become emaciated and are more vulnerable to predation and disease when they are unfed for extended periods, as during the winter. To maintain golden shiners in good condition and improve survival, feeding at a rate of 1 to 2 percent body weight when air temperature exceeds 8°C is recommended (Rowan and Stone 1993, 1994). Feed reduction during the winter may increase mortality of fathead minnows, especially during mild winters when warmer temperatures increase their metabolic rate and caloric needs. Production of large numbers of fathead minnows in good condition in ponds was achieved with a feeding rate of 3 percent body weight per day (using a 32 percent protein feed) from late summer to winter (Ludwig 1996). The rate was reduced to 2 percent of body weight in winter. Feed intake and utilization by baitfish may be reduced by high (30°C) temperatures, and feeding rates on commercial farms may be reduced based on fish behavior.

SUMMARY

Desired market sizes of baitfish are achieved and maintained primarily through manipulation of fish density rather than diet. However, maintaining the health of baitfish after reaching market sizes is challenging, particularly when the fish are held for long periods (e.g., over winter). There is a need for diets and feeding practices that can preserve baitfish health, vigor, and appearance without promoting growth. Dietary effects on body composition of baitfish also are relevant with respect to health, rather than dress-out percentage, as in food fish.

Therefore, nutritional modulation of health rather than growth has the greatest potential to expand the baitfish industry. Additional baitfish nutrition research that incorporates health-monitoring techniques should be conducted. Research techniques commonly used to monitor indices of stress in large fishes (e.g., measurement of serum cortisol and electrolytes) do not yield consistent results in baitfish due to the difficulty of obtaining adequate serum samples from small (<5 g) fish (see Table 5.1). Methods such as tracking mortality of fish following exposure to stress (e.g., low dissolved oxygen) or histological examination of tissues may be more applicable for assessing nutritional effects on the health of baitfish.

TABLE 5.1. Methods of Assessing Diet Effects in Golden Shiners *(Notemigonus crysoleucas)* in Aquaria[1]

Diet Variable	**Method**	**Response Measured**	**Results (Nontraditional Method)**	**Results (Traditional Method)[2]**
Lipid source	Exposure to low dissolved oxygen (D.O.)	Cumulative mortality 24 hours after stressor	Significant differences	Insignificant differences, but same trend as low D.O. test
Lipid source	Crowding stress	Serum cortisol, glucose, and chloride	High values; data highly variable; general stress response evident	Significant differences, but did not reveal fatty acid requirement
Lipid amount	Exposure to low D.O.	Cumulative mortality	No mortality	Significant differences in survival
Ascorbic acid amount	Bacterial challenge (*Cytophaga columnaris)*	Mortality	Inconclusive. Unchallenged fish were also infected with *C. columnaris.*	Significant differences in weight gain or survival, depending on type of diet

1 See text for details.
2 Traditional methods include measurement of weight gain, survival, feed efficiency, and whole-body proximate composition.

REFERENCES

Blazer, V.S. 1992. Nutrition and disease resistance in fish. *Annual Review of Fish Diseases* 2:309-323.

Castell, J.D., R.O. Sinnhuber, J.H. Wales, and J.D. Lee. 1972. Essential fatty acids in the diet of rainbow trout *(Salmo gairdneri):* Growth, feed conversion and some gross deficiency symptoms. *Journal of Nutrition* 102:77-86.

Dabrowski, K., S. Hinterleitner, C. Sturmbauer, N. El-Fiky, and W. Wieser. 1988. Do carp larvae require vitamin C? *Aquaculture* 72:295-306.

Dupree, H.K. 1969. *Influence of Corn Oil and Beef Tallow in Growth of Channel Catfish.* Technical Paper No. 27. Washington, DC: U.S. Fish and Wildlife Service.

Fracalossi, D.M., M.C. Craig-Schmidt, and R.T. Lovell. 1994. Effect of dietary lipid sources on production of leukotriene B by head kidney of channel catfish at different temperatures. *Journal of Aquatic Animal Health* 6:242-250.

Fracalossi, D.M., and R.T. Lovell. 1994. Dietary lipid sources influence responses of channel catfish to challenge with the pathogen *Edwardsiella ictaluri. Aquaculture* 119:287-298.

Gatlin, D.M., III. 1987. Whole-body amino acid composition and comparative aspects of amino acid nutrition of the goldfish, golden shiner and fathead minnow. *Aquaculture* 60:223-229.

Geurden, I., J. Radunz-Neto, and P. Bergot. 1995. Essentiality of dietary phospholipids for carp (*Cyprinus carpio* L.) larvae. *Aquaculture* 131:303-314.

Henderson, R.J. and D.R. Tocher. 1987. The lipid composition and biochemistry of freshwater fish. *Journal of Lipid Research* 26:281-347.

Hertrampf, J.W. 1992. *Feeding Aquatic Animals with Phospholipids. II. Fishes.* Publication #11. Hamburg, Germany: Lucas Meyer (GmbH and Co.), KG.

Kiron, V., H. Fukuda, T. Takeuchi, and T. Watanabe. 1995. Essential fatty acids nutrition and defense mechanisms in rainbow trout *Oncorhynchus mykiss. Comparative Biochemistry and Physiology* 111A:361-367.

Lochmann, R.T. and R. Brown. 1997. Soybean-lecithin supplementation of practical diets for juvenile goldfish *(Carassius auratus). Journal of the American Oil Chemists Society* 74(2):149-152.

Lochmann, R.T. and H. Phillips. 1994a. Dietary protein requirement of juvenile golden shiners *(Notemigonus crysoleucas)* and goldfish *(Carassius auratus)* in aquaria. *Aquaculture* 128:277-285.

Lochmann, R.T. and H. Phillips. 1994b. Vitamin and mineral additions to golden shiners diet tested. *Arkansas Farm Research* 43(3):8-9.

Lochmann, R. and H. Phillips. 1996. Stable isotopic evaluation of the relative assimilation of natural and artificial foods by golden shiners *Notemigonus crysoleucas* in ponds. *Journal of the World Aquaculture Society* 27:168-177.

Lovell, R.T., 1989. *Nutrition and Feeding of Fish.* New York: Van Nostrand Reinhold.

Ludwig, G.M. 1996. Seasonal growth and survival of rosy red and normal-colored fathead minnows receiving different feed rations. *Progressive Fish-Culturist* 58:160-166.

Mazeaud, M.M., F. Mazeaud, and E.M. Donaldson. 1977. Primary and secondary effects of stress in fish: Some new data with a general review. *Transactions of the American Fisheries Society* 106:201-212.

NRC (National Research Council). 1993. *Nutrient Requirements of Fish.* Washington, DC: National Academy Press.

Pickering, A.D. 1981. *Stress and Fish.* New York: Academic Press.

Pozernick, M. and M.D. Wiegand. 1997. Use of canola oil in the feed of larval and juvenile goldfish, *Carassius auratus* (L.). *Aquaculture Research* 28:75-83.

Radunz-Neto, J., G. Corraze, P. Bergot, and S.J. Kaushik. 1996. Estimation of essential fatty acid requirements of common carp larvae using semi-purified artificial diets. *Archives of Animal Nutrition* 49:41- 48.

Rowan, M. and N.M. Stone. 1993. Winter feeding of golden shiners. *Aquaculture Magazine* 19(2):99-102.

Rowan, M. and N.M. Stone. 1994. Winter feeding of golden shiners: Year II. *Arkansas Aquafarming* 12(1):5-6.

Satoh, S., E.E. Poe, and R.P. Wilson. 1989a. Effect of dietary n-3 fatty acids on weight gain and liver polar lipid fatty acid composition of fingerling channel catfish. *Journal of Nutrition* 119:23-28.

Satoh, S., E.E. Poe, and R.P. Wilson. 1989b. Studies on the essential fatty acid requirement of channel catfish, *Ictalurus punctatus. Aquaculture* 79:121-128.

SRAC (Southern Regional Aquaculture Center). 1998. *Dietary Protein and Lipid Requirements of Golden Shiners and Goldfish.* United States Department of Agriculture, Cooperative States Research, Education, and Extension Service. SRAC Publication No. 124.

Stickney, R.R. and J.W. Andrews. 1972. Effects of dietary lipids on growth, feed conversion, lipid and fatty acid composition of channel catfish. *Journal of Nutrition* 102:249-258.

Stone, N., E. Park, L. Dorman, and H. Thomforde. 1997. *Baitfish Culture in Arkansas: Golden Shiners, Goldfish, and Fathead Minnows.* Cooperative Extension Program, University of Arkansas at Pine Bluff, United States Department of Agriculture, and County Governments Cooperating.

Szlaminska, M., A.M. Escafre, and P. Bergot. 1993. Preliminary data on semi-synthetic diets for goldfish (*Carassius auratus* L.) larvae. In S.J. Kaushik and P. Luques (Eds.), *Fish Nutrition in Practice* (pp. 607-612). Paris: Biarritz (France), INRA.

Takeuchi, T., S. Satoh, and T. Watanabe. 1983. Requirement of *Tilapia nilotica* for essential fatty acids. *Bulletin of the Japanese Society of Scientific Fisheries* 49:1127-1134.

Takeuchi, T. and T. Watanabe. 1977. Requirement of carp for essential fatty acids. *Bulletin of the Japanese Society of Scientific Fisheries* 43:541-551.

Thompson, K.D., M.F. Tatner, and R.J. Henderson. 1996. Effects of dietary (n-3) and (n-6) polyunsaturated fatty acid ratio on immune response of Atlantic salmon, *Salmo salar* L. *Aquaculture Nutrition* 2:21-31.

Watanabe, T. 1982. Lipid nutrition in fish. *Comparative Biochemistry and Physiology* 73B:3-15.

Watanabe, T., T. Takeuchi, and C. Ogino. 1975. Effect of dietary methyl linoleate and linolenate on growth of carp-II. *Bulletin of the Japanese Society of Scientific Fisheries* 41:263-269.

Watanabe, T., O. Utsue, I. Koybayashi, and C. Ogino. 1975. Effect of dietary methyl linoleate and linolenate on growth of carp-I. *Bulletin of the Japanese Society of Scientific Fisheries* 41:257-262.

Chapter 6

Nutritional Deficiencies in Commercial Aquaculture: Likelihood, Onset, and Identification

Ronald W. Hardy

INTRODUCTION

Dietary nutrient requirements for major species of cultured fish, for example, channel catfish, *Ictalurus puntatus;* rainbow trout, *Oncorhynchus mykiss;* and tilapia, *Oreochromis aureus,* in the United States are reasonably well known (NRC 1993). Diets for these species are formulated to contain levels of essential nutrients in excess of minimum dietary requirements to account for slight differences in the essential nutrient levels of major diet ingredients, such as fish meal, and to cover any losses associated with diet manufacture and storage. Essential nutrients in fish diets are the following: (1) 10 amino acids; (2) essential fatty acids, for example, n-3 for marine fish and n-6 plus n-3 for freshwater fish; (3) 15 vitamins, including 4 fat-soluble and 11 water-soluble vitamins; and (4) about 10 minerals (Lovell 1998). Fish can obtain most minerals, except phosphorus, from both their rearing water and their diet, and although many minerals are known to be essential for growth and metabolism of fish, the need for a dietary source depends on the mineral and its concentration in the culture water (Lall 1989).

One would think that nutritional deficiencies would be easy to avoid, given that the dietary needs for many species of farmed fish are known or at least estimated. However, the process of diet formulation and diet manufacturing (e.g., grinding, heating, adding moisture, pelleting, and drying) adds an element of uncertainty to the stability and bioavailability of certain essential nutrients, which sometimes results in clinical deficiencies of several nutrients. Substitution of fish meal with grain and oilseed meals can lower the availability of certain nutrients. The trend toward high-energy diet affects the amount of micronutrients needed to support growth and health, plus it increases the chances of nutritional problems associated with lipid oxidation. Thus, nutritional deficiencies still occasionally occur in commercial production, and undetected subclinical deficiencies may be more common, possibly contributing to inefficient fish growth, losses to disease, and baffling problems encountered with attempts to culture new fish species.

Several factors contribute to this situation. First, the published data upon which dietary recommendations are made are somewhat scant; that is, few scientific studies have been conducted, and these relied on different measures of adequate nutrient intake (weight gain, tissue saturation, enzyme activity, bone strength, and so on). Surprisingly, for some major species of cultured fish, such as Atlantic salmon, *Salmo salar,* the scientific literature contains virtually no information on dietary requirements, except for ascorbic acid and a few other vitamins and minerals. Nutritional requirements for Atlantic salmon and some other emerging fish species produced by aquaculture are estimated from data on rainbow trout and Pacific salmon, *Oncorhynchus tshawytscha* and *O. kisutsch,* which are closely related fish species. Information on the nutritional requirements for channel catfish is the most complete among cultured fish. Second, published studies are usually conducted with fingerling fish raised under laboratory conditions, rather than grow-out fish grown in the less-than-ideal environment of commercial aquaculture production. Third, new diet processing technologies, for example, cooking-extrusion, are more abusive to some vitamins than earlier diet processing technologies. Fourth, antagonistic interactions among diet ingredients can cause losses of certain vitamins and render several essential minerals unavailable to fish. Antagonistic interactions among diet ingredients are sometimes difficult to predict and are thus overlooked in diet formulation. Fifth, the advent of high-energy diets has increased fish growth per unit of diet intake, in many cases twice as much as semipurified diets used to determine dietary requirements. This point is critical because the need for essential nutrients is to support fish growth. If fish require twice as much semipurified diet to gain the same amount in a given period as a commercial high-energy diet, the amount of essential nutrients in the high-energy diet will have to be twice that of the semipurified diet. Finally, the forms of vitamins used to supplement diets are evolving, as new products are developed and used. Most fish farmers assume that vitamins are stable during diet storage, but several are not. Generally, diet producers over-supplement these vitamins to ensure adequate levels for three months of storage, but fish farmers sometimes store diets for longer periods, even freezing the diet, which actually accelerates destruction of some nutrients. The consequence of this is inadequate intake of several vitamins by the fish fed the old and/or frozen diet.

LIKELIHOOD OF DEFICIENCIES OF ESSENTIAL NUTRIENTS

Amino Acids

As mentioned earlier, fish, similar to most vertebrates, require ten essential amino acids in their diet. The other ten or so amino acids found in tissue proteins can be synthesized by the cells; thus, they are nonessential (dispensable). Fish require the essential amino acids in proportions more or less similar to the proportions found in their bodies. Thus, fish meal made from whole fish (rather than seafood processing waste) has an amino acid profile similar to the dietary requirements of fish. As diet formulators lower the proportion of fish meal in fish diets, they replace fish meal with rendered products and/or plant proteins, such as soybean meal. These protein sources differ from fish meal in amino acid content and bioavailability. Dietary amino acid deficiencies can result from formulation or production errors, generally by failure to take into account the difference between amino acid content and amino acid bioavailability in these protein sources. However, the consequences of these deficiencies are not catastrophic, for example, slower-than-expected growth and higher diet conversion ratios, rather than fish mortality. Amino acid deficiencies are not an area of concern with respect to fish losses in commercial aquaculture.

Essential Fatty Acids

Conventional wisdom is that marine and freshwater fish living in cold water (10°C) require n-3 highly unsaturated fatty acids (HUFAs) in their diets, while freshwater fish from temperate or warm climates require either a mixture of n-3 and n-6 HUFAs or n-6 HUFAs exclusively (Lovell 1998). In the case of essential fatty acids, deficiencies can result in clinical signs, but inducing a deficiency takes many months, especially in large fish, which can draw on tissue reserves to supply their needs. Diets for salmon, trout, and marine species contain fish meal, which has residual amounts (5 to 9 percent) of fish oil, enough to meet the n-3 HUFA needs of these fish. There are no reported instances of essential fatty acid deficiency in grow-out fish, except for anecdotal accounts that are complicated by the presence of lipid oxidation. Thus, essential fatty acid deficiencies are rare or virtually nonexistent in commercial fish production, excluding spawning marine fish. However, if they did occur, the results would be serious.

Micronutrients

Vitamin and mineral deficiencies, although the easiest problems to avoid in fish diets, are the most common category of deficiencies observed in commercial aquaculture. Deficiencies can result from omission of vitamin or mineral premixes, but more commonly they result from antagonistic interactions with other compounds in diets that lower vitamin levels or lower the bioavailability of some minerals. Deficiencies may also result from incorrect dietary supplementation, either associated with deliberate reductions in vitamin levels added to diets or lack of understanding of the need to increase dietary levels in high-energy diets. Both factors, in combination with faster-growing fish from selection programs, may interact to create deficiencies in commercial production. Anticipating these problems, and recognizing them when they do occur, requires an understanding of how dietary vitamin and mineral requirements are established. This knowledge will facilitate identification of a nutritional problem and allow corrective measures to be taken.

Establishment of dietary vitamin requirements of fish is based upon an interpretation of the best published information concerning minimum amounts needed to prevent deficiency signs and promote rapid growth and optimum health. For many vitamins, various published studies differ in their estimation of dietary requirements for a given fish species. Experts evaluate these studies and choose a required dietary level based upon their best judgement. The NRC (1993) makes dietary recommendations that are always higher than minimum requirements because they contain an additional amount to allow for any losses that might occur in diet manufacture and delivery to fish, in other words, a margin of safety. Vitamin premixes used in diets for many fish species are formulated to supply the entire dietary requirement of the fish. Any contribution to the dietary requirements of the fish from vitamins naturally present in diet ingredients is ignored, although studies are underway to determine the amount and availability of certain vitamins in diet ingredients to channel catfish. One would think that vitamin deficiencies in aquaculture production would never occur. However, they do, in part because our understanding of the dietary vitamin requirements of fish is far from complete. Thus, certain vitamins are potential suspects when nutritional deficiencies are thought to be occurring in commercial aquaculture, namely ascorbic acid, alpha-tocopherol, pantothenic acid, pyridoxine, niacin, and thiamin. Luckily, it is not difficult to identify or rule out a deficiency of any of these vitamins, based upon clinical signs and confirmatory tests.

Dietary mineral deficiencies fall into three general categories: (1) those which affect hard tissue mineralization; (2) those which affect other specific tissues; and (3) those which do not affect specific tissues (see Table 6.1). Because a good portion of the mineral needs of fish can be supplied by rearing water, the list of minerals that are likely to be deficient in commercial aquaculture is relatively short: phosphorus, zinc, iodine, copper, and selenium. Identifying each of these deficiencies is relatively easy.

TABLE 6.1. Categories of Essential Minerals Based Upon Physiological Function

Category/Mineral	Function
1. Bone/scale mineralization	
Calcium	Bone, scale, skin, muscle function
Phosphorus	Bone, scale, skin, phospholipids
Magnesium	Bone, scale, skin, muscle function
2. Specific physiological function	
Iron	Hemoglobin
Selenium	Glutathione peroxidase
Iodine	Thyroid function
Sodium and Potassium	Ionic balance
3. General physiological function	
Copper	Cofactor for enzyme activity
Manganese	Cofactor for enzyme activity
Zinc	Cofactor for enzyme activity

EVOLUTION OF ESSENTIAL NUTRIENT LEVELS IN FISH DIETS

Amino Acids

In general, the amino acid requirements first demonstrated for Pacific salmon (Halver 1972) and catfish (Dupree and Halver 1970) are still in use today. Refinement of dietary requirements for several amino acids for some species, such as arginine in salmonids, has been made since the first published information appeared in the scientific literature (Kaushik 1979; Kaushik et al. 1988; Kim, Kayes, and Amundson 1992; Luzzana, Hardy, and Halver 1998). For the most part, however, in 25 years, little progress has been made in establishing the dietary amino acid requirements of major cultured fish species, with the exception of channel catfish.

Essential Fatty Acids

The first published studies of the essential fatty acid requirements of fish appeared in the 1970s (Castell, Sinnhuber, Lee, et al. 1972; Castell, Sinnhuber, Wales, et al. 1972), including descriptions of clinical deficiency signs in rainbow trout. Trout fingerlings developed deficiency signs within one to three months when placed on a diet devoid of n-3 HUFAs. Since then, numerous studies have shown the importance of n-3 fatty acids to various species of marine fish, and to Pacific salmon. Little information has been published on the n-3 fatty acid requirement of grow-out fish, but substantial evidence exists to demonstrate that as long as the level of n-3 fatty acids in diets is approximately 1 percent, fish grow normally (NRC 1993).

Vitamins

Most early published studies on dietary vitamin needs of fish, especially salmonids, were the result of efforts to increase the survival of salmonids, both in the hatchery and after hatchery release. Thus, they were conducted on fry and fingerling-sized fish and used weight gain and survival (absence of deficiency signs) as the criteria for adequate dietary intake levels. Test diets were less than ideal, resulting in fish growth rates that were much lower and diet conversion ratios that were much higher than those in commercial aquaculture (Hardy 1991). Maximum tissue storage levels were sometimes used as response variables to dietary intake levels, resulting in relatively high requirement estimates, especially when maximum liver storage levels were measured. Given that the initial aim of these efforts was to increase the survival and growth rates of hatchery-raised juvenile salmonids, overestimating the dietary vitamin requirements was a desirable outcome compared to the potential problems associated with erring on the low side. In other words, if the dietary level of supplementation of a given vitamin was two to five times higher than the actual requirement, this was not an economic problem.

Development of commercial fish farming changed researchers' perspectives of the economic consequences of overestimating the dietary vitamin requirements of fish, especially salmon. The use of sophisticated analytical methods for measuring tissue vitamins or metabolically active forms of certain vitamins and development and validation of specific enzyme activity-linked tests that were sensitive to dietary vitamin intake levels led to a lowering of requirement estimates of nearly all of the water-soluble vitamins for salmonids (Woodward 1994). Some commercial diet producers conduct their own proprietary studies to determine minimum levels of vitamin supplementation suitable for commercial fish production. Fish diets, especially for Atlantic salmon, have become more energy dense as a result of higher levels of added fish oil, made possible by the use of cooking-extrusion pelleting technology. The combined effect of these changes has been to lower the level of supplemental vitamins in commercial salmon and trout diets, making the margin for error quite small.

Minerals

Minerals have been the neglected category of micronutrients in fish nutrition research, despite their having been one of the original subjects of fish nutrition research (Gaylord and Marsh 1914). In fact, until the zinc-cataract problem was identified (Ketola 1979), there was little concern about the mineral levels in diets because fish can obtain much of their minerals directly from their culture water, assuming that the water is suitably hard. The main exception to this is phosphorus, which must be supplied by the diet (Lall 1989). The appearance of cataracts in fish, first in rainbow trout and later in Pacific salmon fingerlings, made the mineral nutrition of fish a topic of immediate concern to the aquaculture community. Briefly, cataracts that were characteristic of zinc deficiency appeared in fish fed diets containing zinc at levels well above the reported dietary requirement of 15 to 30 mg Zn/kg diet (Ogino and Yang 1978), suggesting an antagonistic interaction in the diet that lowered the bioavailability of zinc to the fish. Ketola (1979) showed that the condition resulted from diets containing high-ash fish meal; this was corrected by additional zinc supplementation. Later studies showed that high levels of dietary calcium and phosphorus, alone (Hardy and Shearer 1985) or in combination with phytate (Richardson et al. 1985), caused zinc deficiency signs in salmonids, and that this could be prevented by increasing dietary zinc levels. Gatlin and Phillips (1989) reported that channel catfish required substantially higher zinc supplementation in practical catfish diets than the dietary requirement measured with diets containing semipurified diet ingredients. McClain and Gatlin (1988) reported similar findings with tilapia, *Oreochromis aureus.*

Efforts to lower the phosphorus level in fish farm effluent water have resulted in a substantial reduction in total phosphorus content of their diets, bringing the total phosphorus content of diets closer to the available phosphorus level. Rainbow trout, for example, require about 0.6 to 0.7 percent available phosphorus (Ogino and Takeda 1978). Several years ago, rainbow trout diets contained 2 percent or more total phosphorus, mainly because of the high proportion of fish meal in these diets. Of this 2 percent total phosphorus, about half was available phosphorus. Thus, approximately 1.4 percent of the total phosphorus in fish diets was excreted by the fish, either as indigestible phytate or bone phosphorus in the feces (1 percent) or as soluble phosphorus in the urine (0.4 percent). Today, rainbow trout diets contain 1.2 to 1.4 percent total phosphorus, of which about half is available. Thus, the amount of phosphorus excreted by the fish is about one-half of the total dietary phosphorus, mostly as indigestible phosphorus in feces. This improvement has been achieved by restricting high-ash ingredients and increasing the percentage of low-ash rendered products (blood meal, feather meal) and plant protein sources (soybean meal, corn gluten meal) in diets. Fish meal is an excellent source of dietary minerals for fish; thus, the trend toward replacing fish meal with other protein sources will require careful consideration of dietary mineral levels and of possibly antagonistic interactions that might necessitate additional mineral supplementation to fish diets.

ONSET AND IDENTIFICATION OF NUTRITIONAL DEFICIENCIES

Amino acids and essential fatty acids will not be discussed further because their deficiencies are unlikely and/or do not cause fish losses in commercial aquaculture.

Vitamins

Clinical signs of water-soluble vitamin deficiencies, with the exception of vitamin B_{12}, are preceded by loss of appetite (anorexia). The period of time between loss of appetite and appearance of clinical deficiency signs in fish fed semipurified diets varies among vitamins, from two to three days in the case of pyridoxine (vitamin B_6) in rainbow trout fry to several months for postjuvenile trout fed a thiamin-deficient diet (vitamin B_1). Small fish have not yet developed substantial tissue vitamin stores and they grow rapidly; hence, vitamin deficiency signs appear more rapidly in small fish than in larger fish when they are fed a deficient diet. The clinical deficiency signs for vitamins are linked to their physiological functions, and, in some cases, the clinical signs are sufficiently distinct to allow for diagnosis (see Table 6.2). In other cases, the clinical deficiency signs are not sufficiently distinct to permit easy identification (see Table 6.3). Based upon clinical signs of deficiency, diagnostic keys can be constructed, such as the one depicted in Figure 6.1, in which the presence or absence of anemia is the initial categorization used to identify the deficient vitamin.

TABLE 6.2. Primary Clinical Signs of Deficiency of Essential Vitamins for Fish That Are Useful for Diagnosing Deficiency in Commercial Aquaculture

Vitamin	Anorexia	Primary Deficiency Signs
Vitamin A	Yes	Vision problems
Vitamin D	Yes	Impaired bone calcification
Vitamin E	Yes	Anemia, ascites, membrane fragility
Vitamin K	No	Anemia, prolonged prothrombin time
Thiamin	Yes	Hyperirritability, convulsions
Riboflavin	Yes	Lens cataracts
Pyridoxine	Yes	Convulsions, tetany-like paralysis
Pantothenic acid	Yes	Clubbed gills
Ascorbic acid	Yes	Lordosis, scoliosis, hemorrhages

TABLE 6.3. Primary Clinical Signs of Deficiency of Essential Vitamins for Fish That Are Not Useful for Diagnosing Deficiency in Commercial Aquaculture

Vitamin	Anorexia	Primary Deficiency Signs
Niacin	Yes	Skin lesions
Biotin	Yes	Muscle atrophy
Folic acid	Yes	Macrocytic anemia
Vitamin B_{12}	No	Anemia
Inositol	Yes	Reduced phospholipid content
Choline	Yes	Increased liver lipid content

FIGURE 6.1. Sample Key to Vitamin Deficiencies Based Upon Primary Clinical Signs of Deficiency in Salmonids

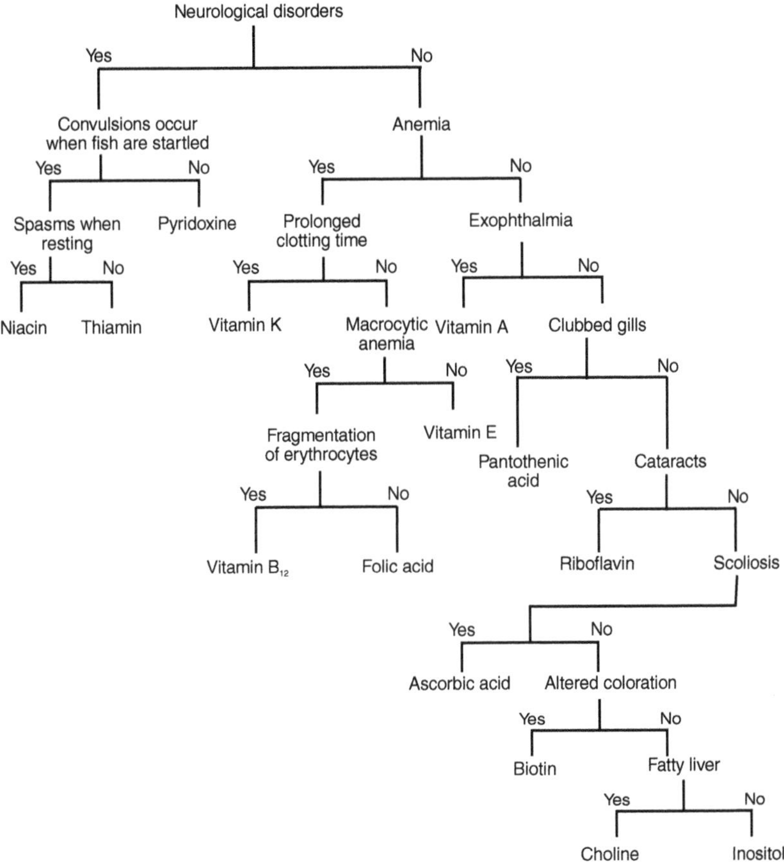

Tissue vitamin stores and turnover rates differ among vitamins, making it possible to categorize vitamins on the basis of the time elapsed from the initiation of a vitamin-deficient diet to the appearance of clinical deficiency signs. A diagnostic key based on published literature on salmon and trout vitamin deficiency studies (with fry and fingerlings fed semipurified diets) is depicted in Figure 6.2. The absolute time between initiation of a diet suspected of being deficient in one or more vitamins and the appearance of clinical deficiency signs depends on many factors, such as body stores, water temperature, growth rate, and fish size. Thus, the number of weeks depicted in Figure 6.2 cannot be taken as absolute values, but rather as values relative to one another, especially when this key is used in a commercial

FIGURE 6.2. Sample Key to Vitamin Deficiencies Based Upon Time of Onset of Primary Deficiency Signs in Juvenile Salmonids

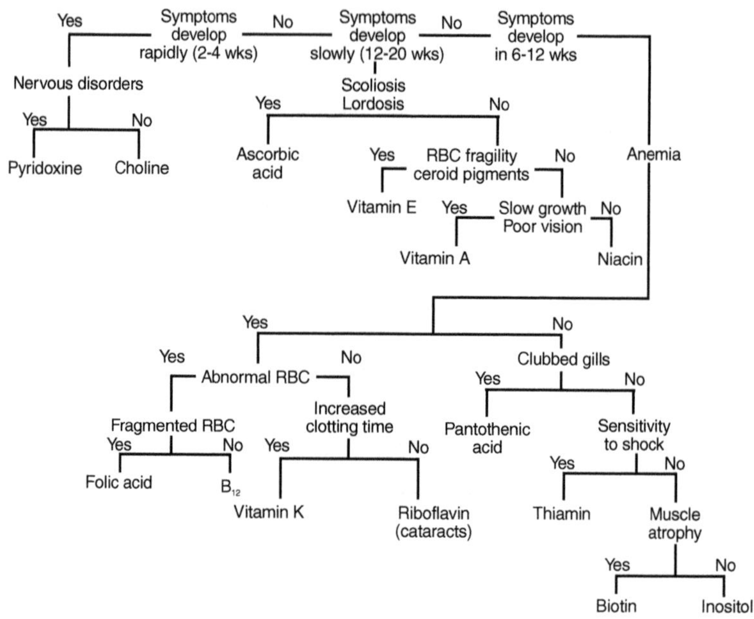

aquaculture setting. In addition, the values shown in Figure 6.2 are based upon the utilization of a diet *completely* deficient in the vitamin in question. In commercial aquaculture, diets are rarely completely deficient, and this difference will affect the number of weeks shown in Figure 6.2.

The main value of this type of key is to focus the farmer or investigator on vitamins that are likely to be depleted, and to suggest what tests might be conducted to identify the nutrient that is deficient, which can then be added to the diet to correct the problem. It also suggests that vitamin deficiencies do not happen overnight but rather take a relatively long period to develop, at least for most vitamins. If a farmer experiences a problem two weeks after receiving a new batch of diet, or after switching diet manufacturers, it is very unlikely that the problem is caused by a vitamin deficiency in the new diet.

Another factor to consider in narrowing down the list of potential deficient vitamins in a diet is the stability of various vitamins during cooking-extrusion and diet storage. Vitamins are produced in forms that are protected from the stresses of diet manufacture by coatings or, in the case of vitamins C and E, by adding chemical moietes at places on the vitamin structure where free radicals react. Vitamins in protected forms are much more stable than pure vitamins, but they are not 100 percent stable. As shown in Table 6.4, crystalline ascorbic acid, vitamin B_{12}, folic acid, biotin, thiamin, vitamin K, and vitamin A are less stable than the other vitamins,

TABLE 6.4. Stability of Vitamins in Steam Pellets and Extruded Pellets After Three Months of Room Temperature Storage

Vitamin	Steam Pellets (%)	Extruded Pellets (%)
Vitamin A (beadlet, cross-linked)	85-95	70-90
Vitamin D (beadlet, cross-linked)	90-100	75-100
Vitamin E (acetate)	90-100	90-100
Vitamin K (MNB)	70-90	40-70
Thiamin	85-100	60-80
Riboflavin	90-100	90-100
Pyridoxine	90-100	80-90
Pantothenic acid	90-100	80-100
Niacin	90-100	90-100
Biotin	90-100	70-90
Folic acid	70-90	50-65
Vitamin B_{12}	60-90	40-80
Inositol	100	100
Choline	100	100
Ascorbic acid, crystalline	30-70	10-30
Ascorbate-2-phosphate	90	90

Source: Gabaudon and Hardy 2000.

with <75 percent of original vitamin activity remaining after extruded pelleting and three months of storage (Gabaudon and Hardy 2000). Luckily, of these vitamins, only ascorbic acid is a concern with respect to vitamin deficiencies likely to be encountered in commercial aquaculture, and ascorbate-2-phosphate, the protected form of ascorbic acid commonly added to fish diets, is relatively stable.

Correctly identifying a vitamin deficiency in commercial aquaculture is not difficult, provided that only one vitamin deficiency exists at a given time. The presence or absence of anemia can generally be used to distinguish between vitamin E and ascorbic acid deficiency. In addition, evidence of broken back, scoliosis or lordosis, shortened opercules, or flexible mandible can be used to make a presumptive diagnosis of ascorbic acid deficiency in most fish. For vitamin E deficiency, measuring erythrocyte fragility by placing erythrocytes in vials containing a salt solution ranging from 1.0 to 0.2 percent in 0.1 percent increments and looking for evidence of cell rupture can be used to make a presumptive diagnosis (Draper and Csallany 1969). Thiamin and pyridoxine deficiencies result in neurological disorders that are evident in the fish. Thiamin-deficient fish are subject to what appear to be convulsions when startled by a sudden splash on the surface of the water or a blow to the side of a tank. They startle and continue to swim, as if escaping from a predator, crashing into the sides of tanks and thrashing on the surface. Pyridoxine-deficient fish look normal one minute and seem paralyzed the next, sinking in the water column, often tail first, with only their gills moving; they look frozen. Pantothenic acid deficiency causes gill hyperplasia, meaning that the gill lamella grow and fuse together, restricting respiration. Fish gasp, especially when stressed, and die with gaping mouths. Niacin-deficient fish have skin lesions and muscle spasms when resting.

Minerals

In practical fish culture, phosphorus deficiency is the only likely deficiency to affect bone and scale mineralization. Deficiencies of calcium and magnesium, the other main constituents of bone, are relatively difficult to create and unlikely to occur. Phosphorus in fish diets is mainly supplied as bone, either fish meal, poultry by-product meal, or meat-and-bone meal. About two-thirds of the phosphorus in plant-derived diet ingredients is present as phytate, the storage form of phosphorus in seeds (wheat, corn, soybeans, etc.). Phytate is indigestible to monogastric animals; thus, phosphorus in phytate is unavailable. In fish diets containing mainly plant-derived diet ingredients, such as catfish or tilapia diets, phosphorus is supplemented as dicalcium phosphate to meet the dietary needs of the fish (Lovell 1998).

Phosphorus deficiencies occur when the amount of available phosphorus in a diet is lower than the dietary requirement of the fish. The bioavailability of phosphorus in common diet ingredients has been measured (Riche and Brown 1996; Sugiura et al. 1998), but it varies with calcium content in fish and animal meals. With regulations in Europe mandating that salmonid diets contain no more than 1 percent phosphorus and have a feed conversion ratio of 1:1, there is little margin for error with respect to ensuring a sufficient level of available phosphorus. The trend in salmon and trout diets of replacing fish meal with soybean meal, corn gluten meal, and other plant-derived ingredients makes it difficult to formulate diets containing 1.0 percent phosphorus and 0.6 percent available phosphorus. When high-fat diets are used, the dietary requirement for available phosphorus is likely 0.75 percent or higher. One study reports that Atlantic salmon fry require 1 percent available phosphorus in their diet (Asgard and Shearer 1997). Phosphorus in diet ingredients varies in availability to salmonids (see Table 6.5), and if too many changes are made in diets over time (high-fat diets, replacement of fish meal with plant-derived protein sources) without consideration of the effects of these changes on phosphorus level and availability, phosphorus deficiency can result. Luckily, phosphorus deficiency in fish rarely results in mortality; rather, fish simply stop eating and growing. Their bones are poorly calcified and become flexible, most noticeably the jaw and operculum cover. This occurs as phosphorus is withdrawn from hard tissue to meet the metabolic demands for blood constituents, cellular energy generation (ATP), and phospholipids used in cellular and subcellular membranes. Fish

TABLE 6.5. Apparent Availability of Phosphorus in Diet Ingredients Determined with Rainbow Trout and Coho Salmon

Ingredient	Rainbow Trout (%)	Coho Salmon (%)
Herring meal	44.5	57.3
Anchovy meal	50.4	47.4
Menhaden meal	36.5	40.4
Deboned whitefish meal	46.8	54.7
Poultry by-product meal	63.5	67.7
Feather meal	61.7	75.4
Soybean meal	22.0	28.4
Wheat gluten meal	74.7	56.9
Corn gluten meal	8.5	15.8
Wheat middlings	55.3	41.0
Wheat flour (diet grade)	47.0	50.1

Source: Sugiura et al. 1998.

can continue to grow for extended periods when fed a phosphorus-deficient diet, utilizing their body stores to supply metabolic needs. Eventually, however, their reserves are used up and clinical signs of deficiency appear. There are no visible signs of phosphorus deficiency that fish farmers can look for prior to the appearance of clinical deficiency signs, for example, anorexia and slow fish growth. Phosphorus deficiency can be confirmed by examining bone flexibility, especially the jaw, looking for check marks on rib bones, and measuring whole-body or bone phosphorus content (Shearer and Hardy 1987).

Zinc is the other mineral that can become deficient in fish diets, mainly through antagonistic interactions with other diet ingredients, such as phytate and high-ash levels (high calcium and phosphorus). Zinc deficiency results in bilateral lens cataracts, which are irreversible. Other nutritional deficiencies can cause cataracts, including riboflavin and methionine, but zinc cataracts are more likely to occur in commercial aquaculture. As is the case with phosphorus, prior to the appearance of cataracts, fish appear normal. Onset appears to be sudden because cataracts are difficult to see without taking fish out of the water. Once they are noticed on a farm, most likely a large number of fish will be affected. The actual onset of cataracts is progressive, with the degree of lens opacity becoming higher over a period of time.

Iodine deficiency causes thyroid hyperplasia (goiter) in fish. Fish differ in their susceptibility to iodine deficiency. Iodine is included in trace mineral supplements to trout and salmon diets, and deficiency is extremely rare but easy to diagnose.

Copper and selenium deficiencies are unlikely to cause acute deficiency signs or fish loss. Their deficiencies are more likely to be subclinical, causing fish to perform poorly and, possibly, to have eroded fins, in the case of copper (Barrows and Lellis 1999). Some evidence suggests that selenium deficiency in fish increases susceptibility to disease (Thorarinsson et al. 1994).

OTHER NUTRITIONAL PROBLEMS

Lipid oxidation creates problems both in the diet and in the fish. Oxidation of lipids during diet storage can lower levels of ascorbic acid (vitamin C) and alpha-tocopherol (vitamin E), both vitamins being sacrificial antioxidants. Diets are normally supplemented with protected forms of these vitamins, which prevents their oxidation during diet storage, but it also prevents them from serving as antioxidants in the diet. When fish eat diets containing oxidizing lipids, a range of pathological problems can occur. These problems are associated with (1) the products of oxidation, namely, aldehydes, ketones, and other potentially toxic compounds, and (2) free radicals that can overwhelm the capacity of cells to detoxify them. The primary health effects associated with adding oxidizing fish oil to fish diets are liver degeneration, anemia, and spleen abnormalities (Smith 1979; Hung, Cho, and Slinger 1980; Moccia et al. 1984). Oxidizing fish oil also causes pathological abnormalities in fish by destroying unprotected tocopherol and ascorbic acid in tissues (Bell and Cowey 1985). In catfish and trout, many of the pathological effects associated with a diet containing oxidizing fish oil can be prevented by increasing the dietary intake of tocopherol (Murai and Andrews 1974; Moccia et al. 1984). However, acute toxicity is not always prevented by increasing dietary tocopherol intake, especially in small fish that have not accumulated substantial tissue stores of tocopherol. Growth rates of Atlantic and coho salmon fry are reported to be inversely proportional to the level of oxidation in starter diets, with reduced growth being observed at levels of oxidation that do reduce growth in larger fish (Ketola, Smith, and Kindschi 1989). There is also some suggestion in the literature that there may be species differences in sensitivity to oxidizing diets. The link between tocopherol deficiency signs and those associated with consumption of oxidizing lipids is weak. If oxidizing lipids caused pathological problems in fish solely by reducing tissue tocopherol levels, one might assume that signs of tocopherol deficiency, such as muscular dystrophy, exudative diathesis, and depigmentation, would result from consumption of diets containing oxidizing lipids. However, this is not the case.

Other nutritional problems associated with diets include the presence of mold toxins and contaminants arising from the use of contaminated diet ingredients. Luckily, the latter rarely occurs, but when it does, the consequences are serious. Fish species differ in their sensitivity to mold toxins, such as aflatoxins. Rainbow trout are more sensitive than coho salmon. Catfish are much less sensitive. Chronic ingestion of low levels of aflatoxins causes liver tumors, while ingestion of high levels causes death (Hendricks and Bailey 1989).

Gizzerosine is a toxic compound found in fish meals produced from scombroid fishes, such as tuna and mackerel. It is a conjugate of the amino acid lysine and histamine, and it causes fatal bleeding lesions in the gizzards of birds. Its toxicity to rainbow trout has been studied, and while it appears to cause stomach swelling, it does not affect growth or cause mortality (Fairgrieve et al. 1994).

IMPORTANT POINTS FOR FISH FARMERS

Nutritional deficiencies are relatively rare in commercial aquaculture, but they do occur. In recent years, deficiencies of ascorbic acid, vitamin E, phosphorus, and zinc have occurred in commercial aquaculture in the United States and abroad. Pyridoxine deficiency was recently observed in wild cutthroat trout fry being fed commercial trout starter diet in the author's laboratory. Nutritional toxicity resulting from lipid oxidation in diets has also occurred in recent years in commercial aquaculture in the United States. Other nutritional deficiencies may occur on a subclinical basis and escape detection but influence fish growth rates, diet conversion ratios, and susceptibility to infectious disease.

Fish farmers must be alert to loss of appetite in their fish. Although this is not useful in identifying a specific nutritional deficiency and does not occur in some nutritional deficiencies, it is easy to notice and usually precedes a clinical deficiency by weeks or months. Fish should be examined for cataracts, soft bones or broken backs, and fin erosion whenever they are handled in the normal course of fish farming. If losses occur and a nutritional deficiency is suspected, farmers should save a sample of the diet for later analysis of vitamin levels and degree of lipid oxidation. An independent professional should check fish for anemia, soft bones, gill hyperplasia, and gross liver abnormalities. All of these tests can be easily and rapidly done. To confirm a vitamin or mineral deficiency, tissue nutrient or metabolite levels should be measured. Luckily, if vitamin and mineral deficiencies are detected early, they can be reversed, or at least limited, in the fish population.

REFERENCES

Asgard, T. and K.D. Shearer. 1997. Dietary phosphorus requirement of juvenile Atlantic salmon, *Salmo salar* L. *Aquaculture Nutrition* 3:17-23.

Barrows, F.T. and W.A. Lellis. 1999. The effect of dietary protein and lipid source on dorsal fin erosion in rainbow trout, *Oncorhynchus mykiss*. *Aquaculture* 180:167-175.

Bell, J.G. and C.B. Cowey. 1985. Roles of vitamin E and selenium in the prevention of pathologies related to fatty acid oxidation in salmonids. In C.B. Cowey, A.M. Mackie, and J.G. Bell (Eds.), *Nutrition and Feeding in Fish* (pp. 333-347). London: Academic Press.

Castell, J.D., R.O. Sinnhuber, D.J. Lee, and J.H. Wales. 1972. Essential fatty acids in the diet of rainbow trout *(Salmo gairdneri):* Physiological symptoms of EFA deficiency. *Journal of Nutrition* 102:87-92.

Castell, J.D., R.O. Sinnhuber, J.H. Wales, and D.J. Lee. 1972. Essential fatty acids in the diet of rainbow trout *(Salmo gairdneri):* Growth, feed conversion and some gross deficiency symptoms. *Journal of Nutrition* 102:77-85.

Draper, H.H. and A.S. Csallany. 1969. A simplified hemolysis test for vitamin E deficiency. *Journal of Nutrition* 98:390-394.

Dupree, H.K. and J.E. Halver. 1970. Amino acids essential for the growth of channel catfish, *Ictalurus punctatus. Transactions of the American Fisheries Society* 99:90-92.

Fairgrieve, W.T., M.S. Myers, R.W. Hardy, and F.M. Dong. 1994. Gastric abnormalities in rainbow trout *(Oncorhynchus mykiss)* fed amine-supplemented diets or chicken gizzard-erosion-positive fish meal. *Aquaculture* 127:219-232.

Gabaudon, J. and R.W. Hardy. 2000. Vitamin Sources for Fish Feeds, In R.R. Stickney (Ed.), *Encyclopedia of Aquaculture* (pp. 961-964). New York: John Wiley & Sons.

Gatlin, D.M.I. and H.F. Phillips. 1989. Dietary calcium, phytate and zinc interactions in channel catfish. *Aquaculture* 79:259-266.

Gaylord, H.R. and M.C. Marsh, 1914. *Carcinoma of the Thyroid in the Salmonoid Fishes,* Bulletin 32. Washington, DC: U.S. Bureau of Fisheries.

Halver, J.E. 1972. *Fish Nutrition.* New York: Academic Press.

Hardy, R.W. 1991. Pacific salmon, *Oncorhynchus* spp. In R.P.Wilson (Ed.), *Handbook of Nutrient Requirements of Finfish* (pp. 105-121). Boca Raton, FL: CRC.

Hardy, R.W. and K.D. Shearer. 1985. Effect of dietary calcium phosphate and zinc supplementation on whole body zinc concentration of rainbow trout *(Salmo gairdneri). Canadian Journal of Fisheries and Aquatic Sciences* 42:181-184.

Hendricks, J.D. and G.S. Bailey. 1989. Adventitious toxins, In J.E.Halver (Ed.), *Fish Nutrition* Second Edition (pp. 605-651). New York: Academic Press.

Hung, S.S.O., C.Y. Cho, and S.J. Slinger. 1980. Measurement of oxidation in fish oil and its effect on vitamin E nutrition of rainbow trout *(Salmo gairdneri). Canadian Journal of Fisheries and Aquatic Sciences* 37:1248-1253.

Kaushik, S. 1979. Application of a biochemical method for the estimation of amino acid needs in fish: Quantitative arginine requirements of rainbow trout in different salinities. In K. Tiews and J.E. Halver (Eds.), *Finfish Nutrition and Fishfeed Technology* (pp. 197-207). Berlin: Heenemann GmbH.

Kaushik, S.J., B. Fauconneau, L. Terrier, and J. Gras. 1988. Arginine requirement and status assessed by different biochemical indices in rainbow trout (*Salmo gairdneri* R.). *Aquaculture* 70:75-95.

Ketola, H.G. 1979. Influence of dietary zinc on cataracts in rainbow trout *(Salmo gairdneri). Journal of Nutrition* 109:965-969.

Ketola, H.G., C.E. Smith, and G.A. Kindschi. 1989. Influence of diet and oxidative rancidity on fry of Atlantic and coho salmon. *Aquaculture* 79:417-423.

Kim, K.I., T.B. Kayes, and C.H. Amundson. 1992. Requirements for lysine and arginine by rainbow trout *(Oncorhynchus mykiss). Aquaculture* 106:333-344.

Lall, S.P. 1989. The Minerals. In J.E. Halver (Ed.), *Fish Nutrition* (pp. 219-257). New York: Academic Press.

Lovell, R.T. 1998. *Nutrition and Dieting of Fish,* Second Edition. Boston: Kluwer Academic Publishers.

Luzzana, U., R.W. Hardy, and J.E. Halver. 1998. Dietary arginine requirement of fingerling coho salmon *(Oncorhynchus kisutch). Aquaculture* 163:137-150.

McClain, W.R. and D.M. Gatlin. 1988. Dietary zinc requirement of *Oreochromis aureus* and effects of dietary calcium and phytate on zinc bioavailability. *Journal of the World Aquaculture Society* 19:103-108.

Moccia, R.D., S.S.O. Hung, S.J. Slinger, and H.W. Ferguson. 1984. Effect of oxidized fish oil, vitamin E and ethoxyquin on the histopathology and haematology of rainbow trout, *Salmo gairdneri* Richardson. *Journal of Fish Diseases* 7:269-282.

Murai, T. and J.W. Andrews. 1974. Interactions of dietary α-tocopherol, oxidized menhaden oil and ethoxyquin on channel catfish *(Ictalurus punctatus). Journal of Nutrition* 104:1416-1431.

NRC (National Research Council). 1993. *Nutrient Requirements of Fish.* Washington, DC: National Academy Press.

Ogino, C. and H. Takeda. 1978. Requirements of rainbow trout for dietary calcium and phosphorus. *Bulletin Japanese Society of Scientific Fisheries* 44:1019-1022.

Ogino, C., and G.Y. Yang. 1978. Requirement of rainbow trout for dietary zinc. *Bulletin Japanese Society of Scientific Fisheries* 44:1015-1018.

Richardson, N.L., D.A. Higgs, R.M. Beames, and J.R. McBride. 1985. Influence of dietary calcium, phosphorus, zinc and sodium phytate level on cataract incidence, growth and histopathology in juvenile chinook salmon *(Oncorhynchus tshawytscha). Journal of Nutrition* 115:553-567.

Riche, M. and P.B. Brown. 1996. Availability of phosphorus from feedstuffs fed to rainbow trout, *Oncorhynchus mykiss. Aquaculture* 142:269-282.

Shearer, K.D. and R.W. Hardy. 1987. Phosphorus deficiency in rainbow trout fed a diet containing deboned fillet scrap. *Progressive Fish-Culturist* 49:192-197.

Smith, C.E. 1979. The prevention of liver lipoid degeneration and microcytic anaemia in rainbow trout *Salmo gairdneri* Richardson fed rancid diets: A preliminary report. *Journal of Fish Diseases.* 2:429-437.

Sugiura, S.H., F.M. Dong, C.K. Rathbone, and R.W. Hardy. 1998. Apparent protein digestibility and mineral availabilities in various feed ingredients for salmonids. *Aquaculture* 159:177-200.

Thorarinsson, R., M. Landolt, D.G. Elliott, R.J. Pascho, and R.W. Hardy. 1994. Effect of dietary vitamin E and selenium on growth, survival and the prevalence of *Renibacterium salmoninarum* infection in chinook salmon *(Oncorhynchus tshawytscha). Aquaculture* 121:343-358.

Woodward, B. 1994. Dietary vitamin requirements of cultured young fish, with emphasis on quantitative estimates for salmonids. *Aquaculture* 124:133-168.

Chapter 7

Immunity and Disease Resistance in Fish

Craig A. Shoemaker
Phillip H. Klesius
Chhorn Lim

INTRODUCTION

Cultured finfish and shellfish accounted for over 25 percent of world aquatic animal production, or close to 100 million metric tons, in 1996 (FAO 1998). As fish increase as a protein source throughout the world and natural fish stocks decline, culture practices will intensify and environmental conditions in culture systems will deteriorate. Deterioration of the culture environment will lead to more problems with disease, not only because of the increased number of animals in a limited and confined space, but also because of the influence of poor environmental conditions such as water quality on the fish immune system. The result will be an increased and continued negative effect of disease on production.

The state of being immune is defined as the inherited ability to resist infection. Immunity is the result of the recognition of nonself or a foreign agent, with the subsequent response and memory in vertebrate animals. The response includes expansion of cells for the immune response, expression of the cells and molecules (e.g., antibody), and, finally, the coordination of the response by regulatory substances. Disease resistance is the innate (natural) defense mechanisms of an animal against foreign invaders. Figure 7.1 is a schematic representation of the result of the response to a pathogen by fish. The study of fish immunity and disease resistance is relatively young in comparison to the study of mammalian immunity. Most early research (1960s) on fish immunology focused on the comparative aspect of the immune system with fish and other species (i.e., vertebrates and invertebrates). More recently (1970 to present), research has focused on understanding how the fish immune system responds to foreign agents or how innate resistance can be selected for by breeding to produce stocks of fish with superior

FIGURE 7.1. Schematic Representation of the Response of a Fish Following an Encounter with a Pathogen

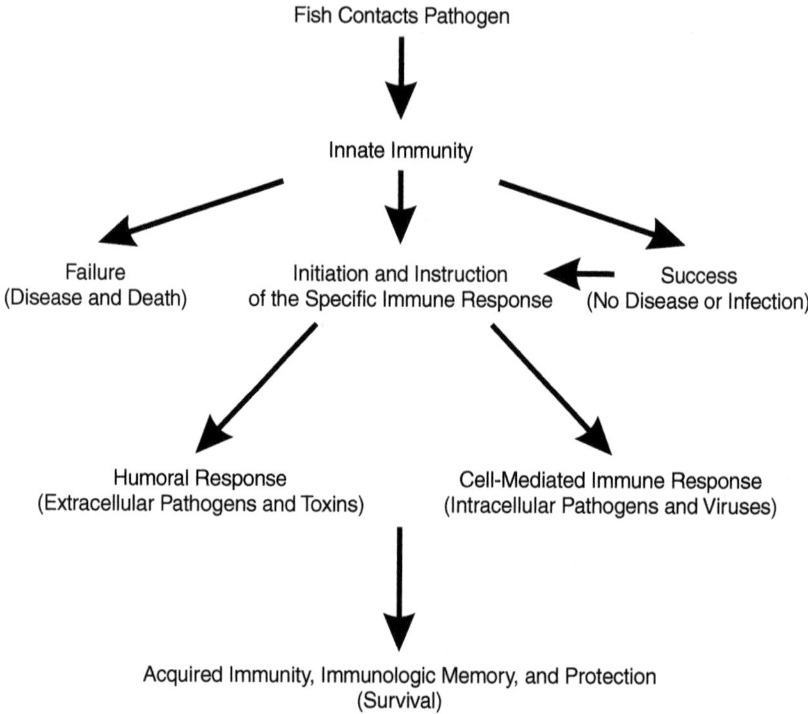

disease resistance. This latter research has led to the basic understanding of how a fish recognizes and responds to pathogens. Future research is needed to better characterize specific cell types and cytokines involved in the immune response, as is the case with warm-blooded animals. Experimentation in these areas will allow for a better understanding of disease resistance and acquired immunity in fish. The final area that requires emphasis is the influence of stress on the overall physiology of fish, which directly influences natural resistance and acquired immunity.

THE IMMUNE TISSUES AND ORGANS OF FISH

The most important immunocompetent organs and tissues of fish include the thymus, kidney (anterior and posterior), spleen, and liver. The immune tissues in these organs are not as well defined or organized as those in mammals (Manning 1994). Evidence suggests that the thymus is responsible for the development of T-lymphocytes, as in other jawed vertebrates. However, much of the data supporting this is indirect evidence, obtained either by immunizing with T-dependent antigens (Ellsaesser, Bly, and Clem 1988) or by using monoclonal antibodies as cell surface markers (Passer et al. 1996) and functional in vitro assays.

The kidney is important in hematopoiesis and immunity in fish. Blood cell differentiation occurs here instead of in the bone marrow, as in mammals. Early in development, the entire kidney is involved in production of immune cells and the early immune responses. As the fish matures, the anterior kidney becomes the most important site of blood cell formation and immune functions, whereas the posterior kidney is primarily involved in blood filtration and/or urinary functions. Blood supply to the kidney is via a renal portal system. The blood flow through the kidney is slow, and exposure to antigens occurs. There appears to be a concentration of melanomacrophage aggregates or immune cells (i.e., believed similar in function to germinal centers in spleen and lymph nodes of birds and mammals [Agius 1985]) in the anterior kidney of most teleost fish. These melanomacrophage centers are aggregates of reticular cells, macrophages, lymphocytes, and plasma cells; they may be involved in antigen trapping and may play a role in immunologic memory (Secombes, Manning, and Ellis 1982).

The spleen is a secondary immune organ in fish believed to be involved in immune reactivity and blood cell formation (Manning 1994). Most fish spleens are not distinctly organized into red and white pulp, as in mammals, but white and red pulp are identifiable. Lymphocyte and macrophages are present in the spleen of fish. Most macrophages are arranged in melanomacrophage centers in the spleen, and it is believed that they are primarily responsible for the breakdown of erythrocytes. Van Muiswinkel, Lamers, and Roumbout (1991) suggest that these melanomacrophage centers are involved in antigenic stimulation, and that this stimulation may persist for about a year, thus suggesting a role in immunologic memory. Manning (1994) suggests that the antigenic stimulation is not organized into thymus-dependent and -independent areas but rather occurs throughout the organ.

The liver was included under this section because, in mammals, it is responsible for production of components of the complement cascade and acute-phase proteins (e.g., C-reactive protein), which are important in the natural resistance of the animal. Fletcher (1981) suggests that the liver of fish plays a similar role. However, research to support this claim is lacking.

The mucus and skin of fish act as natural barriers to foreign substances and disease agents. The mucus has been demonstrated to contain molecules with immune actions. These substances include lysozyme (Fletcher and White 1973; Ourth 1980), complement, and natural antibodies or immunoglobulins (Fletcher and Grant 1969; Di Conza and Halliday 1971; Zilberg and Klesius 1997). Most research on the presence of immunoglobulin or antibody in the mucus suggests that mucus immunoglobulin is not a result of the transduction of immunoglobulin from the serum (Fletcher and Grant 1969; Lobb and Clem 1981; Peleterio and Richards 1988; Rombout et al. 1993; Zilberg and Klesius 1997). Specific antibodies were also reported to be present in the mucus of channel catfish, *Ictalurus punctatus* (Lobb 1987), and rainbow trout, *Oncorhynchus mykiss* (St. Louis-Cormier, Osterland, and Anderson 1984). Zilberg and Klesius (1997) demonstrated that mucus immunoglobulin in channel catfish was elevated following exposure to bacteria, but they were unable to detect specific antibodies in the mucus. Most studies have been done using different antigens and/or routes of administration; thus, the question of the presence of a specific antibody in the mucus remains.

INNATE IMMUNITY AND DISEASE RESISTANCE

Nonspecific Immune Cells

The nonspecific immune cells in fish consist of monocytes/macrophages, neutrophils (or granulocytes), and nonspecific cytotoxic cells or natural killer cells (see Table 7.1). Monocytes and/or tissue macrophages are probably the single most important cell in the immune response of fish. Not only are they important in the production of cytokines (Clem et al. 1985), but they also are the primary cells involved in phagocytosis and the killing of pathogens upon first recognition and subsequent infection (Shoemaker, Klesius, and Plumb 1997). Vallejo, Miller, and Clem (1992) also suggest the macrophage as being the primary antigen-presenting cell in teleosts, thus linking the nonspecific and acquired immune responses.

Neutrophils (granulocytes) are the primary cells involved in the initial stages of inflamation (12 to 24 hours) in fish (Manning 1994). The function of the neutrophil may be cytokine production to recruit immune cells to the area of damage or infection. In channel catfish, the neutrophil is phagocytic but appears to kill bacteria by extracellular mechanisms rather than via intracellular mechanisms (Ellis 1981; Waterstrat, Ainsworth, and Capley 1991). The role neutrophils play in immunity probably varies with species of fish.

TABLE 7.1. Fish Nonspecific Immune Cells and Their Functional Characteristics

Cell Type	Functional Characteristics
Monocyte/macrophage	Phagocytosis, cytokine production, intracellular killing, antigen processing and presentation
Neutrophil/granulocyte	Inflammation, cytokine production, extracellular killing, phagocytosis
Nonspecific cytotoxic cells/ natural killer cells	Recognition and target cell lysis

Nonspecific cytotoxic cells are present in teleost fish. These cells are closely related in function to the mammalian natural killer cells. Evans and Jaso-Friedman (1992) provide an excellent review of nonspecific cytotoxic cells in channel catfish. Nonspecific cytotoxic function in the lysis of target cells following receptor binding and subsequent signaling for the lytic cycle to occur and destroy the target. In fish, these cells appear to be important in parasitic (Evans and Gratzek 1989) and viral (Hogan et al. 1996) immunity.

Phagocytosis

Phagocytosis occurs in fish and is the most primitive defense mechanism. The initial step in phagocytosis is the movement of the immune cell (monocyte or macrophage) in response to the foreign agent. The movement is by chemokinesis (nondirectional movement of the phagocyte) or chemotaxis (directional movement of the phagocyte). Weeks-Perkins and Ellis (1995) and Klesius and Sealey (1996) were among the first to demonstrate that fish macrophages possess the ability to move by chemokinesis or chemotaxis in response to a bacterial antigen either in vitro or in vivo. After movement in response to the foreign agent, attachment occurs. Ainsworth (1993) demonstrated that attachment occurs via lectins and is enhanced by opsonization. Shoemaker, Klesius, and Plumb (1997) showed that opsonization with specific antibody against *E. ictaluri* enhanced attachment, as indirectly measured by an increase in the killing of *Edwardsiella ictaluri* by channel catfish macrophages obtained from previously infected fish. The next step in phagocytosis is engulfment of the foreign agent. Engulfment is simply moving the foreign agent into the cell with subsequent phagosome formation. Killing and digestion of the foreign agent is the final step of phagocytosis.

In fish, destruction or killing can occur by oxygen-dependent or oxygen-independent mechanisms. Oxygen-independent mechanisms involve low pH, lysozyme, lactoferrin, and proteolytic and hydrolytic enzymes. Oxygen-dependent mechanisms are well defined in salmonids, and the pathways are becoming better defined in other species.

Nonspecific Humoral Molecules of Fish

Most nonspecific humoral molecules involved in the natural resistance of fish are presented in Table 7.2. Lectins are important for nonspecific binding to sugars located on the surface of bacteria and pathogens, resulting in precipitation and agglutination reactions. The lectins are glycoproteins that bind by noncovalent reactions to the sugar portion of molecules. Sharon and Lis (1993) suggest lectins are also involved in cell recognition and binding, thus playing an important role in cellular communication as well as defensive actions.

TABLE 7.2. Suggested Nonspecific Humoral Molecules and Their Mode of Action in Serum of Fish

Molecule	Composition	Mode of Action
Lectins	Specific sugar binding proteins	Recognition, precipitation, agglutination, opsonin activity
Lytic enzymes	Catalytic proteins lysozyme, complement components	Hemolytic and antibacterial activities
Transferrin/lactoferrin	Glycoprotein	Iron binding
Ceruloplasmin	Acute-phase protein	Copper binding
C-reactive protein	Acute-phase protein	Opsonization or activation of complement
Interferon	Protein	Aid in resistance to viral infection

Lytic enzymes such as lysozyme and members of the complement cascade have been described in fish. Lysozyme is an antibacterial molecule that cleaves the 1-4-ß-linkages between N-acetylmuramic acid and N-acetylglucosamine in the cell wall of bacteria. Alexander (1985) suggests that lysozyme may be effective in killing Gram-positive bacteria, whereas Gram-negative bacteria require complement. Members of the complement cascade may be lytic enzymes by themselves but more often are enzymes that lead to the formation of new products when the pathway is activated. The molecules also are important in opsonization of target cells and attraction and activation of cells important in inflammation.

Acute-phase proteins (serum proteins involved in nonspecific defense) appear in the serum and brains of animals following tissue injury or infection (Gordon 1976). Two such proteins are C-reactive protein and ceruloplasmin, both of which have been identified in fish serum. Ceruloplasmin is responsible for the binding of copper and was shown to be elevated in the presence of cadmium by Syed, Coombs, and Keir (1979). C-reactive protein was found to be elevated in response to elevated cortisol levels (stress) (Wingfield and Grimm 1977) and endotoxin stimulation (White et al. 1981). Kiron et al. (1995) also suggested that nutrition (protein level) influenced levels of C-reactive protein. Research suggests that these proteins are probably produced in response to stress and play a role in natural resistance to infection.

Complement

Complement was initially described as a biological substance that complemented the action of antibody. Actually, antibody or foreign substances are needed to activate the complement cascade. Complement was first described around 1900 as a serum component having bactericidal activity in mammals.

It was not until the 1960s or 1970s that complement was studied in fish (Sakai 1992). Research now suggests that the complement system appears to be present throughout the entire vertebrate phylum, as well as in invertebrate species studied to date (Whaley 1988). Complement is an important humoral defense mechanism not only in a specific way (i.e., interaction with a specific antibody) but also in a nonspecific manner (i.e., interaction with surface molecules of bacteria, viruses, and parasites). In mammals, complement consists of 20 or more chemically distinct serum proteins and glycoproteins that exhibit enzymatic properties (Kuby 1992). Researchers have demonstrated that each of the components reacts in an enzymatic cascade, generating products that are able to clear antigenic molecules and immune complexes and participate in the processes of inflammation and phagocytosis by macrophages and neutrophils.

Fish complement in general can be activated through the classical pathway (i.e., specific immunoglobulin or IgM) or the alternative pathway (i.e., bacterial cell wall components, viral components, or surface molecules of parasites). Although both pathways exist, Sakai (1992) suggests that characterization of the components is incomplete for fish. In rainbow trout serum, complement components have been isolated and characterized as similar to mammalian C3 and C5 (Nonaka et al. 1981; Nonaka et al. 1984). Complement components in fish have been demonstrated to be important for opsonization (Nonaka 1985) and bactericidal activity by the formation of membrane attack complexes (Jenkins et al. 1991).

HUMORAL IMMUNITY

Humoral immunity is defined as the antibody response to foreign antigens. Fish possess B-cells (surface Ig-positive cells) that are considered to be similar to the mammalian B-cells. The surface immunoglobulin of B-cells serves as the receptor for antigen recognition and is of the same specificity to the antibody molecule that will be produced (Janeway and Travers 1994). After antigen binds the surface immunoglobulin, the B-cells proliferate and begin producing antibody. Kaattari (1992) provides an excellent review of B-cell function and antibody response in fish. Fish antibody has been described as a tetrameric IgM-like molecule (i.e., structurally similar to mammalian IgM). The antibody molecule in fish is made up of eight heavy chains (mu) and eight light chains (lambda). Wilson and Warr (1992) state that genes responsible for immunoglobulin production are believed structurally similar in all vertebrates and point out that this includes the V_H, D, J_H, and C exons for heavy chain and V_L, J_L, and C_L exons for light chains. Immunologic memory has been demonstrated to occur in fish (Arkoosh and Kaattari 1991). An interesting difference between primary and memory responses of fish and mammals is that fish do not switch to class G immunoglobulin as do mammals. The memory response of fish is still IgM, as was the primary response, and there appear to be conflicting reports as to affinity maturation of fish immunoglobulin. Fish IgM possesses eight antigen binding sites, and while affinity at each site may be relatively low, the molecule has a greater effective binding strength or avidity because of the presence of the multiple binding sites. Fish IgM is functionally similar to mammalian IgM, in that it is capable of specifically binding epitopes on the surface of bacteria, viruses, and toxins. Fish IgM is a potent activator of complement and a very efficient opsonin and agglutinin.

CELL-MEDIATED IMMUNITY

Cell-mediated immunity is important in elimination of intracellular pathogens (e.g., intracellular bacteria, parasites, and viruses). The cell-mediated immune response is well defined in mammals and is becoming better understood in fish. Cell-mediated immunity relies on contact of the foreign invader with the subsequent presentation of an antigen having the correct major histocompatability complex (MHC II) to T-helper cells in mammals (Kuby 1992). The nature of MHC molecules in fish is now being elucidated (Vallejo, Miller, and Clem 1992), and molecular cloning of MHC I and II in fish has been carried out for teleost fish (Grimholt et al. 1993; Glamann 1995; Godwin et al. 1997). It is being assumed that the function of MHC is similar to that in mammals; however, as Grimholt and Lie (1998) point out, no studies on fish have established the function of fish MHC molecules. Once T-helper cells are stimulated, they produce cytokines that result in stimulation of effector cells (cytotoxic lymphocytes) or macrophages. The cytokines not only stimulate the aforementioned cells but also function in recruitment of new cells to the area and in activation of the newly recruited cells. For defense against intracellular bacteria, it appears that the recruitment and subsequent activation of macrophages, which enhance their intracellular killing ability, result in the elimination of the pathogen. Cellular immunity via the macrophage has recently been suggested to be an important mechanism for immunity to *E. ictaluri* in channel catfish by Shoemaker, Klesius, and Plumb (1997), Shoemaker and Klesius (1997), and Klesius and Shoemaker (1999). Many of the cytokines involved in cellular immunity in fish are unknown and need further study. Secombes, Hardie, and Daniels (1996) describe cytokines as simple polypeptides or glycoproteins that act as signaling molecules of the immune system.

FACTORS INFLUENCING INNATE RESISTANCE AND ACQUIRED IMMUNITY OF FISH

Many factors may contribute to innate resistance and acquired immunity in fish (see Table 7.3). A number of fish species have been shown to possess genetic variation in disease resistance to pathogens. More recently, Wolters and Johnson (1994) and Wolters, Wise, and Klesius (1996) demonstated genetic

TABLE 7.3. Factors That May Influence Disease Resistance and Immune Response of Fish

General	Specific
Genetics	Individuals may exhibit differences in innate resistance and acquired immunity
Environment	Temperature, season, photoperiod
Stress	Water quality, pollution, density, handling and transport, breeding cycles
Nutrition	Feed quality and quantity, availability of micronutrients, addition of immunostimulants, antinutritional factors contained in some feed ingredients
Fish	Age, species or strains, individuals
Pathogen	Exposure levels, type (parasite, bacterial, viral), serovar, virulence

variability in innate resistance of channel catfish to *E. ictaluri.* Genetic variation to innate resistance may allow for selection of stocks or lines of fish with enhanced natural resistance; however, innate immune mechanisms can be overwhelmed (Klesius and Shoemaker, unpublished observation). A better use of genetic variability may be to look at the acquired immune response following vaccination. Wilkie and Mallard (1999) demonstrated that selection for high and low immune response was possible in hogs and that high immune responders responded better to vaccination. This is an important area that needs to be addressed with respect to fish vaccination.

Environmental factors and stress also influence natural disease resistance and acquired immune function in fish. Environmental factors include, but are not limited to, temperature, season, and photoperiod. Bly and Clem (1991) demonstrated in vitro immunosuppression in channel catfish, following exposure to low water temperatures. Stress may include water quality, pollution, density, handling and transport stress (cultured fish), and breeding cycles. Many of these stressors are "perceived" by people and may or may not have an influence on the fish. More work is needed to determine what the factors (stressors) are and how they affect natural resistance and immunity. Many experiments have been conducted to determine in vitro responses of immune cells following the stress, but a better approach may be to determine the effect of stress on disease resistance or acquired immunity against specific pathogens (i.e., in vivo).

Nutrition has long been known to be important in natural resistance and immunity to infection in humans. Recently, much research has accumulated on the influence of nutrition on fish health. Important factors have been feed quality and quantity, availability of micronutrients, addition of vitamins, and the determination of antinutritional factors contained in some feed ingredients. One important area of research has been in the use of feed additives that enhance or stimulate innate immune mechanisms in fish: the immunostimulants. One such additive is beta-glucan. In fish, beta-glucan has been demonstrated to enhance immune cell function associated with antibacterial function (Ainsworth, Mao, and Boyle 1994; Duncan and Klesius 1996). However, these researchers were unable to demonstrate protection to a specific pathogen upon challenge of immunostimulant-fed fish. Enhancement of nonspecific immune function may not result in increased disease resistance. A better approach may be to combine these additives (immunostimulants) in combination with vaccination to help activate or initiate the acquired immune response via innate mechanisms.

The fish themselves (i.e., species, strain or line, individual) as well as their age or size may also influence innate and acquired immunity. Genetic variation in the innate immune response and antibody response has been demonstrated at the species level (Wolters, Wise, and Klesius 1996). Recently, Balfry, Shariff, and Iwama (1997) demonstrated differences in nonspecific immunity in strains of Nile tilapia, *Oreochromis niloticus.* Age and size was shown to influence serum Ig levels in catfish (Klesius 1990). Future work in these areas may lead to strains or lines of fish with enhanced immune mechanisms and, thus, greater survival following infection.

The pathogen also is an important factor influencing innate resistance and acquired immunity. Other factors include exposure level (Klesius and Shoemaker, unpublished observation), type (e.g., parasite, bacterial, or viral), serotype, or serovar. Type of pathogen, such as parasite, bacterium, or virus, influences the nature of the immune response. For example, intracellular bacteria require cell-mediated immunity, whereas extracellular bacteria require humoral immunity for successful defense against the pathogen. Serotype or serovar (which may be an indicator of virulence) of pathogen also influences immunity. Klesius and Shoemaker (1997) recently demonstrated that the isolate of *E. ictaluri* influenced survival following vaccination with live bacteria (i.e., acquired immunity). They suggested that the variability in protective ability induced by *E. ictaluri* isolate(s) was dependent on antigenic heterogeneity.

CONCLUSION

Fish immunologists have made great strides toward the understanding of the immune mechanisms of fish in the past 30 to 40 years. Future studies on fish immunology will undoubtedly be driven by the increase in aquaculture throughout the world. Studies will most likely concentrate on the immune response following vaccination. A continued effort on characterization of the cell network and the regulatory substances (cytokines) involved in natural resistance and acquired immunity is needed. The final areas of research will involve the influence of environmental pollution on disease resistance and immunity in fish as well as the influence of stress on the physiology of the fish and how this affects innate and acquired immunity.

REFERENCES

Agius, C. 1985. The melano-macrophage centres of fish. In M.J. Manning and M.F. Tatner (Eds.), *Fish Immunology* (pp. 85-105). London: Academic Press.

Ainsworth, A.J. 1993. Carbohydrate and lectin interactions with *Edwardsiella ictaluri* and channel catfish, *Ictalurus punctatus* (Rafinesque), anterior kidney leukocytes and hepatocytes. *Journal of Fish Diseases* 16:449-459.

Ainsworth, A.J., C.P. Mao, and C.R. Boyle. 1994. Immune response enhancement in channel catfish, *Ictalurus punctatus,* using beta-glucan from *Schizophyllum commune.* In J.S. Stolen and T.C. Fletcher (Eds.), *Modulators of Fish Immune Responses* (pp. 67-81). Fairhaven, NJ: SOS Publications.

Alexander, J.B. 1985. Non-immunoglobulin humoral defense mechanisms in fish. In M.J. Manning and M.F. Tatner (Eds.), *Fish Immunology* (pp. 133-140). London: Academic Press.

Arkoosh, M.R. and S.L. Kaattari. 1991. Development of immunological memory in rainbow trout *(Oncorhynchus mykiss).* I. An immunochemical and cellular analysis of the B-cell response. *Developmental and Comparative Immunology* 15:279-293.

Balfry, S.K., M. Shariff, and G.K. Iwama. 1997. Strain differences in non-specific immunity of tilapia *Oreochromis niloticus* following challenge with *Vibrio parahaemolyticus. Diseases of Aquatic Organisms* 30:77-80.

Bly, J.E. and L.W. Clem. 1991. Temperature-mediated processess in teleost immunity: In vitro immunosuppression induced by in vivo low temperature in channel catfish. *Veterinary Immunology and Immunopathology* 28:365-377.

Clem, L.W., R.C. Sizemore, C.F. Ellsaesser, and N.W. Miller. 1985. Monocytes as accessory cells in fish immune responses. *Developmental and Comparative Immunology* 9:803-809.

Di Conza, J.J. and W.J. Halliday. 1971. Relationship of catfish and serum antibodies to immunoglobulin in mucus secretions. *Australian Journal of Experimental Biology and Medical Science* 49:517-519.

Duncan, P.L. and P.H. Klesius. 1996. Dietary immunostimulants enhance nonspecific immune responses in channel catfish but not resistance to *Edwardsiella ictaluri. Journal of Aquatic Animal Health* 8:241-248.

Ellis, A.E. 1981. Nonspecific defense mechanisms in fish and their role in disease processes. *Development and Biological Standards* 49:337-352.

Ellsaesser, C.F., J.E. Bly, and L.W. Clem. 1988. Phylogeny of lymphocyte heterogeneity: The thymus in channel catfish. *Developmental and Comparative Immunology* 12:787-799.

Evans, D.L. and J.B. Gratzek. 1989. Immune defense mechanisms in fish to protozoan and helminth infections. *American Zoologist* 29:409-418.

Evans, D.L. and L. Jaso-Friedman. 1992. Nonspecific cytotoxic cells as effectors of immunity in fish. In M. Faisal and F.M. Hetrick (Eds.), *Annual Review of Fish Diseases,* Volume 2 (pp. 109-121). New York: Pergamon Press.

FAO (Food and Agriculture Organization) Fisheries Department. 1998. *Trends in Global Aquaculture Production: 1984-1996.* Circular No. 815, Rome, Italy: FAO.

Fletcher, T.C. 1981. Non-antibody molecules and the defense mechanisms of fish. In A.D. Pickering (Ed.), *Stress and Fish* (pp. 171-183). New York: Academic Press.

Fletcher, T.C. and P.T. Grant. 1969. Immunoglobulins in the serum and mucus of the plaice *(Pleuronectes platessa). Biochemical Journal* 115:65.

Fletcher, T.C. and A. White. 1973. Lysozyme activity in the plaice *(Pleuronectes platessa). Experientia* 29:1283-1285.

Glamann, J. 1995. Complete coding sequence of rainbow trout MHC II beta chain. *Scandinavian Journal of Immunology* 14:365-372.

Godwin, U.B., A. Antao, M.R. Wilson, V.G. Chinchar, N.W. Miller, L.W. Clem, and T.J. McConnel. 1997. MHC class II beta genes in the channel catfish *(Ictalurus punctatus). Developmental and Comparative Immunology* 21:13-23.

Gordon, A.H. 1976. The acute phase plasma proteins. In R. Bianchi, G. Mariani and A.S. McFarlane (Eds.), *Plasma Protein Turnover* (pp. 381-394). London: Macmillan Press Ltd.

Grimholt, U., I. Hordvik, V.M. Fosse, I. Olsaker, C. Endressen, and O. Lie. 1993. Molecular cloning of major histocompatibility complex class I cDNAs from Atlantic salmon *(Salmo salar). Immunogenetics* 37:469-473.

Grimholt, U. and O. Lie. 1998. The major histocompatability complex in fish. *Revue Scientifique et Technique, Office International des Epizootics* 17:121-127.

Hogan, R.J., T.B. Stuge, L.W. Clem, N.W. Miller, and V.G. Chinchar. 1996. Antiviral cytotoxic cells in the channel catfish. *Developmental and Comparative Immunology* 20:115-127.

Janeway, C.A. Jr. and P. Travers. 1994. *Immunobiology: The Immune System in Health and Disease.* New York: Current Biology Ltd., Garland Publishing Inc.

Jenkins, J.A., R. Rosell, D.D. Ourth, and L.B. Coons. 1991. Electron microscopy of bactericidal effects produced by the alternative complement pathway of channel catfish. *Journal of Aquatic Animal Health* 3:16-22.

Kaattari, S.L. 1992. Fish B lymphocytes: Defining their form and function. In M. Faisal and F.M. Hetrick (Eds.), *Annual Review of Fish Diseases,* Volume 2 (pp. 161-180). New York: Pergamon Press.

Kiron, V., T. Watanabe, H. Fukuda, N. Okamoto, and T. Tekeuchi. 1995. Protein nutrition and defense mechanisms in rainbow trout *(Oncorhynchus mykiss). Comparative Biochemistry and Physiology* 111:351-359.

Klesius, P.H. 1990. Effect of size and temperature on the quantity of immunoglobulin in channel catfish, *Ictalurus punctatus. Veterinary Immunology and Immunopathology* 24:187-195.

Klesius, P.H. and W.M. Sealey. 1996. Chemotactic and chemokinetic responses of channel catfish macrophages to exoantigen from *Edwardsiella ictaluri. Journal of Aquatic Animal Health* 8:314-318.

Klesius, P.H. and C.A. Shoemaker. 1997. Heterologous isolates challenge of channel catfish, *Ictalurus punctatus,* immune to *Edwardsiella ictaluri. Aquaculture* 57:147-155.

Klesius, P.H. and C.A. Shoemaker. 1999. Development and use of a modified live *Edwardsiella ictaluri* vaccine against enteric septicemia of catfish. In R.D. Schultz

(Ed.), *Veterinary Vaccines and Diagnostics* (pp. 523-537). Advances in Veterinary Medicine, Volume 41. San Diego, CA: Academic Press.

Kuby, J. 1992. *Immunology.* New York: W.H. Freeman and Co.

Lobb, C.J. 1987. Secretory immunity induced in catfish, *Ictalurus punctatus,* following bath immunization. *Developmental and Comparative Immunology* 11:727-738.

Lobb, C.J. and W. Clem. 1981. The metabolic relationship of the immunoglobulins in fish serum, cutaneous mucus and bile. *Journal of Immunology* 127:1525-1529.

Manning, M.J. 1994. Fishes. In R.J. Turner (Ed.), *Immunology: A comparative approach* (pp. 69-100). Chichester, England: John Wiley and Sons, Ltd.

Nonaka, M. 1985. Evolution of complement system. *Developmental and Comparative Immunology* 9:377.

Nonaka, M., N. Iwaki, C. Nakai, M. Nozaki, T. Kaidoh, M. Nonaka, S. Natsumme-Sakai, and M. Takahashi. 1984. Purification of a major serum protein of rainbow trout *(Salmo gairdneri),* homologous to the third complement component of mammalian complement. *Journal of Biological Chemistry* 259:6327-6333.

Nonaka, M., N. Yamaguchi, S. Natsuume-Sakai, and M. Takahashi. 1981. The complement system of rainbow trout *(Salmo gairdneri).* 1. Identification of the serum lytic system homologous to mammalian complement. *Journal of Immunology* 126:1489-1494.

Ourth, D.D. 1980. Secretory IgM, lysozyme and lymphocytes in the skin and mucus of channel catfish, *Ictalurus punctatus. Developmental and Comparative Immunology* 4:65-74.

Passer, B.J., C.H. Chen, N.W. Miller, and M.D. Cooper. 1996. Identification of a T lineage antigen in the catfish. *Developmental and Comparative Immunology* 20:441-450.

Peleterio, M.C. and R.H. Richards. 1988. Immunocytochemical studies on immunoglobulin-containing cells in the epidermis of rainbow trout *Salmo gairdneri,* Richardson: Influence of bath vaccination. *Journal of Fish Biology* 32:845-858.

Rombout, J.H.W.M., N. Taverne, M. van de Kamp, and A.J. Taverne-Thiele. 1993. Differences in mucus and serum immunoglobulin of carp (*Cyprinus carpio* L.). *Developmental and Comparative Immunology* 17:309-317.

Sakai, D.K. 1992. Repertoire of complement in immunological defense mechanisms of fish. In M. Faisal and F.M. Hetrick (Eds.), *Annual Review of Fish Diseases,* Volume 2 (pp. 223-247). New York: Pergamon Press.

Secombes, C.J., L.J. Hardie, and G. Daniels. 1996. Cytokines in fish: An update. *Fish and Shellfish Immunology* 6:291-304.

Secombes, C.J., M.J. Manning, and A.E. Ellis. 1982. The effect of primary and secondary immunization on the lymphoid tissues of the carp, *Cyprinus carpio* L. *Journal of Experimental Zoology* 220:277-287.

Sharon, N. and H. Lis. 1993. Carbohydrates in cell recognition. *Scientific American* 268(1):82-89.

Shoemaker, C.A. and P.H. Klesius. 1997. Protective immunity against enteric septicaemia in channel catfish, *Ictalurus punctatus* (Rafinesque), following controlled exposure to *Edwardsiella ictaluri. Journal of Fish Diseases* 20:361-368.

Shoemaker, C.A., P.H. Klesius, and J.A. Plumb. 1997. Killing of *Edwardsiella ictaluri* by macrophages from channel catfish immune and susceptible to enteric septicemia of catfish. *Veterinary Immunology and Immunopathology* 58:181-190.

St. Louis-Cormier, E.A., C.K. Osterland, and D.P. Anderson. 1984. Evidence for a cutaneous secretory immune system in rainbow trout *(Salmo gairdneri). Developmental and Comparative Endocrinology* 8:71-80.

Syed, A., T.L. Coombs, and H.M. Keir. 1979. Effects of cadmium on copper-dependent enzymes in plaice, *Pleuronectes platessa. Biochemistry Society Transactions* 7:711-713.

Vallejo, A.N., N.W. Miller, and L.W. Clem. 1992. Antigen processing and presentation in teleost immune responses. In M. Faisal and F.M. Hetrick (Eds.), *Annual Review of Fish Diseases,* Volume 2 (pp. 73-89). New York: Pergamon Press.

van Muiswinkel, W.B., C.H.J. Lamers, and J.H.W.M. Roumbout. 1991. Structural and functional aspects of the spleen in bony fish. *Research Immunology* 142:362-366.

Waterstrat, P.R., A.J. Ainsworth, and G. Capley. 1991. In vitro responses of channel catfish, *Ictalurus punctatus,* neutrophils to *Edwardsiella ictaluri. Developmental and Comparative Immunology* 15:53-63.

Weeks-Perkins, B.A. and A.E. Ellis. 1995. Chemotactic responses of Atlantic salmon *(Salmo salar)* macrophages to virulent and attenuated strains of *Aeromonas salmonicida. Fish and Shellfish Immunology* 5:313-323.

Whaley, K. 1988. The complement system. In K. Whaley (Ed.), *Complement in Health and Disease* (pp. 1-35). Boston, MA: MTP Press Ltd.

White, A., T.C. Fletcher, M.B. Pepsy, and B.A. Baldo. 1981. The effect of inflammatory agents on C-reactive protein and serum amyloid P-component levels in plaice (*Pleuronectes platessa* L.) serum. *Comparative Biochemistry and Physiology* 69:325-329.

Wilkie, B.N. and B.A. Mallard. 1999. Genetic effects on vaccination. In R.D. Schultz (Ed.), Veterinary Vaccines and Diagnostics. *Advances in Veterinary Medicine,* Volume 41 (pp. 39-50). San Diego, CA: Academic Press.

Wilson, M.R. and G.W. Warr. 1992. Fish immunoglobulins and the genes that encode them. In M. Faisal and F.M. Hetrick (Eds.), *Annual Review of Fish Diseases,* Volume 2 (pp. 201-221). New York: Pergamon Press.

Wingfield, J.C. and A.S. Grimm. 1977. Seasonal changes in plasma cortisol, testosterone and oestradiol-17 beta in the plaice, *Pleuronectes platessa* L. *General and Comparative Endocrinology* 31:1-11.

Wolters, W.R. and M.R. Johnson. 1994. Enteric septicemia resistance in blue catfish and three channel catfish strains. *Journal of Aquatic Animal Health* 6:329-334.

Wolters, W.R., D.J. Wise, and P.H. Klesius. 1996. Survival and antibody response of channel catfish, blue catfish and channel catfish female × blue catfish male hybrids after exposure to *Edwardsiella ictaluri. Journal of Aquatic Animal Health* 8:249-254.

Zilberg, D. and P.H. Klesius. 1997. Quantification of immunoglobulin in the serum and mucus of channel catfish at different ages and following infection with *Edwardsiella ictaluri. Veterinary Immunology and Immunopathology* 58:171-180.

Chapter 8

Dietary Ascorbic Acid Requirement for Growth and Health in Fish

Meng H. Li
Edwin H. Robinson

INTRODUCTION

Since most fish lack the ability to biosynthesize ascorbic acid (AA), or if they have the capability to synthesize the vitamin, the quantity produced is insufficient to meet metabolic needs, AA must be provided in the diet. Also, since AA is an essential dietary component that functions in numerous metabolic processes, considerable research has been conducted concerning its role in the nutrition of fish. Diets devoid of AA result in prominent deficiency signs in fish. Most notable are spinal deformities, particularly scoliosis and lordosis. Often abnormalities of the cartilage of the eye, gill, gill opercula, and fins are observed in AA-deficient fish, as well as reduced appetite, slow growth, internal and external hemorrhage, fin erosion, and anemia. The minimum amount of AA required in the diet needed for normal growth and health of fish generally ranges from 10 to 50 mg/kg of diet. The requirement varies among fish species, but intraspecies differences such as fish strain, size, and age also affect the dietary requirement. The amount of AA that must be added to the diet for normal function is also affected by the abundance of natural food organisms found in pond water and by the form of the vitamin that is added to the diet. Considering that ingredients typically used in commercial fish feeds are essentially devoid of AA and further considering that processing these ingredients into fish feed pellets requires conditions of high pressure, heat, and moisture, AA must be added, not only

---

The manuscript is approved for publication as Journal Article No. J-9435 of the Mississippi Agricultural and Forestry Experiment Station (MAFES), Mississippi State University. This project is supported under MAFES Project Number MIS-0822.

in a quantity sufficient for normal growth and health, but also to compensate for losses of the vitamin during feed processing. The role of AA in the nutrition and health of fish is examined in detail herein.

DEFICIENCY SIGNS

Most fish require dietary AA for normal growth and functions. Lack of AA in the diet causes numerous deficiency signs (see Table 8.1). Although AA deficiency signs vary with species, the most prominent deficiency sign is broken-back syndrome (i.e. scoliosis and lordosis), which is caused by reduced synthesis of collagen (a component of the bone), gill-support cartilage, skin, and fins. Studies have demonstrated that lack of dietary AA impaired collagen formation in rainbow trout, *Oncorhynchus mykiss* (Sato et al. 1978a); channel catfish, *Ictalurus punctatus* (Wilson and Poe 1973; Lim and Lovell 1978; Wilson, Poe, and Robinson 1989; El Naggar and Lovell 1991; Mustin and Lovell 1992); and African catfish, *Clarias gariepinus* (Eya 1996). Deformed opercula and gill arches are common AA deficiency signs that are also related to reduced collagen formation. Wilson and Poe (1973) found that serum alkaline phosphatase activity was lower in channel catfish fed AA-free diets. Serum alkaline phosphatase level generally reflects osteoblastic activity (Gould and Shwachman 1942). Other common deficiency signs include anorexia, lethargy, slow growth, dark skin coloration, internal and external hemorrhages, fin erosion, anemia, and increased mortality.

Some fish do not appear to require AA in their diets. Ascorbic acid deficiency signs and dietary requirements have not been demonstrated in white sturgeon, *Acipenser transmontanus* (Hung 1991). Sufficient activity of L-gulonolactone oxidase, a key enzyme for de novo synthesis of AA, was found in white sturgeon. In common carp, *Cyprinus carpio,* no AA deficiency signs were found in fish raised from 0.25 g to 3.5 g and from 35 g to 350 g (Sato et al. 1978b), possibly because distinct L-gulonolactone oxidase activity was present in hepatopancreas in common carp (Yamamoto, Sato, and Ikeda 1978) so that the fish were able to synthesize an adequate amount of AA from glucose and gulonolactone (Ikeda, Sato, and Kimura 1964). However, Dabrowski et al. (1988) found that common carp fry fed formulated dry diets were significantly smaller than those fed live *Artemia nauplii* and developed typical AA deficiency signs (lordosis, caudal fin erosion, deformed opercula, and gill arches). They suggested that common carp required a dietary source of AA during early larval development, possibly because transformation of the dehydro-form was limited at this stage or the dehydro-form and the reduced form of AA were transported differently by the intestine. Gouillou-Coustans, Bergot, and Kaushik (1998) reported that common carp fry require 45 mg AA/kg of diet for normal growth and to prevent signs of deficiency.

TABLE 8.1. Ascorbic Acid Deficiency Signs in Various Species of Fish

Species	**Deficiency signs**	**Reference**
African catfish, *Clarias gariepinus*	Broken skull, hemorrhages and tissue erosion in fins, snout, and operculum	Eya 1996
Atlantic salmon, *Salmo salar*	Scoliosis, lordosis, hemorrhages along spinal cord and in the muscle	Lall et al. (1990) and Sandnes et al. (1992)
Ayu, *Plecoglossus altivelis*	Loss of appetite, exophthalmia, hemorrhage in eyes and fins, congestion of the back and head, erosions in gill opercula and lower jaw	Aoe (1980)
Blue tilapia, *Tilapia aurea*	Scoliosis, hemorrhage in fins, mouth, and swim bladder, short gill lamellae	Stickney et al. (1984)
Channel catfish, *Ictalurus punctatus*	Scoliosis, lordosis, dark skin coloration, fin erosion, hemorrhage around mouth and fins, reduced hematocrit and bone collagen concentration	Lim and Lovell (1978)
Coho salmon, *Oncorhynchus kisutch*	Scoliosis, lordosis, failure in wound repair	Halver et al. (1969)
Common carp, *Cyprinus carpio*	No deficiency signs found	Sato et al. (1978b)
	Lordosis, caudal fin erosion, deformed opercula and gill arches	Dabrowski et al. (1988)
European sea bass, *Scophthalmus maximus*	No deficiency signs found	Merchie et al. (1996)
Gilthead sea bream, *Sparus aurata*	Scale loss, depigmentation, internal and external hemorrhages, glomerulonephritis	Alexis et al. (1997)
Grass carp, *Ctenopharyngodon idella*	Hemorrhage in the eyes and fins	Lin (1991)
Hybrid tilapia, *Tilapia nilotica*♀ × T. *aurea* ♂	Reduced hematocrit, loss of pigmentation on skin, skin lesions, loss of scales, hemorrhage in fins and skin	Shiau and Jan (1992)
Japanese eel, *Anguilla japonica*	Hemorrhage in fins, head and skin, low jaw erosion	Arai (1972)
Mayan cichlid, *Cichlasoma urophthalmus*	Scoliosis, dark coloration, short opercula, hemorrhage in the eyes, head, and fins, skin and fin erosion, loss of scale, exophthalmia, swollen abdomen, loss of pigmentation	Chavez de Martinez (1990)
Nile tilapia, *T. nilotica*	Scoliosis, lordosis, dark skin pigmentation, petechial hemorrhage, short opercula with exposed gills, erratic swimming behavior	Abdelghany (1996)

TABLE 8.1 *(continued)*

Species	Deficiency signs	Reference
Pacu, *Piaractus mesopotamicus*	Hyperplasia, hypertrophy, and dysplasia of bone cartilage of gill filaments, twisted gill lamellae	Martins (1995)
Rainbow trout, *O. mykiss*	Scoliosis, lordosis, distortion of gill filamental cartilage, failure in wound repair	Halver et al. (1969)
	Impaired iron metabolism	Hilton et al. (1978)
Red drum, *Sciaenops ocellatus*	Scoliosis, lordosis, hemorrhage, loss of equilibrium	Aguirre and Gatlin (1999)
Sunshine bass, *Morone chrysops*♀ × *M. saxatilis*♂	Abnormalities in isthmus cartilage formation	Sealey and Gatlin (1999)
Turbot, *Scophthalmus maximus*	Hypertyrosinemia, renal granulomatous nodules	Coustans et al. (1990)
White sturgeon, *Acipenser transmontanus*	No deficiency signs found	Hung (1991)
Yellowtail, *Seriola lalandi*	Scoliosis, dark coloration, hemorrhage in body surface, hypochromic anemia	Shimeno (1991)

Note: Anorexia, reduced weight gain, and increased mortality, which are common vitamin deficiency signs, are excluded.

Under laboratory conditions, AA deficiency signs usually become obvious at 8 to 12 weeks in channel catfish fed AA-free diets (Lim and Lovell 1978; Wilson, Poe, and Robinson 1989; El Naggar and Lovell 1991; Li, Johnson, and Robinson 1993). Nile tilapia, *Tilapia nilotica* (Soliman, Jauncey, and Roberts 1986b), and hybrid tilapia, *T. nilotica* ♀ × *T. aurea* ♂ (Shiau and Jan 1992) exhibited AA deficiency signs at six weeks in fish fed an AA-free diet. Rainbow trout fry fed an AA-free diet for nine weeks showed typical deficiency signs (Matusiewicz et al. 1994), whereas rainbow trout fingerlings did not develop deficiency signs after fed an AA-free diet for 18 weeks (Matusiewicz et al. 1995). The length of time required to induce deficiency signs varies with species, size of fish, tissue storage, and environmental conditions.

DIETARY REQUIREMENT

With few exceptions, the AA requirement for maximum weight gain of fish is similar to that for preventing deficiency signs. Lim and Lovell (1978) reported that channel catfish required 30 mg AA/kg in their diet for normal growth but required 60 mg AA/kg to prevent all deficiency signs. Chavez de Martinez (1990) found that Mayan cichlid, *Cichlasoma urophthalmus*, require 40 mg AA/kg in their diet for maximum growth but require 110 mg AA/kg to prevent deficiency signs.

Reported AA requirement values for maximum weight gain and for prevention of deficiency signs vary among fish species, ranging from <10 to 122 mg/kg (see Table 8.2). Most of the requirement values are in the range of 10 to 50 mg/kg. The dietary AA requirement is dependent upon metabolic function and size or age of fish. A dietary AA level of 50 mg/kg was sufficient for normal growth and bone development in coho salmon, *Oncorhynchus kisutch*, but 400 mg/kg was required for maximum wound healing

TABLE 8.2. Ascorbic Acid Requirement for Maximum Weight Gain and Prevention of Deficiency Signs in Various Species of Fish

Species	Requirement (mg/kg)	AA source	Initial fish Size (g/fish)	Reference
African catfish, *Clarias gariepinus*	45	ECAA	19.9	Eya (1996)
Atlantic salmon, *Salmo salar*	50	ECAA	Fry	Lall et al. (1990)
	10-20	AMP	0.16	Sandnes et al. (1992)
Ayu, *Plecoglossus altivelis*	R	AA	Fry	Aoe (1980)
Blue tilapia, *Tilapia aurea*	50	AA	2	Stickney et al. (1984)
Catfish, *Clarias batrachus*	69	AA	1.5	Mishra and Mukhopadhyay (1996)
Channel catfish, *Ictalurus punctatus*	25	AA	14	Andrews and Murai (1975)
	50	AA	2	Andrews and Murai (1975)
	60	AA	2.3	Lim and Lovell (1978)
	11	APP	13	El Naggar and Lovell (1991)
	15	APP	5.5	Robinson (1992)
Coho salmon, *Oncorhynchus kisutch*	50	AA	0.4	Halver et al. (1969)
Common carp, *Cyprinus carpio*	NR	AA	0.3, 35	Sato et al. (1978b)
	R		0.001	Dabrowski et al. (1988)
	45	APP	Fry	Gouillou-Coustans et al. (1998)
European sea bass, *Scophthalmus maximus*	20	APP	Fry	Merchie et al. (1996)

TABLE 8.2 *(continued)*

Species	Requirement (mg/kg)	AA source	Initial fish Size (g/fish)	Reference
Gilthead sea bream, *Sparus aurata*	R	AA	0.5	Alexis et al. (1997)
	<25	APP	9	Henrique et al. (1998)
Grass carp, *Ctenopharyngodon idella*	R	AA	Fingerling	Lin (1991)
Hybrid tilapia, *Tilapia nilotica* ♀ × *T. aurea* ♂	79	AA	1.1	Shiau and Jan (1992)
Japanese eel, *Anguilla japonica*	R	AA	1-3	Arai (1972)
Japanese flounder, *Paralichthys olivaceus*	60-100	AMP	3, 43	Teshima et al. (1993)
Mexican cichlid, *Cichlasoma urophthalmus*	40	AA	0.16	Chavez de Martinez (1990)
Nile tilapia, *T. nilotica*	50	AS APP	0.6	Abdelghany (1996)
Pacu, *Piaractus mesopotamicus*	139	AP	8.6	Martins (1995)
Rainbow trout, *O. mykiss*	50-100	AA	0.3	Halver et al. (1969)
	40	ECAA	6.7	Hilton et al. (1978)
	<10	AMP	3.0	Cho and Cowey (1993)
Red drum, *Sciaenops ocellatus*	15	APP	3.6	Aguirre and Gatlin (1999)
Sunshine bass, *Morone chrysops* ♀ × *M. saxatilis* ♂	22	APP	0.6	Sealey and Gatlin (1999)
Turbot, *Scophthalmus maximus*	R		Fingerling	Coustans et al. (1990)
White sturgeon, *Acipenser transmontanus*	NR	AA	10-15	Hung (1991)
Yellowtail, *Seriola lalandi*	122		Fingerling	Shimeno (1991)

Note: The following denotes AA sources:

AA = crystalline L-ascorbic acid
ECAA = ethylcellulose-coated ascorbic acid
APP = ascorbyl-2-polyphosphate
AS = ascorbyl-2-sulfate
AP = ascorbyl-6-palmitate
NR = not required for normal growth
R = required but no requirement value determined

(Halver, Ashley, and Smith 1969). Similarly, Lim and Lovell (1978) found that a dietary AA concentration of 30 mg/kg was adequate for maximum weight gain and bone formation, but 60 mg/kg was required for prevention of all deficiency signs and wound repair in channel catfish. Andrews and Murai (1975) reported that channel catfish raised from 2 to 7 g required 50 mg AA/kg for maximum weight gain, while for fish raised from 14 to 100 g, 25 mg/kg was sufficient. Li and Lovell (1985) also demonstrated that dietary AA requirement for channel catfish decreased with size. They found that fish raised from 10 to 150 g required 30 mg/kg of diet for maximum weight gain, while for fish raised from 3 to 19 g, the requirement was 60 mg/kg.

It should be noted that all requirements listed in Table 8.2 were from controlled studies using small fingerlings or fry. There is evidence that some fish raised in earthen ponds do not appear to require a dietary AA source for maximum weight gain. Robinson, Li, and Oberle (1998) demonstrated that supplemental AA in practical diets may be unnecessary for pond-raised channel catfish from large fingerling to marketable size. Two experiments were conducted in which channel catfish fingerlings (initial sizes: 118 and 33 g/fish, respectively) were stocked into 0.04 ha earthen ponds at a rate of 24,700 fish/ha and fed practical diets with or without supplemental AA for a growing season. Results showed no significant differences in weight gain, feed conversion efficiency, survival, or hematocrit between fish fed diets with and without supplemental AA. No deficiency signs were observed in fish fed the diet without supplemental AA. Apparently channel catfish were able to meet their AA requirement from natural food organisms or residual amounts of AA inherent in dietary ingredients, although levels of ascorbic acid inherent in dietary ingredients are generally considered insignificant. Natural food organisms such as phytoplankton and zooplankton are rich in AA (Merchie et al. 1995; Robinson and Li 1996). However, no quantitative data on contribution of natural food organisms to AA nutrition of channel catfish in intensive culture are available. Supplemental AA is not necessary in pond-raised hybrid tilapia *T. aurea* ♀ × *T. nilotica* ♂ (Nitzan, Angeoni, and Gur 1996).

REPRODUCTION

Early work with mammals indicated that the AA concentration in the ovary varies with the reproductive cycle, suggesting a possible role of the vitamin in reproduction (Lutwak-Mann 1958). An increase in the AA concentration in ovarian tissue and a decrease in other tissues during maturation of the ovary have been reported in several species, including rainbow trout (Hilton et al. 1979); Atlantic cod, *Gadus morhua* (Sandnes and Braekkan 1981); crucian carp, *Carassius carassius* (Seymour 1981); and Arctic char, *Salvelinus alpinus* (Dabrowski 1991). Studies with rainbow trout indicated that dietary supplementation of AA for broodfish improved reproduction performance. Sandnes et al. (1984) reported that a supplementation level of

115 mg AA/kg increased egg hatchability, which was significantly correlated to the AA concentration in the egg. They suggested that broodfish should be fed adequate amounts of the vitamin to ensure eggs contained more than 20 μg AA/kg. Waagbo, Thorsen, and Sandnes (1989) demonstrated that dietary AA had a role in vitellogenesis of rainbow trout. Blom and Dabrowski (1995) found that fecundity, total egg mass, and survival of eyed-stage embryos increased with increasing dietary AA levels. They suggested that an amount eight times higher than the National Research Council (1993) recommended dietary level of 50 mg AA/kg for rainbow trout was necessary for optimizing tissue levels and maximizing reproduction success. In another study, Blom and Dabrowski (1996) fed female rainbow trout broodfish diets containing 0 or 360 mg AA/kg for 10 months, and the fry from these females were fed diets with 20 or 500 mg AA/kg for up to 4.5 months. They found that fry with high maternal AA that were fed a high AA diet had higher survival than fry in other groups. Survival of fry with low maternal AA fed a high AA diet was the same as that of those with high maternal AA fed a low AA diet, suggesting that maternal AA status affects survival of the fry. Dabrowski and Ciereszko (1996) demonstrated that dietary AA supplementation protects against male infertility in rainbow trout. They found that fertilization rate and hatchability were correlated to AA concentrations in seminal plasma, which were affected by dietary AA levels.

Dietary AA supplementation in broodfish has also been shown to improve egg hatchability and survival of fry in Mozambique tilapia, *T. mossambica* (Soliman, Jauncey, and Roberts 1986a). Mangor-Jensen et al. (1994) evaluated the effects of dietary AA on egg quality and fry survival of Atlantic cod. They fed broodfish diets containing 0, 50, or 500 mg supplemental AA/kg for three months. Egg strength measured by chorion rigidity and the internal pressure was lower in fish fed the diet containing 500 mg AA/kg than in those fed the lower AA diets. However, dietary AA levels did not affect weight of egg mass, neutral buoyancy of the egg, fertilization rate, and survival of the fry.

It seems clear that dietary AA plays an important role in fish reproduction. However, based on results from these studies it is difficult to draw a conclusion that broodfish require a higher level of dietary AA than the levels recommended by the National Research Council (1993). Most of these studies just described evaluated effects of diets with and without supplemental AA on the reproduction performance. In studies that evaluated various levels of dietary AA, data were inconclusive with regard to the quantitative requirement of AA for reproduction success of broodfish. Feeding AA-free diets to broodfish does not appear to affect weight gain and survival of the broodfish itself, indicating that the dietary AA requirement of broodfish for normal growth and health is relatively low (Waagbo, Thorsen, and Sandnes 1989; Mangor-Jensen et al. 1994; Blom and Dabrowski 1995; Dabrowski and Ciereszko 1996).

IMMUNE RESPONSE AND DISEASE RESISTANCE

There has been considerable interest concerning the effect of dietary AA on immune response and disease resistance in fish. For channel catfish, published data seem to agree that under laboratory conditions immune response of the fish is depressed if an AA-free diet is fed. However, the results from studies in which high levels of AA were fed to channel catfish to improve disease resistance are contradictory. Durve and Lovell (1982) reported that a dietary AA level of 150 mg/kg increased disease resistance of channel catfish against *Edwardsiella tarda* as compared to dietary AA levels of 60 mg/kg and below at 23°C but not at 33°C. They explained that the requirement for resistance to infection is probably higher at low temperatures because of lower natural resistance of the fish. Two studies demonstrated that high levels of dietary AA (1,000 to 3,000 mg/kg of diet) decreased mortality of channel catfish that were experimentally exposed to *E. ictaluri* (Li and Lovell 1985; Liu et al. 1989). Other studies with channel catfish showed that low levels of dietary AA (25 to 50 mg/kg) were as effective as megadose levels of the vitamin (Floyd 1987; Li, Johnson, and Robinson 1993; Li, Wise, and Robinson 1998). Johnson and Ainsworth (1991) found no difference in percentage phagocytosis, phygocytic index, and bactericidal capacities of neutrophils in the anterior kidney of channel catfish fed diets containing 100 or 1,000 mg AA/kg.

Research on the effect of dietary AA on immune response and disease resistance of salmonids has also yielded inconsistent results. Navarre and Halver (1989) reported that diets containing 500 mg AA/kg and above improved disease resistance and antibody production in rainbow trout infected with *Vibrio anguillarum.* Anggawati-Satyabudhy, Grant, and Halver (1990) found that rainbow trout fed a diet containing 320 mg AA/kg had a higher survival rate after exposure to an infectious hematopoietic necrosis virus than fish fed lower AA diets, but antibody production was not different among fish fed different diets. Hardie, Fletcher, and Secombes (1991) demonstrated a significant effect of high dietary AA (2,750 mg/kg) on serum complement activity of Atlantic salmon as compared to fish fed a dietary AA level of 50 or 310 mg/kg. After challenge with *Aeromonas salmonicida,* fish fed 50 mg AA/kg had a higher mortality than fish fed the two higher AA levels. Nonspecific defense mechanisms such as respiratory burst activity and erythrophagocytosis were not affected by dietary AA levels. Waagbo et al. (1993) reported that Atlantic salmon fed a diet containing 4,000 mg AA/kg had a higher survival rate than fish fed lower AA diets after exposure to *A. salmonicida.* They also found that serum complement activity, lysozyme activity of the anterior kidney, and postvaccination antibody production were higher in fish fed the diet containing a megadose of AA. Verlhac and Gabaudan (1994) studied the influence of dietary AA on the immune response of rainbow trout and Atlantic salmon. They found that fish fed a diet containing 1,000 mg AA/kg had a higher leucocyte count, mitogen-induced proliferation of lymphocytes, and natural cytotoxicity than fish fed a diet containing 60 mg AA/kg. However, Bell, Higgs, and Traxler (1984) found no differences in disease resistance and antibody production of sockeye salmon, *O. nerka,* against *Renibacterium salmoninarum* between fish fed diets containing 10 and 100 mg AA/kg. Lall et al. (1990) reported that elevated dietary AA concentrations up to 2,000 mg/kg did not improve disease resistance and antibody titer in Atlantic salmon infected with *A. salmonicida* or *V. anguillarum.* Erdal et al. (1991) found no significant effect of high levels of dietary L-ascorbate (2,980 mg/kg) on disease resistance and antibody titer in Atlantic salmon infected with *V. salmonicida* or *Yersina ruckeri;* however, fish fed elevated dietary levels of L-ascorbate-2-sulfate (4,770 mg/kg) had increased antibody production.

There are several reports on the effects of dietary AA on the immune responses of other species. Roberts, Davis, and Pulsford (1995) evaluated the effect of AA supplementation on the nonspecific immunity in juvenile turbot, *Scophthalmus maximus.* A basal diet was supplemented with 300; 1,000; or 2,000 mg calcium ascorbate/kg. Results showed that serum lysozyme and phagocytic capacity of kidney and spleen cells were positively correlated with dietary AA concentration. Total serum protein and differential leucocyte counts were not affected by dietary supplemental AA. However, Merchie et al. (1996) found no immunostimulative effect of high supplemental AA levels for turbot infected with *V. anguillarum.* Nitzan, Angeoni, and Gur (1996) found that dietary AA supplementation at a level of 458 mg/kg had no significant effect on disease resistance in pond-raised hybrid tilapia *T. aurea* ♀ × T. *nilotica*♂ .

Interactions of AA and other vitamins have also been studied. A study with channel catfish, in which fish were fed a combination of dietary levels of AA (0, 20, 200 mg/kg) and folic acid (0, 0.4, 4.0 mg/kg), showed that after the fish were infected with *E. ictaluri,* maximum survival and antibody production were achieved when the fish were fed diets containing high levels of both vitamins or one vitamin at a high level (Duncan and Lovell 1994). Wahli et al. (1998) studied the effect of combined AA and α-tocopherol on nonspecific immunity and disease resistance of rainbow trout. Diets containing a combination of 0, 30, or 2,000 mg AA/kg and 0, 30, or 800 mg α-tocopherol/kg were fed for eight months. They found that the combination of high dietary doses of both AA and α-tocopherol significantly stimulated proliferation of lymphocytes and macrophage oxidative burst activity when compared with fish fed low levels of both vitamins. Maximum survival after infection with hemorrhagic septicemia virus was achieved when fish were fed a diet containing both vitamins at the highest level or one at the highest level. A similar trend followed when fish were infected with the bacterial pathogen *Y. ruckeri.*

Although many studies have been conducted on the effect of dietary AA on immune response and disease resistance, the efficacy of using megadoses of dietary AA to improve immune response is still unresolved for some species. However, megadoses of AA are not recommended for channel catfish diets. The reasons for different results from the different studies described here are not clear. The differences may have been caused by several factors, including species, strain, size, and history of fish studied; pathogens; methods and pathogenicity of infection; duration of feeding; length of mortality observation period; or other differences in experimental design.

STRESS RESPONSE

Ascorbic acid appears to have a role in the synthesis of corticosteroids. Feeding high levels of AA has been proposed to be beneficial for reducing the effects of physiological stress in fish (Jaffa 1989; Hardie, Fletcher, and Secombes 1991). However, Dabrowska et al. (1991) found no consistent differences in plasma cortisol concentration after short-term confinement stress in common carp fed diets containing 84 to 3,815 mg AA/kg. Research with channel catfish also showed no correlation between dietary AA and serum cortisol concentration after a two-hour confinement stress in aquarium-raised (Li, Wise, and Robinson 1998) or pond-raised fish (Davis et al. 1998). Also, response of Atlantic salmon to confinement stress as measured by plasma glucose concentration was not affected by dietary AA levels (White et al. 1993). It should be noted that the stress applied in these studies was a short-term, acute stress. We are unaware of research on the effect of dietary AA on serum cortisol concentration of fish caused by long-term chronic stress.

Several studies have evaluated the effect of dietary AA on the tolerance of fish to environmental stressors. Mazik, Brandt, and Tomasso (1987) demonstrated that channel catfish fed an AA-free diet had a lower tolerance to ammonia and died of hypoxia at higher dissolved oxygen levels than fish fed diets containing 78 or 390 mg AA/kg, but there were no differences in tolerance of ammonia and low oxygen between fish fed two diets containing AA. Wise, Tomasso, and Brandt (1988) indicated that a megadose of dietary AA (7,950 mg/kg) reduced nitrite-induced methemoglobinemia in channel catfish as compared to that of fish fed a diet containing 64 mg AA/kg. This may be due to the fact that AA reduces nitrite in the blood because of its antioxidant properties and is not directly related to a reduction in stress. Tolerance of Japanese parrot fish, *Oplegnathus fasciatus,* and spotted parrot fish, *O. punctatus,* to low dissolved oxygen appeared to be improved as dietary AA increased (Ishibashi et al. 1992). Gilthead seabream, *Sparus aurata,* fed an AA-free diet showed a higher concentration of plasma glucose three hours after hypoxia stress and a higher concentration of plasma cortisol nine hours after hypoxia stress than fish fed diets containing AA (Henrique et al. 1998). Waagbo and Sandnes (1996) evaluated the effects of dietary AA on parr-smolt transformation of Atlantic salmon and found that dietary AA levels up to 1,000 mg/kg did not affect mortality, serum chloride and cortisol concentrations, or hematological parameters during a challenge test in seawater.

ASSESSMENT OF STATUS

Assessment of the AA status of animals usually relates to measurement of ascorbate in various tissues. No single method of assessment has been developed that is without fault. Chemical methods used to assay for AA have generally been troublesome and often were not adequate for measurement of the various components of tissue ascorbate. Newer methods of AA analysis based on the use of high-performance liquid chromatography (HPLC) are generally improvements over the older methods and allow for separation of tissue ascorbate into components (Wang et al. 1988). Several studies have been conducted to assess body ascorbate reserves in various animals as a tool in predicting AA status. The best tissue for evaluating AA status in fish is debatable. Levels of serum ascorbate do not appear to be a good indicator of the AA status of fish because serum ascorbate reflects recent dietary intake rather than indicating tissue reserves. As a result, other tissues have been studied as indicators of AA reserves in fish. Halver (1972) suggested that anterior kidney ascorbate levels reflected AA status in fish. Other researchers (Hilton, Cho, and Slinger 1978) considered liver ascorbate levels to be a better indicator of AA status. The anterior kidney is small, making AA assay on individual fish problematic. Also, in a study with channel catfish, anterior kidney ascorbate concentrations were highly variable, whereas liver ascorbate concentrations were relatively consistent and thus appeared to be a better indicator of AA status (Lim and Lovell 1978). These researchers suggested that a liver ascorbate concentration below 30 μg/g was indicative of an AA deficiency. In rainbow trout, liver ascorbate concentrations of 20 μg/g or below were considered to be an indicator of poor AA status (Hilton, Cho, and Slinger 1978). Recent studies with channel catfish indicated that a liver ascorbate concentration below 30 μg/g does not necessarily reflect an AA deficiency (Robinson 1992). Fish fed diets containing 45 and 60 mg AA/kg, which were adequate to prevent deficiency signs, had liver ascorbate levels of 16.5 and 18.2 μg/g, respectively. Fish fed levels of AA as low as 15 mg/kg diet had liver ascorbate levels of 4.3 μg/g, but external signs of AA deficiency were not evident. However, mild subclinical signs of AA deficiency were apparent. Some researchers have suggested that vertebral collagen is a good early subclinical indicator for AA deficiency in channel catfish (Lim and Lovell 1978). These researchers suggested that a vertebral collagen level of 25 percent or below was indicative of AA deficiency in channel catfish. Decreases in vertebral collagen appear prior to the occurrence of clinical signs or decrease in alkaline phosphatase activity in channel catfish fed AA-deficient diets. The ratio of hydroxyproline to proline in bone and skin collagen has also been suggested as a sensitive indicator of AA deficiency in fish (Sato et al. 1978a).

Various measurements have been used to evaluate the AA status of fish; however, none has been completely satisfactory. There are good arguments for measuring ascorbate concentrations in various tissues as an indicator of AA status, but there is not enough evidence to recommend a particular technique that will consistently predict AA status in fish. Because of the ease of obtaining sufficient tissue for assay and since new, relatively simple assays for AA are available, liver ascorbate concentrations may turn out to be the method of choice to use in assessing AA status in fish. However, further data are needed to establish levels of liver ascorbate that are reliable indicators of AA deficiency. These values may be difficult to establish because of the many factors that influence tissue AA concentrations.

TISSUE STORAGE

Tissue levels of AA in fish generally reflect dietary AA levels, and the requirement for tissue saturation is much higher than that for normal growth and preventing deficiency signs (Lim and Lovell 1978; Dabrowski et al. 1990; Robinson 1992; Sandnes, Torrissen, and Waagbo 1992; Li, Johnson, and Robinson 1993; Matusiewicz et al. 1995; Merchie et al. 1996; Gouillou-Coustans, Bergot, and Kaushik 1998). It has been reported that a dietary AA level of 500 to 1,000 mg/kg is needed for AA saturation in liver and anterior kidney of channel catfish (Li, Johnson, and Robinson 1993; Li and Robinson 1994). Dabrowski et al. (1990) reported that juvenile rainbow trout fed diets containing 0 to 500 mg AA/kg for six months required 20 mg AA/kg to prevent all deficiency signs and 264 mg AA/kg for maintaining whole body AA saturation, respectively. Similarly, Matusiewicz et al. (1995) found that weight gain of juvenile rainbow trout fed diets containing 0 to 1,280 mg AA/kg for four months was not different among dietary treatments, but 360 mg AA/kg was required for AA saturation in liver and kidney tissues. In a study with common carp larvae, Gouillou-Coustans, Bergot, and Kaushik (1998) demonstrated that 45 mg AA/kg was required for normal growth, but 270 mg AA/kg was required for maximum body AA concentration.

The relationship between tissue AA levels and fish health is not well understood. It has been speculated that tissue saturation of AA may alleviate the effect of environmental pollutants (Halver 1985), protect fish against disease infections (Lovell 1989), or guarantee reproduction success in broodfish (Soliman, Jauncey, and Roberts 1986a; Blom and Dabrowski 1995, 1996; Dabrowski and Ciereszko 1996). However, other studies mentioned in previous sections do not support these hypotheses.

USE IN COMMERCIAL AQUACULTURE

Stability During Feed Manufacture and Storage

A major consideration when formulating fish diets is that AA is extremely sensitive to oxygen and dietary components that accelerate oxidation, such as fats and trace minerals, particularly when exposed to conditions of high moisture, heat, and pressure encountered during the manufacture of most aquatic animal diets. In a typical steam-pelleting process, steam is added to the ground mash to increase the moisture level to 15 to 18 percent, and temperatures are in the range of 70 to 85°C. These conditions are conducive to destruction of the vitamin; however, many aquatic animal diets are prepared by extrusion, in which milling conditions are more rigorous—moisture and temperature may reach 25 percent and 150°C, respectively, resulting in high losses of AA (Robinson 1992). In addition, extruded diets are commonly dried at up to about 130°C for periods of up to 30 to 40 minutes, which may result in additional losses of the vitamin. Losses of crystalline AA from channel catfish diets may be as high as 55 percent during steam pelleting or up to 70 percent during extrusion and drying (see Table 8.3).

TABLE 8.3. Stability of Ascorbic Acid in Fish Feeds During Feed Manufacture

Source	Feed Type	Manufacturing Process	Percent Loss	Reference
AA	trout	cold pelleting	90	Hilton et al. (1977)
AA	trout	warm pelleting	29	Skelbaek et al. (1990)
AA	catfish	steam pelleting	23-34	Lovell and Lim (1978)
ECAA	catfish	steam pelleting	10-24	Lovell and Lim (1978)
AA	catfish	extrusion	55-69	Lovell and Lim (1978)
ECAA	catfish	extrusion	40-55	Lovell and Lim (1978)
FCAA	catfish	extrusion	32-53	Robinson (1992)
FCAA	catfish	extrusion	43	Robinson (1992)
FCAA	catfish	extrusion	52	Robinson (1992)
AA	trout	steam pelleted	57	Grant et al. (1989)
APP	trout	steam pelleted	5	Grant et al. (1989)
ECAA	trout	extrusion	68	Gadient (1991)
ECAA	trout	extrusion	58	Gadient et al. (1992)
ECAA	trout	extrusion	60	Gadient et al. (1992)
APP	trout	extrusion	7	Gadient (1991)
APP	trout	extrusion	4	Gadient et al. (1992)
ECAA	catfish	extrusion	61	Robinson et al. (1989)
APP	catfish	extrusion	17	Robinson et al. (1989)
AA	fish	cold pelleted	35	Soliman et al. (1987)
NaAA	fish	cold pelleted	39	Soliman et al. (1987)
GCAA	fish	cold pelleted	12	Soliman et al. (1987)
AS	fish	cold pelleted	4	Soliman et al. (1987)

AA = crystalline L-ascorbic acid
ECAA = ethylcellulose-coated ascorbic acid
FCAA = fat-coated ascorbic acid
APP = ascorbyl-2-polyphosphate
NaAA = sodium ascorbic acid
GCAA = glyceride-coated ascorbic acid
AS = ascorbyl-2-sulfate
AP = ascorbyl-6-palmitate

Losses during storage of diets containing crystalline AA may reach 100 percent, depending on storage conditions (see Table 8.4). Losses of AA are accelerated as moisture increases in the finished diet. Losses of the vitamin increased dramatically in channel catfish diets as diet moisture levels increased from about 5 percent to 10 to 15 percent (Robinson 1992). Losses of AA in moist salmonid diets were rapid, with essentially all AA activity lost within one hour (Grant et al. 1989). Freezing diet samples minimizes losses of the vitamin during storage (Hilton, Cho, and Slinger 1977; Grant et al. 1989), but freezing is not a feasible approach to diet storage in commercial aquaculture.

Various methods have been used to stabilize AA. One of the most common methods has been to coat AA with ethylcellulose or fat. Coated products are somewhat more stable than crystalline AA; however, the protective coating may be damaged during feed manufacture, and losses of AA can still be considerable (see Tables 8.3 and 8.4). Another approach to avoid losses of the vitamin has been to add AA after pelleting or extrusion and

TABLE 8.4. Stability of AA Sources During Storage

Source	**Temper-ature**	**Feed Type**	**% Loss**				**Reference**
				4		**12 (weeks)**	
AA	25	trout-P		55		74	Grant et al. (1989)
APP	25	trout-P		12		25	Grant et al. (1989)
AA	40	trout-P		87		95	Grant et al. (1989)
APP	40	trout-P		29		32	Grant et al. (1989)
ECAA	25	trout-E		—		85	Gadient (1991)
APP	25	trout-E		—		11	Gadient (1991)
ECAA	—	trout-E		74		87	Gadient et al. (1992)
ECAA	—	trout-E		70		—	Gadient et al. (1992)
APP	—	trout-E		1		0	Gadient et al. (1992)
ECAA	—	catfish-E		78		—	Robinson (1992)
APP	—	catfish-E		0		—	Robinson (1992)
			1	**12**	**24**	**48 (hours)**	
AA	—	moist-P	58	100	—	—	Grant et al. (1989
APP	—	moist-P	0	4	24	—	Grant et al. (1989)
ECAA	—	moist-P	99	—	100	—	Grant et al. (1989)
APP	—	moist-P	0	—	4	16	Schai (1991)
APP	—	moist-P	0	—	0	2	Schai (1991)

AA = crystalline L-ascorbic acid
ECAA = ethylcellulose-coated ascorbic acid
APP = ascorbyl-2-polyphosphate
P = pelleted feed
E = extruded feed

drying by spraying the diet pellet with AA suspended in oil (Lovell 1984). This method is problematic because it is difficult to keep the vitamin evenly suspended in the oil, some of the vitamin is found in the diet dust and thus may be unavailable to the fish, and sprayer nozzles may clog, resulting in uneven distribution of the vitamin. The most successful method to stabilize AA has been to chemically bind the C-2 carbon of AA to a chemical moiety such as a fatty acid, sulfate, or phosphate. Esterification with a sulfate or phosphate group appears to be particularly effective in stabilizing the vitamin, thus increasing its retention during diet manufacture and storage (see Tables 8.3 and 8.4). Losses of L-ascorbyl-2-sulfate during feed processing ranges from 4 to 28 percent (Soliman, Jauncey, and Roberts 1987; Schuep, Marmet, and Studer 1989). Losses of L-ascorbyl-2-polyphosphate during manufacture of fish diets have been reported to be from 5 to 17 percent (Grant et al. 1989; Robinson, Brent, and Crabtree 1989; Gadient, Fenster, and Latscha 1992). The polyphosphate compound has been shown to be 4 to 55 times more stable than crystalline AA at 25°C and 13 to 100 times more stable at 40°C (Grant et al. 1989). Another phosphorylated ascorbic acid, L-ascorbyl-2-phosphate magnesium salt, also appears to be stable with losses during manufacture of around 10 to 20 percent and losses during storage for 20 months of about 30 percent (Shigueno and Itoh 1988).

Bioavailability of Stabilized Forms

Several factors must be considered when stabilizing AA either by coating or chemical alteration, but a major consideration is that the stable product must be biologically available. Coated products should be easily digested to release AA for absorption. Research has shown that the ethylcellulose AA (ECAA) is readily available (Murai, Andrews, and Bauernfeind 1978; Robinson 1992) and is equivalent to crystalline AA in biopotency in fish (Murai, Andrew, and Bauernfeind 1978). Fat-coated AA is also available to fish and appears to be biologically equivalent to crystalline AA (Robinson 1992).

Derivatives of AA that have been stabilized chemically require enzymatic degradation to release the active AA. Sulfate transferase is required to release AA from L-ascorbyl-2-sulfate (AS) (Tucker and Halver 1984), and phosphatase is required to release the vitamin from the phosphorylated AA compounds (Liao and Seib 1990; Dabrowski, Matusiewicz, and Blom 1994). L-ascorbate-2-sulfate has been shown to have antiscorbutic activity in rainbow trout (Halver et al. 1975; Dabrowski et al. 1990); channel catfish (Murai, Andrew, and Bauernfeind 1978; Brandt, Deyoe, and Seib 1985; Wilson, Poe, and Robinson 1989; El Naggar and Lovell 1991); and Nile tilapia (Soliman, Jauncey, and Roberts 1986b; Abdelghany 1996). It has been reported that the antiscorbutic activity of AS was equivalent to crystalline AA on an equimolar basis in rainbow trout (Halver et al. 1975) and Nile tilapia (Abdelghany 1996); however, the antiscorbutic activity of AS is apparently not equivalent to crystalline AA or certain other stable AA derivatives for some fish, including rainbow trout (Murai, Andrew, and Bauernfeind 1978; Tsujimura et al. 1978; Soliman, Jauncey, and Roberts 1986b; Dabrowski et al. 1990; El Naggar and Lovell 1991).

L-ascorbate-2-sulfate has been shown to have only one-fourth the activity of crystalline AA in one study with channel catfish (Murai, Andrews, and Bauernfeind 1978). In another study with channel catfish, the AS was reported to have only about 5 percent the activity of crystalline AA or L-ascorbyl-2-monophosphate based on maximum growth (El Naggar and Lovell 1991). However, Wilson, Poe, and Robinson (1989) found no differences in weight gain of channel catfish fed diets containing 100 mg AA supplied by either ECAA or AS, but fish fed the AS diet had a lower carcass AA than fish fed the ECAA diet. In vitro studies of channel catfish intestine have shown that AS is poorly absorbed (Buddington et al. 1993), which would account for its poor antiscorbutic activity in that species. Also, it has been reported to be poorly absorbed by rainbow trout (Fenster 1990; Dabrowski and Kock 1989) and common carp (Fenster 1990; Dabrowski 1990).

In contrast to AS, phosphate derivatives of AA have been shown to have antiscorbutic activity equivalent to that of the parent compound in various animals including several fish species. The magnesium and sodium salt of L-ascorbyl-2-monophosphate has been shown to have equal antiscorbutic activity as crystalline AA in channel catfish (El Naggar and Lovell 1991; Mustin and Lovell 1992) and rainbow trout (Dabrowski et al. 1996). Similarly, L-ascorbyl-2-polyphosphate (APP) has been shown to have activity equal to that of crystalline or coated AA in channel catfish (Robinson, Brent, and Crabtree 1989; Wilson, Poe, and Robinson 1989); rainbow trout (Grant et al. 1989; Sato, Miyasaki, and Yoshinaka 1991; Matusiewicz et al. 1995); Nile tilapia (Abdelghany 1996); fathead minnow, *Pimephales promelas* (Grant et al. 1989); and guppy, *Poecilia reticulata* (Grant et al. 1989). Based on the efficacy of phosphate derivatives as a substitute for AA in fish diets, it is clear that for the species studied, AA is released from the phosphorylated compound and absorbed. This has been shown to be true of monophosphate derivatives of AA in in vitro studies of channel catfish intestine (Buddington et al. 1993). This work implied that the bioavailability of monophosphate derivatives of AA is a two-stage process in which intestinal hydrolysis removes the phosphate group and the released AA is absorbed via carrier-mediated processes. Further, it was suggested that the affinity of AA carriers was higher for phosphorylated forms of AA and that less AA was oxidized prior to uptake, which allowed more AA to be transported. These results would account for the observation that tissue AA levels are often higher in fish fed phosphorylated forms of AA as compared to those of fish fed equimolar concentrations of AA supplied by crystalline and coated forms (Grant et al. 1989; Wilson, Poe, and Robinson 1989; El Naggar and Lovell 1991; Abdelghany 1996).

Supplementation

For many years, crystalline AA and ethylcellulose-coated AA were the most common economical sources of the vitamin available for use in aquatic animal diets. Ethylcellulose-coated AA has been the predominant form of the vitamin used in fish diets for the past several years. However, as previously discussed, it is relatively unstable during diet manufacture and storage. In general, losses during diet manufacture were 50 percent or greater. Additional losses were encountered during storage. Most commercially available coated products appear to be similar in stability and bioavailability, whereas the phosphorylated products offer the advantage of stability during diet manufacture and storage—their disadvantage is cost. Phosphorylated products may cost four to five times more than coated AA products on an equal activity basis.

There is no one best choice in choosing the source of AA for use in all fish diets. Economics is certainly an important factor; however, one must consider several factors in making a choice. The overriding consideration should be to provide the proper amount of AA activity to the target species. To do so, diet processing methods and the length of time the diet will be stored prior to use must be considered. If, as in the channel catfish industry in the southeastern United States, diet is stored only for a day or so, it may be more economical to overfortify with the less expensive AA sources to account for losses during manufacture than to use a more expensive yet stable source. If, however, the diet is to be stored for extended periods of time or the feed contains a high level of moisture, a more stable form of the vitamin may be the best choice. Of the commercial forms of AA currently available, the phosphorylated forms are the most stable and, thus, a more accurate means to provide a targeted amount of the vitamin.

REFERENCES

Abdelghany, A. 1996. Growth response of Nile tilapia *Oreochromis niloticus* to dietary L-ascorbic acid, L-ascorbyl-2-sulfate, and L-ascorbyl-2-polyphosphate. *Journal of the World Aquaculture Society* 27:449-455.

Aguirre, P. and D.M. Gatlin, III. 1999. Dietary vitamin C requirement of red drum *(Sciaenops ocellatus). Aquaculture Nutrition* 5:247-249.

Alexis, M.N., K.K. Karanikolas, and R.H. Richards. 1997. Pathological findings owing to the lack of ascorbic acid in cultured gilthead bream (*Sparus aurata* L.). *Aquaculture* 151:209-218.

Andrews, J.W. and T. Murai. 1975. Studies on vitamin C requirements of channel catfish. *Journal of Nutrition* 105:557-561.

Anggawati-Satyabudhy, A.M.A., B.F. Grant, and J.E. Halver. 1990. Effect of L-ascorbyl-2-phosphates (AsPP) on growth and immunoresistance of rainbow trout to infectious hematopoietic necrosis (IHN) virus. In M. Takeda and T. Watanabe (Eds.), *Proceedings of the Third International Symposium on Feeding and Nutrition in Fish,* Toba, Japan, August 28 to September 1, 1989 (pp. 411-426). Tokyo, Japan: Japan Translation Center.

Aoe, H. 1980. Vitamins. In C. Ogino (Ed.), *Nutrition and Diets in Fish* (pp. 186-196). Tokyo, Japan: Koseisha-Koseikaku.

Arai, S. 1972. Qualitative requirement of young eels, *Anguilla japonica,* for water soluble vitamins and their deficiency symptoms. *Bulletin of Freshwater Research Laboratories* (Tokyo, Japan) 22:69-83.

Bell, G.R., D.A. Higgs, and G.S. Traxler. 1984. The effect of dietary ascorbate, zinc, and manganese on the development of experimentally induced bacterial kidney disease in sockeye salmon *(Oncorhynchus nerka). Aquaculture* 36:293-311.

Blom, J.H. and K. Dabrowski. 1995. Reproductive success of female rainbow trout *(Oncorhynchus mykiss)* in response to graded dietary ascorbyl monophosphate levels. *Biology of Reproduction* 52:1073-1080.

Blom, J.H. and K. Dabrowski. 1996. Ascorbic acid metabolism in fish: Is there a maternal effect on the progeny? *Aquaculture* 147:215-224.

Brandt, T.M., C.W. Deyoe, and P.A. Seib. 1985. Alternate sources of vitamin C for channel catfish. *Progressive Fish-Culturist* 47:55-59.

Buddington, R.K., A.A. Puchal, K.L. Houpe, and W.J. Diehl. 1993. Hydrolysis and absorption of two monophosphate derivatives of ascorbic acid by channel catfish *Ictalurus punctatus* intestine. *Aquaculture* 114:317-326.

Chavez de Martinez, M.C. 1990. Vitamin C requirement of the Mexican native cichlid *Cichlasoma urophthalmus* (Gunther). *Aquaculture* 86:409-416.

Cho, C.Y. and C.W. Cowey. 1993. Utilization of monophosphate esters of ascorbic acid by rainbow trout *(Oncorhynchus mykiss).* In S.J. Kanshik and P. Luquet (Eds.), *Fish Nutrition in Practice* (pp. 149-156). Paris, France: Institut National de la Recherch Agronomique.

Coustans, M.F., J. Guillaume, R. Metailler, and O. Dugornay. 1990. Effect of an ascorbic acid deficiency on tyrosinemia and renal granulomatous disease in turbot *(Pisces scophthalmidae).* Interaction with a slight polyhypovitaminosis. *Comparative Biochemistry and Physiology* 97:145-155.

Dabrowska, H., K. Dabrowski, K. Meyer-Burgdorff, W. Hanke, and K.D. Gunther. 1991. The effect of large doses of vitamin C and magnesium on stress responses in common carp *(Cyprinus carpio). Comparative Biochemistry and Physiology* 89B:539-545.

Dabrowski, K. 1990. Absorption of ascorbic acid and ascorbic sulfate and ascorbate metabolism in stomachless fish, common carp. *Journal of Comparative Physiology B* 160:549-561.

Dabrowski, K. 1991. Ascorbic acid status in high-mountain char (*Salvelinus alpinus* L.) related to the reproductive cycle. *Environmental Biology of Fishes* 31:213-217.

Dabrowski, K. and A. Ciereszko. 1996. Ascorbic acid protects against male infertility in a teleost fish. *Experimentia* (Switzerland) 52:97-100.

Dabrowski, K., N. El-Fiky, G. Kock, M. Frigg, and W. Wieser. 1990. Requirement and utilization of ascorbic acid and ascorbic sulfate in juvenile rainbow trout. *Aquaculture* 91:317-337.

Dabrowski, K., S. Hinterleitner, C. Sturmbauer, N. El-Fiky, and W. Wieser. 1988. Do carp larvae require vitamin C? *Aquaculture* 72:295-306.

Dabrowski, K. and G. Kock. 1989. Absorption of ascorbic acid and ascorbic sulfate and their interaction with minerals in the digestive tract of rainbow trout *(Oncorhynchus mykiss). Canadian Journal of Fisheries and Aquatic Sciences* 46:1952-1957.

Dabrowski, K., M. Matusiewicz, and J.H. Blom. 1994. Hydrolysis, absorption and bioavailability of ascorbic acid esters in fish. *Aquaculture* 124:169-192.

Dabrowski, K., K. Matusiewicz, M. Matusiewicz, P.P. Hoppe, and J. Ebeling. 1996. Bioavailability of vitamin C from two ascorbyl monophosphate esters in rainbow trout, *Oncorhynchus mykiss* (Walbaum). *Aquaculture Nutrition* 2:3-10.

Davis, K.B., B.A. Simco, M. Li, and E. Robinson. 1998. Effect of reduction of supplemental dietary vitamins on the stress response of channel catfish *(Ictalurus punctatus). Journal of the World Aquaculture Society* 29:319-324.

Duncan, P.L. and R.T. Lovell. 1994. Influence of vitamin C on the folate requirement of channel catfish, *Ictalurus punctatus,* for growth, hematopoiesis, and resistance to *Edwardsiella ictaluri* infection. *Aquaculture* 127:233-244.

Durve, V.S. and R.T. Lovell. 1982. Vitamin C and disease resistance in channel catfish *(Ictalurus punctatus). Canadian Journal of Fisheries and Aquatic Sciences* 39:948-951.

El Naggar, G.O. and R.T. Lovell. 1991. L-ascorbyl-2-monophosphate has equal antiscorbutic activity as L-ascorbic acid but L-ascorbyl-2-sulfate is inferior to L-ascorbic acid for channel catfish. *Journal of Nutrition* 121:1622-1626.

Erdal, J.I., O. Evensen, O.K. Kaurstad, A. Lillehaug, R. Solbakken, and K. Thorud. 1991. Relationship between diet and immune response in Atlantic salmon (*Salmo Salar* L.) after feeding various levels of ascorbic acid and omega-3 fatty acids. *Aquaculture* 98:363-379.

Eya, J. 1996. "Broken-skull disease" in African catfish *Claria gariepinus* is related to a dietary deficiency of ascorbic acid. *Journal of the World Aquaculture Society* 27:493-498.

Fenster, R. 1990. Ascorbyl-2-sulfate, an alternative source of vitamin C for fish? In M. Takeda and T. Watanabe (Eds.), *Proceedings of the Third International Symposium on Feeding and Nutrition in Fish,* Toba, Japan, August 28 to September 1, 1989 (pp. 166-171). Tokyo, Japan: Japan Translation Center.

Floyd, R. 1987. Field efficacy of vitamin C for prevention of enteric septicemia of channel catfish. *Proceedings of International Association of Aquatic Animal Medicine* 18:181-183.

Gadient, M. 1991. Trial Ex 91.4. Rovimix Stay-C, Technical Dossier. Hoffman La Roche, Inc., Nutley, New Jersey.

Gadient, M., R. Fenster, and T. Latscha. 1992. Vitamin stability in aquaculture feeds. *Fish Farmer,* January/February: 27-28.

Gouillou-Coustans, M.F., P. Bergot, and S.J. Kaushik. 1998. Dietary ascorbic acid needs for common carp *(Cyprinus carpio)* larvae. *Aquaculture* 453-461.

Gould, B.S. and H. Shwachman. 1942. Bone and tissue phosphatase in experimental scurvy and studies on the source of serum phosphatase. *American Journal of Physiology* 161:135:485.

Grant, B.F, P.A. Seib, M. Liao, and K.E. Corpron. 1989. Polyphosphorylated L-ascorbic acid: A stable form of vitamin C for aquaculture feeds. *Journal of the World Aquaculture Society* 20:143-157.

Halver, J.E. 1972. The role of ascorbic acid in fish disease and tissue repair. *Bulletin of Japanese Society of Scientific Fisheries* 38:79-92.

Halver, J.E. 1985. Recent advances in vitamin nutrition and metabolism. In C.B. Cowey, A.M. Mackie and J.G. Bell (Eds.), *Nutrition and Feeding in Fish* (pp. 415-429). New York: Academic Press.

Halver, J.E., L.M. Ashley, and R.R. Smith. 1969. Ascorbic acid requirement of coho salmon and rainbow trout. *Transactions of the American Fisheries Society* 98:762-771.

Halver, J.E., R.R. Smith, B.M. Tolbert, and E.M Baker. 1975. Utilization of ascorbic acid in fish. *Annals of the New York Academy of Science* 258:81-102.

Hardie, L.J., T.C. Fletcher, and C.J. Secombes. 1991. The effect of dietary vitamin C on the immune response of the Atlantic salmon (*Salmo salar* L.). *Aquaculture* 95:201-214.

Henrique, M.M.F., E.F. Gomes, M.F. Gouillou-Coustans, A. Oliva-Teles, and S.J. Davies. 1998. Influence of supplementation of practical diets with vitamin C on growth and response to hypoxic stress of seabream, *Sparus aurata. Aquaculture* 161:415-426.

Hilton, J.W., C.Y. Cho, R.G. Brown, and S.J. Slinger. 1979. The synthesis, half-life and distribution of ascorbic acid in rainbow trout. *Comparative Biochemistry and Physiology* 63A:447-453.

Hilton, J.W. C.Y. Cho, and S.J. Slinger. 1977. Factors affecting the stability of supplemental ascorbic acid in practical trout diets. *Journal of Fisheries Research Board of Canada* 34:683-687.

Hilton, J.W., C.Y. Cho, and S.J. Slinger. 1978. Effect of graded levels of supplemental ascorbic acid in practical diets fed to rainbow trout *(Salmo gairdneri). Journal of Fisheries Research Board of Canada* 35:431-436.

Hung, S.S.O. 1991. Sturgeon, *Acipenser* spp. In R.P. Wilson (Ed.), *Handbook of Nutrient Requirements of Finfish* (pp. 153-160). Boca Raton, FL: CRC Press.

Ikeda, S., M. Sato, and R. Kimura. 1964. Biochemical studies on L-ascorbic acid in aquatic animals. III. Biosynthesis of L-ascorbic acid by carp. *Bulletin of the Japanese Society of Scientific Fisheries* 30:365-369.

Ishibashi, Y., K. Kato, S. Ikeda, O. Murata, T. Nasu, and H. Kumai. 1992. Effect of dietary ascorbic acid on the tolerance for low oxygen stress in fish. *Bulletin of the Japanese Society of Scientific Fisheries* 58:1555.

Jaffa, M. 1989. Vitamin C can curb those stress associated losses. *Fish Farmer* 12:18-19.

Johnson, M.R. and A.J. Ainsworth. 1991. An elevated dietary level of ascorbic acid fails to influence the response of anterior kidney neutrophils to *Edwardsiella ictaluri* in channel catfish. *Journal of Aquatic Animal Health* 3:266-273.

Lall, S.P., G. Oliver, D.E.M. Weerakoon, and J.A. Hines. 1990. The effect of vitamin C deficiency and excess on immune response in Atlantic salmon (*Salmo salar* L.). In M. Takeda and T. Watanabe (Eds.), *Proceedings of the Third International*

Symposium on Feeding and Nutrition in Fish. Tokyo, Japan, August 28 to September 1, 1989 (pp. 427-441). Tokyo, Japan: Japan Translation Center.

Li, M.H., M.R. Johnson, and E.H. Robinson. 1993. Elevated dietary vitamin C concentrations did not improve resistance of channel catfish, *Ictalurus punctatus,* against *Edwardsiella ictaluri* infection. *Aquaculture* 117:303-312.

Li, M.H. and E.H. Robinson. 1994. Effect of dietary vitamin C on tissue vitamin C concentration in channel catfish, *Ictalurus punctatus,* and clearance rate at two temperatures—A preliminary investigation. *Journal of Applied Aquaculture* 4(2):59-71.

Li, M.H., D.J. Wise, and E.H. Robinson. 1998. Effect of dietary vitamin C on weight gain, tissue ascorbate concentration, stress response, and disease resistance of channel catfish *Ictalurus punctatus. Journal of the World Aquaculture Society* 29:1-8.

Li, Y. and R.T. Lovell. 1985. Elevated levels of dietary ascorbic acid increase immune response in channel catfish. *Journal of Nutrition* 115:123-131.

Liao, M.L. and P.A. Seib. 1990. A stable form of vitamin C: L-ascorbate-triphosphate. Synthesis, isolation, and properties. *Journal of Agricultural Food Chemistry* 38:355-366.

Lim, C. and R.T. Lovell. 1978. Pathology of the vitamin C deficiency syndrome in channel catfish *(Ictalurus punctatus). Journal of Nutrition* 108:1137-1146.

Lin, D. 1991. Grass carp, *Ctenopharyngodon idella.* In R.P. Wilson (Ed.), *Handbook of Nutrient Requirements of Finfish* (pp. 89-96). Boca Raton, FL: CRC Press.

Liu, P.R., J.A. Plumb, M. Guerin, and R.T. Lovell. 1989. Effect of megalevels of dietary vitamin C on the immune response of channel catfish *Ictalurus punctatus* in ponds. *Disease of Aquatic Organisms* 7:191-194.

Lovell, R.T. 1984. Ascorbic acid metabolism in fish. In I. Wegger, F.J. Tagwerker and J. Moustgaard (Eds.), *Ascorbic Acid Metabolism in Domestic Animals* (pp. 196-205). Copenhagen: The Royal Danish Agricultural Society.

Lovell, R.T. and C. Lim. 1978. Vitamin C in pond diets for channel catfish. *Transactions of the American Fisheries Society* 107:321-325.

Lovell, T. 1989. *Nutrition and Feeding of Fish.* New York: Van Nostrand Reinhold.

Lutwak-Mann, C. 1958. The dependence of gonad function upon vitamins and other nutritional factors. In R.S. Harris, G.F. Marrian, and K.V. Thimann (Eds.) *Vitamins and Hormones.* Volume 16 (pp. 35-75). New York: Academic Press.

Mangor-Jensen, A, J.C. Holm, G. Rosenlund, O. Lie, and K. Sandnes. 1994. Effects of dietary vitamin C on maturation and egg quality of cod *Gadus morhua* L. *Journal of the World Aquaculture Society* 25:30-40.

Martins, M.L. 1995. Effect of ascorbic acid deficiency on the growth, gill filament lesions and behavior of pacu fry *Piaractus mesopotamicus. Brazilian Journal of Medical and Biological Research* 28:563-568.

Matusiewicz, M., K. Dabrowski, L. Volker, and K. Matusiewicz. 1994. Regulation of saturation and depletion of ascorbic acid in rainbow trout. *Journal of Nutrition and Biochemistry* 5:204-212.

Matusiewicz, M., K. Dabrowski, L. Volker, and K. Matusiewicz. 1995. Ascorbate polyphosphate is a bioavailable vitamin C source in juvenile rainbow trout: Tissue saturation and compartmentalization model. *Journal of Nutrition* 125:3055-3061.

Mazik, P.M., T.M. Brandt, and J.R. Tomasso. 1987. Effects of dietary vitamin C on growth, caudal fin development, and tolerance of aquaculture-related stressors in channel catfish. *Progressive Fish-Culturist* 49:13-16.

Merchie, G., P. Lavens, P. Dhert, M. Dehasque, H. Nellis, A. De Leenheer, and P. Sorgeloos. 1995. Variation of ascorbic acid content in different live food organisms. *Aquaculture* 134:325-337.

Merchie, G, P. Lavens, P. Dhert, M.G.U. Gomez, H. Nellis, A. De Leenheer, and P. Sorgeloos. 1996. Dietary ascorbic acid requirements during the hatchery production of turbot larvae. *Journal of Fisheries Biology* 49:573-583.

Mishra S. and P.K. Mukhopadhyay. 1996. Ascorbic acid requirement of catfish fry *Clarias batrachus* (Linn.). *Indian Journal of Fisheries* 43:157-162.

Murai, T., J.W. Andrews, and J.C. Bauernfeind. 1978. Use of L-ascorbic acid, ethocel-coated ascorbic acid and ascorbate-2-sulfate in diets for channel catfish, *Ictalurus punctatus. Journal of Nutrition* 108:1761-1776.

Mustin, W.G., and R.T. Lovell. 1992. Na-L-ascorbyl-2-monophosphate as a source of vitamin C for channel catfish. *Aquaculture* 105:95-100.

National Research Council. 1993. *Nutritional Requirements of Fish.* Washington, DC: National Academy Press.

Navarre, O., and J.E. Halver. 1989. Disease resistance and humoral antibody production in rainbow trout fed high levels of vitamin C. *Aquaculture* 79:207-221.

Nitzan, S., H. Angeoni, and N. Gur. 1996. Effects of ascorbic acid polyphosphate (AAPP) enrichment on growth, survival and disease resistance of hybrid tilapia. *Israeli Journal of Aquaculture (Bamidgeh)* 48(3):133-141.

Roberts, M.L., S.J. Davis, and A.L. Pulsford. 1995. The influence of ascorbic acid (vitamin C) on non-specific immunity in the turbot *Scophthalmus maximus. Fish and Shellfish Immunology* 5(1):27-38.

Robinson, E.H. 1992. *Vitamin C Studies with Catfish: Requirements, Biological Activity and Stability.* Technical Bulletin 182. Mississippi State: Mississippi Agricultural and Forestry Experiment Station.

Robinson, E.H., J.R. Brent, and J.T. Crabtree. 1989. AsPP, an ascorbic acid, resists oxidation in fish feed. *Feedstuffs* 61(40):64-66.

Robinson, E.H. and M.H. Li. 1996. *A Practical Guide to Nutrition, Feed, and Feeding of Catfish* (Revised). Bulletin 1041. Mississippi State: Mississippi Agricultural and Forestry Experiment Station.

Robinson, E.H., M.H. Li, and D. Oberle. 1998. *Catfish Vitamin Nutrition.* Bulletin 1078. Mississippi State: Mississippi Agricultural and Forestry Experiment Station.

Sandnes, K. and O.R. Braekkan. 1981. Ascorbic acid and the reproductive cycle of ovaries in cod *(Gadus morhua). Comparative Biochemistry and Physiology* 70A:545-551.

Sandnes, K., O. Torrissen, and R. Waagbo. 1992. The minimum dietary requirement of vitamin C in Atlantic salmon *(Salmo salar)* fry using Ca ascorbate-2-monophosphate as dietary source. *Fish Physiology and Biochemistry* 10:315-319.

Sandnes, K., Y. Ulgenes, O.R. Braekkan, and F. Utne. 1984. The effect of ascorbic acid supplementation in broodstock feed on reproduction of rainbow trout *(Salmo gairdneri). Aquaculture* 43:167-177.

Sato, M., T. Miyasaki, and R. Yoshinaka. 1991. Utilization of L-ascorbyl 2-phosphate in rainbow trout as a dietary vitamin C source. *Bulletin of the Japanese Society of Scientific Fisheries* 57:1923-1926.

Sato, M., R. Yoshinaka, Y. Yamamoto, and S. Ikeda. 1978a. Dietary ascorbic acid requirement of rainbow trout for growth and collagen formation. *Bulletin of the Japanese Society of Scientific Fisheries* 44:1029-1035.

Sato, M., R. Yoshinaka, Y. Yamamoto, and S. Ikeda. 1978b. Nonessentiality of ascorbic acid in the diet of carp. *Bulletin of the Japanese Society of Scientific Fisheries* 44:1151-1156.

Schai, E. 1991. Rovimix Stay-C. Technical Dossier. Enclosure 3, 1992. Vitamins and Fine Chemicals Division, Hoffmann-La Roche, Nutley, New Jersey.

Schuep, W., J. Marmet, and W. Studer. 1989. Stability of ascorbyl-2-sulfate in trout feed measured by HPLC. *Aquaculture* 79:249-258.

Sealey, W.M. and D.M. Gatlin, III. 1999. Dietary vitamin C requirement of hybrid striped bass (*Morone chrysops* ♀ × *M. saxatilis* ♂). *Journal of the World Aquaculture Society* 30:297-301.

Seymour, E.A. 1981. Gonadal ascorbic acid and change in level with ovarian development in the crucian carp, *Carassius carassius* (L.). *Comparative Biochemistry and Physiology* 75A:541-543.

Shiau, S.Y. and F.L. Jan. 1992. Dietary Ascorbic acid requirement of juvenile tilapia *Oreochromis niloticus* × *O. aurea. Bulletin of the Japanese Society of Scientific Fisheries* 58:671-675.

Shigueno, K. and S. Itoh. 1988. Use of Mg-L-ascorbyl-2-phosphate as vitamin C source in shrimp diets. *Journal of the World Aquaculture Society* 19:169-175.

Shimeno, S. 1991. Yellowtail, *Seriola quinqueradiata.* In R.P. Wilson (Ed.) *Handbook of Nutrient Requirements of Finfish* (pp. 181-191). Boca Raton, FL: CRC Press.

Skelbaek, T., N.G. Anderson, M. Winning, and S. Westergaard. 1990. Stability in feed and bioavailability to rainbow trout of two ascorbic acid forms. *Aquaculture* 84:335-343.

Soliman, A.K, K. Jauncey, and R.J. Roberts. 1986a. The effect of dietary ascorbic acid supplementation on hatchability, survival rate and fry performance in *Oreochromis mossambicus. Aquaculture* 59:197-208.

Soliman, A.K, K. Jauncey, and R.J. Roberts. 1986b. The effect of varying forms of dietary ascorbic acid on the nutrition of juvenile tilapia *(Oreochromis niloticus). Aquaculture* 52:1-10.

Soliman, A.K., K. Jauncey, and R.J. Roberts. 1987. Stability of L-ascorbic acid (vitamin C) and its forms in fish feeds during processing, storage and leaching. *Aquaculture* 60:73-83.

Stickney, R.R., R.B. McGeachin, D.H. Lewis, J. Marks, A. Riggs, R.F. Sis, E.H. Robinson, and W. Wurts. 1984. Response of *Tilapia aurea* to dietary vitamin C. *Journal of the World Mariculture Society* 15:179-185

Teshima, S.I., A. Kanazawa, S. Koshio, and S. Itoh. 1993. L-ascorbyl-2-polyphosphate-Mg as vitamin C source for the Japanese flounder *(Paralichthys olivaceus)*. In S.J. Kanshik and P. Luquet (Eds.) *Fish Nutrition in Practice* (pp. 149-156). Paris: Institut National de la Recherch Agronomique.

Tsujimura, M., H. Yoshikawa, T. Hasagawa, T. Suzuki, T. Kaisai, T. Suwa, and S. Kitamura. 1978. Studies on the vitamin C activity of ascorbic acid-2-sulfate on the feeding test of new born rainbow trout. *Vitamins* (Japan) 52:35-44.

Tucker, B.W. and J.E. Halver. 1984. Ascorbate-2-sulfate metabolism in fish. *Nutrition Review* 42:173-179.

Verlhac, V. and J. Gabaudan. 1994. Influence of vitamin C on the immune system of salmonids. *Aquaculture and Fishery Management* 25:21-36.

Waagbo, R., J. Glette, E. Raa-Nilsen, and K. Sandnes. 1993. Dietary vitamin C, immunity and disease resistance in Atlantic salmon *(Salmo salar)*. *Fish Physiology and Biochemistry* 12:61-73.

Waagbo, R. and K. Sandnes. 1996. Effect of dietary vitamin C on growth and parr-smolt transformation in Atlantic salmon, *Salmo salar* L. *Aquaculture Nutrition* 2(2):65-69.

Waagbo, R., T. Thorsen, and K. Sandnes. 1989. Role of dietary ascorbic acid in vitellogenesis in rainbow trout *(Salmo gairdneri)*. *Aquaculture* 80:301-314.

Wahli, T., V. Verlhac, J. Gabaudan, W. Schuep, and W. Meier. 1998. Influence of combined vitamin C and E on non-specific immunity and disease resistance of rainbow trout, *Oncorhynchus mykiss* (Walbaum). *Journal of Fish Disease* 21:127-137.

Wang, X., M. Liao, T. Hung, and P. A. Seib. 1988. Liquid chromatographic determination of L-ascorbate-2-polyphosphate in fish feeds by enzymatic release of L-ascorbate. *Journal of Association of Official Analytical Chemists* 71:1158-1161.

White, T.A., T.C. Fletcher, D.F. Houlihan, and C.J. Secombes. 1993. The effect of stress on the immune response of Atlantic salmon (*Salmo salar* L.) fed diets containing different amounts of vitamin C. *Aquaculture* 114:1-18.

Wilson, R.P. and W.E. Poe. 1973. Impaired collagen formation in the scorbutic channel catfish. *Journal of Nutrition* 103:1359-1364.

Wilson, R.P., W.E. Poe, and E.H. Robinson. 1989. Evaluation of L-ascorbyl-2-polyphosphate (AsPP) as a dietary ascorbic acid source for channel catfish. *Aquaculture* 81:129-136.

Wise, D.J., J.R. Tomasso, and T.M. Brandt. 1988. Ascorbic acid inhibition of nitrite-induced methemoglobinemia in channel catfish. *Progressive Fish-Culturist* 50:77-80.

Yamamoto, Y., M. Sato, and S. Ikeda. 1978. Existence of L-gulonolactone oxidase in some teleosts. *Bulletin of the Japanese Society of Scientific Fisheries* 44:775-779.

Chapter 9

Dietary Iron and Fish Health

Chhorn Lim
Phillip H. Klesius
Craig A. Shoemaker

INTRODUCTION

Iron is a trace mineral of fundamental importance for most higher animals, including fish, because of its functions in oxidation-reduction activity and oxygen transport. In biological systems, iron can exist in the ferrous (Fe^{2+}) or ferric (Fe^{3+}) state, and this permits iron to donate or accept electrons and thus participate in the oxidation-reduction reactions, including those involved in oxygen transport. It occurs in the animal body as a component of the respiratory pigment (heme compounds), such as hemoglobin in red blood cells and myoglobin in muscle, as well as the heme enzymes such as peroxidase, catalase, and cytochromes. The remainder of the iron in the body is found in nonheme compounds such as transferrin (siderophilin), lactoferrin, ferritin, and hemosiderin (Harper 1973; Kaneko 1980). Transferrin, which is the principal protein carrier of iron in the blood, plays an important role in iron metabolism. Ferritin and hemosiderin are iron storage compounds found in the liver, spleen, and bone marrow.

Fish can absorb soluble iron from water across the gill membrane (Roeder and Roeder 1966); however, intestinal mucosa is considered the major site of iron absorption (Lall 1989). Diet is considered the major source of iron because of low concentrations of soluble iron in natural waters (NRC 1993). The dietary requirements of iron for optimum growth and prevention of various deficiency signs have been demonstrated for several fish species, including red sea bream, *Chrysophrys major* (Sakamoto and Yone 1976, 1978b); yellowtail, *Seriola quinqueradiata* (Ikeda, Ozaki, and Uematsu 1973); common carp, *Cyprinus carpio* (Sakamoto and Yone 1978a); eel, *Anguilla japonica* (Nose and Arai 1979); Atlantic salmon, *Salmo salar* (Lall and Hines 1987; Andersen, Maage, and Julshamn 1996);

and channel catfish, *Ictalurus punctatus* (Gatlin and Wilson 1986; Lim, Sealey, and Klesius 1996).

Iron also is an important micronutrient that has been shown to affect immune system function and host defense against infection. Either a deficiency or an excess of iron can compromise the immune system (Beisel 1982; Bhaskaram 1988). From the point of view of infection, Bullen, Rogers, and Griffiths (1978) suggested that a clear distinction must be made between the amount of iron present in the body fluids and its availability to pathogenic bacteria. An enormous amount of research has examined the relationships between iron status and the host immune response and susceptibility to infectious disease in humans and terrestrial animals. However, few studies have evaluated the effect of dietary iron on immune response and disease resistance in fish. This chapter reviews available information on iron as it relates to fish health. Bioavailability of various sources of iron and factors affecting iron absorption are also presented.

DEFICIENCY AND TOXICITY SIGNS

Iron deficiency is not a common problem in fish culture because commercial diets normally contain certain amounts of fish meal and/or animal proteins that are rich sources of iron. Moreover, commercial fish diets are routinely supplemented with a trace mineral mix containing sufficient amounts of iron to meet fish requirements. However, iron deficiency has been produced in several fish species cultured under laboratory conditions and fed low-iron diets (see Table 9.1). Iron deficiency causes hypochromic microcytic anemia, which is characterized by decreased hemoglobin, hematocrit, mean corpuscular volume, and mean corpuscular hemoglobin. Iron deficiency anemia has been reported in Atlantic salmon (Andersen, Maage, and Julshamn 1996); brook trout, *Salvelinus fontinalis* (Kawatsu 1972); channel catfish (Gatlin and Wilson 1986; Lim, Sealey, and Klesius 1996; Lim and Klesius 1997); common carp (Sakamoto and Yone 1978a); eel (Nose and Arai 1976); and red sea bream (Sakamoto and Yone 1978b). Decreased appetite, feeding activity, and feed efficiency, along with reduced growth and reduced serum iron and transferrin saturation in the blood have also been observed in channel catfish fed iron-deficient diets (Gatlin and Wilson 1986; Lim, Sealey, and Klesius 1996; Lim and Klesius 1997). However, studies with red sea bream (Sakamoto and Yone 1978b), common carp (Sakamoto and Yone 1978a) and Atlantic salmon (Andersen, Maage, and Julshamn 1996), did not indicate the adverse effect of iron deficiency on growth and feed efficiency. A significant reduction in survival and an increase in the level of total serum iron-binding capacity are also characteristic signs of iron deficiency anemia of channel catfish, when fish were fed an iron-deficient diet for 17 weeks (Lim and Klesius 1997).

TABLE 9.1. Iron Deficiency Signs in Certain Species of Fish

Species	**Deficiency Signs**	**Reference**
Atlantic salmon, *Salmo salar*	Reduced hemoglobin, hematocrit, mean corpuscular hemoglobin, and mean corpuscular volume.	Andersen et al. (1996)
Brook trout, *Salvelinus fontinalis*	Decreased growth, red blood cell count, hematocrit, hemoglobin, mean corpuscular volume, mean corpuscular hemoglobin, and mean corpuscular hemoglobin concentration.	Kawatsu (1972)
Channel catfish, *Ictalurus punctatus*	Reduced appetite, growth, feed efficiency, red blood cell count, hemoglobin, hematocrit, plasma iron, and transferrin saturation.	Gatlin and Wilson (1986)
	Decreased appetite and growth, and reduced hemoglobin, hematocrit, mean corpuscular hemoglobin, and mean corpuscular volume.	Lim et al. (1996)
	Reduced growth, feed efficiency, hematocrit, total cell count, serum iron and transferrin saturation, and increased mortality and total serum iron binding capacity.	Lim and Klesius (1997)
Common carp, *Cyprinus carpio*	Reduced hemoglobin, hematocrit, mean corpuscular hemoglobin, mean corpuscular volume, and mean corpuscular hemoglobin concentration.	Sakamoto and Yone (1978a)
Eel, *Anguilla japonica*	Decreased red blood cell count, hemoglobin, hematocrit, mean corpuscular hemoglobin, mean corpuscular volume, and mean corpuscular hemoglobin concentration.	Nose and Arai (1976)
Red sea bream, *Pagrus major*	Decreased hemoglobin, hematocrit, mean corpuscular hemoglobin, mean corpuscular volume, mean serum iron and corpuscular hemoglobin concentration, serum iron and iron saturation index, and increased percentage of immature erythrocyte.	Sakamoto and Yone (1978b)
Yellowtail, *Seriola lalandi*	Reduced growth, red blood cell count, hemoglobin, and serum iron and protein level.	Ikeda et al. (1973)

Excessive levels of dietary iron can be toxic to fish. Toxicity of dietary iron was observed in rainbow trout, *Oncorhynchus mykiss,* fed greater than 1,380 mg iron (as ferrous sulfate)/kg diet (Desjardins, Hicks, and Hilton 1987). The major signs of iron toxicity include reduced growth, poor feed efficiency, diet refusal, high mortality, diarrhea, and histopathological damage to liver cells (Desjardins 1985). Moreover, the catalytic effect of iron on lipid peroxidation or diet rancidity may have also contributed to the toxicity of high levels of dietary iron. Iron is one of the primary metals involved in lipid oxidation, and ferrous iron is a more potent catalyst of lipid peroxidation than ferric iron (Chvapil, Aronson, and Peng 1974; Lee et al. 1981). Ferrous iron catalyzes the formation of hydroperoxides and free radical peroxides by providing a free radical initiator in the presence of unsaturated fatty acids and oxygen. Fish diets normally contain high levels of unsaturated fats. Thus, high levels of iron supplementation may be harmful to fish due to the adverse effect of diet rancidity, which leads to decreased diet intake, destruction of fatty acids and other nutrients, the presence of toxic constituents such as hydroperoxides, and an alteration of intestinal function. Desjardins, Hicks, and Hilton (1987) reported that when diets are highly rancid (based on the concentration of malonaldehyde), the toxic levels of dietary iron for rainbow trout appears to range from 86 to 265 mg iron (as ferrous sulfate)/kg diet. Baker, Martin, and Davies (1997) observed that feeding African catfish, *Clarias gariepinus,* a fish meal-based diet containing 6,354.4 mg iron (supplied by iron sulfate)/kg diet for five weeks at a rate of 2 percent body weight/day resulted in suppressed growth and increased oxidative stress. They reported increased concentrations of the lipid peroxidation product (malondialdehyde) in the liver and heart and a depletion of antioxidant vitamin E in the liver of these fish.

DIETARY REQUIREMENT

The total dietary iron requirements have been determined for certain species of fish (see Table 9.2). Total dietary levels of iron required to maintain optimum hematological values and iron status were 30 mg iron/kg diet for channel catfish (Gatlin and Wilson 1986; Lim, Sealey, and Klesius 1996); 60 to 100 mg/kg of diet for Atlantic salmon (Lall and Hines 1987; Andersen, Maage, and Julshamn 1996); 150 mg/kg diet for red sea bream (Sakamoto and Yone 1978a); and 170 mg/kg diet for eel (Nose and Arai 1976). In rainbow trout, supplementation of semipurified diet with 60 mg iron/kg diet is adequate (Desjardins 1985). Iron has also been demonstrated to be a dietary essential for other species, such as brook trout, common carp, and yellowtail, but the requirement values for these species have not been determined.

TABLE 9.2. Iron Requirement of Certain Species of Fish

Species	Requirements (mg/kg diet)	Reference
Atlantic salmon, *Salmo salar*	60	Lall and Hines (1987)
	60-100	Andersen et al. (1996)
Channel catfish, *Ictalurus punctatus*	30	Gatlin and Wilson (1986) and Lim et al. (1996)
Eel, *Anguilla japonica*	170	Nose and Arai (1976)
Rainbow trout, *Oncorhynchus mykiss*	60	Desjardins (1985)
Red sea bream, *Pagrus major*	150	Sakamoto and Yone (1978b)

BIOAVAILABILITY

Relatively little information is available on the absorption and metabolism of iron in fish. Lall (1989) suggested that the mechanisms of iron absorption from the digestive tract, storage, and excretion in fish may be similar to those in other vertebrates. Fish can absorb soluble iron from the water across the gill membrane, as evidenced by the increased growth and hemoglobin level in swordtail, *Xiphophorus helleri,* and platyfish, *Xiphophorus maculatus* (Roeder and Roeder 1966). However, intestinal mucosa is considered the major site of iron absorption and diet is the major source of iron for metabolic purposes due to low concentrations of soluble iron in natural waters (Lall 1989; NRC 1993). The absorbed iron is bound to transferrin or siderophilin, a specific protein that serves as the principal carrier of iron in the blood. Transferrin occurs in the blood of all vertebrate species (Underwood 1977). In rainbow trout, iron injected intraperitoneally was absorbed from the peritoneal cavity and stored in the liver, spleen, and head kidney (Walker and Fromm 1976). There is little or no detectable iron lost in urine and feces of either normal or iron-deficient fish.

The nutritional value of iron sources depends not only upon their concentration in the diet ingredient but also upon the amount absorbed and metabolized—bioavailability to the animal. Intrinsic (such as size, age, and nutritional status) and extrinsic (such as the chemical form of iron and dietary factors) factors affect iron bioavailability to fish (Paripatananont and Lovell 1997). Dietary factors are elements or compounds that may bind to iron to form insoluble precipitates, or that have chelation effects which may positively or negatively affect iron absorption (O'Dell 1984). It has been reported that trace minerals chelated with organic compounds, such as amino acids or peptides, have higher bioavailability for various animals than inorganic forms. Chelation protects the mineral element from forming insoluble compounds or complexes in the digestive tract and facilitates the absorption of the mineral across membranes (Ashmead and Zunino 1992). Ashmead (1992) suggested that the mineral chelate is absorbed from the gut intact, delivered to various parts of the body, and degraded at the site where the element is needed. Paripatananont and Lovell (1997) reported that the type of diet (purified and practical diet) and source of iron affect the net absorption of iron by channel catfish. Net absorption of iron was significantly higher for iron proteinate than for ferrous sulfate heptahydrate. Net absorption of inorganic iron was significantly higher in the purified diet than in the practical diet, but net absorption of chelated iron did not differ among the diets. Lim, Sealey, and Klesius (1996), however, found that in purified diets ferrous sulfate heptahydrate was equally effective as ferric methionine complex for prevention of iron deficiency anemia in fingerling channel catfish. Sakamoto and Yone (1979) showed that ferrous chloride and ferric chloride were equally effective in preventing anemia in red sea bream, but a higher concentration was required when ferric citrate was used. For Atlantic salmon, Andersen and colleagues (1997) found that heme iron is the most available, iron sulfate is moderately available, and metallic iron is the least available. Iron from ferric oxide (Fe_2O_3) is not available or has very low availability, as is metallic iron for Atlantic salmon (Maage and Sveier 1998). Sugiura et al. (1998) showed that absorption of iron in fish meals and plant ingredients by coho salmon, *Oncorhynchus kisutch,* and rainbow trout was very low, whereas absorption of iron from wheat gluten meal was high, but the iron content in wheat gluten was very low. Iron in feather meal is highly available for these species because heme iron and/or feather meal is devoid of antagonistic substances that interfere with iron absorption. Thus, the requirement values reported may vary depending on the composition of the diets and the source of dietary iron.

IMMUNE RESPONSE AND DISEASE RESISTANCE

Iron is one of the most important micronutrients because of its effect on immune system functions and host defense against infections (Beisel 1982; Bhaskaram 1988). Iron is as essential a nutrient for microorganisms as it is for fish and other animals. Certain bacteria require iron for growth and replication, and for the production and release of certain exotoxins. Some bacteria synthesize siderospores, which have the ability to withdraw, solubilize, and chelate ferric iron to achieve their growth. However, during an infectious process in warm-blooded animals, the availability of iron in the body fluid to the invading microorganisms is restricted by the rapid sequestration of iron in tissue storage forms, and the ability of iron-binding proteins (such as transferrin and lactoferrin) to bind and withhold the circulating iron from the siderospores of the invading organisms (Weinberg 1974; Beisel 1982). To reinforce this withholding process, infected hosts promptly become hypoferremic by halting intestinal assimilation of iron, and by increasing liver storage of the metal (Weinberg 1974). Excessive iron in the body fluid (hyperferremic) may overwhelm the iron-binding ability and increase susceptibility of the host to the invading pathogens. This, however, does not suggest that iron deficiency resulting in anemia provides protection against infectious diseases (Berger 1996). Earlier research with warm-blooded animals indicated that iron-deficient or anemic animals are much more susceptible to infections than those with adequate iron. However, mild iron deficiency provided protection against infection, whereas excessive iron enhanced infectious illness (Sherman 1992). Berger (1996) indicated that iron is an interesting trace mineral, in that either a deficiency or an excess can adversely affect the immune system.

Studies with fish appear to indicate that there were interactions between the dietary level of iron and the immune response and disease resistance. However, no conclusive evidence has been established. Sealey, Lim, and Klesius (1997) reported that iron deficiency did not significantly suppress antibody titer of channel catfish in response to formalin-killed *Edwardsiella ictaluri.* Engulfment of bacteria by macrophages as measured by chemiluminescence was likewise not depressed by iron deficiency. It appeared, however, that maximum phagocytic engulfment of opsonized *E. ictaluri* by macrophages was observed in fish fed the 60 mg supplemental iron/kg of diet from either iron sulfate or iron methionine. They also suggested that antibody titer and engulfment of bacteria by macrophages were generally unaffected by the source of dietary iron. Macrophage chemotaxis, on the other hand, was significantly affected by dietary levels of iron. Chemotactic response of macrophages to *E. ictaluri* exoantigen expressed in terms of chemotactic index or ratio was significantly suppressed for fish fed the diet with no iron supplementation (Sealey, Lim, and Klesius 1997; Lim and Klesius 1997). The chemotactic ratio also decreased when the group fed the iron-replete diet (30 mg total iron/kg diet) was switched to an iron-deficient diet for four weeks. Suppression of macrophage chemotaxis, however, was reversed when the iron-deficient catfish were fed an iron-replete diet (Lim and Klesius 1997). In another study with channel catfish, Lim et al. (2000) obtained higher mean macrophage chemotaxis in fish fed a diet containing 30 mg iron/kg as compared to those fed an iron-unsupplemented diet or the diet supplemented with 300 mg iron/kg. The suppression of macrophage chemotactic response, however, was more pronounced in fish fed the iron-deficient diets than in those fed iron-excessive diets.

It is generally believed that anemic animals are more susceptible to infection than those with adequate iron. Sealey, Lim, and Klesius (1997) observed increased mortality of iron-deficient channel catfish following challenge by bath immersion with *E. ictaluri.* More recent studies, however, showed that dietary iron did not protect against mortality of channel catfish from *E. ictaluri,* but the average number of days to which the first mortality occurred after *E. ictaluri* challenge was significantly earlier for fish fed iron-deficient diets (Lim and Klesius 1997; Lim et al. 2000). They indicated that the early onset of mortality of fish fed the iron-deficient diet could be due to the synergistic effect of iron deficiency and *E. ictaluri* infection. Sealey, Lim, and Klesius (1997) observed that supplementation of diet with 180 mg iron/kg from iron sulfate, but not from iron methionine, resulted in increased mortality of channel catfish challenged with *E. ictaluri.* They suggested that a total dietary iron level of 30 mg/kg required for optimum growth and prevention of anemia in channel catfish may be sufficient for optimum immune response and resistance to *E. ictaluri.* Nakai and colleagues (1987) reported that increased availability of free iron following intramuscular injection of ferric ammonium citrate significantly increased the virulence of *Vibrio anguillarum* infection in eels and ayu, *Plecoglossus altivelis,* although the rate of infection was more severe in eels than in ayu. With Atlantic salmon, it has been reported that fish fed a low-iron diet were somewhat protected against *V. anguillarum* (S. P. Lall, Institute of Marine Biosciences, National Research Council, Halifax, Canada, personal communication). Ravndal and colleagues (1994) also observed a significant association between a high concentration of serum iron and mortality of Atlantic salmon infected with *V. anguillarum.* Thus, in fish, as in warm- blooded animals, a delicate balance exists between the need for iron for host defense mechanisms and the need for iron for microbial growth (Sherman, 1992).

CONCLUSION

Iron has been shown to be dietary essential for several fish species, but the quantitative iron requirement for optimum growth and prevention of deficiency signs has been determined only for selected species. Based on the requirement values reported and the iron content of feed ingredients, most practical diets for fish should contain adequate iron levels without supplementation. However, because the absorption of iron in most feedstuffs is very low, and the presence of other dietary components that may inhibit iron absorption, practical fish diets may require iron supplementation. The level of supplemental iron used is of critical importance because iron is a nutrient that can have profound effects on the immune response and resistance of hosts to infectious diseases. In fish, as in homeotherms, a delicate balance exists between the need for iron for host defense mechanisms and the need for iron to sustain microbial growth. A deficiency or excess of iron could compromise the immune system and, thus, resistance to infection. Currently, available information on the role of iron in immune function and disease resistance in fish is scanty and conflicting. Since iron deficiency seems to predispose fish to infection, a level of iron adequate to meet the requirements for growth and prevention of deficiency signs should be used in fish diets.

REFERENCES

Andersen, F., M. Lorentzen, R. Waagbo, and A. Maage. 1997. Bioavailability and interactions with other micronutrients of three dietary iron sources in Atlantic salmon, *Salmo salar,* smolts. *Aquaculture Nutrition* 3:239-246.

Andersen, F., A. Maage, and K. Julshamn. 1996. An estimation of dietary iron requirement of Atlantic salmon, *Salmo salar* L., parr. *Aquaculture Nutrition* 2:41-47.

Ashmead, H.D. 1992. *The roles of amino acid chelates in animal nutrition.* Park Ridge, NJ: Noyes Publications.

Ashmead, H.D. and H. Zunino. 1992. Factors which affect the intestinal absorption of minerals. In H.D. Ashmead (Ed.), *The Roles of Amino Acid Chelates in Animal Nutrition* (pp. 221-246). Park Ridge, NJ: Noyes Publications.

Baker, R.T.M., P. Martin, and S.D. Davies. 1997. Ingestion of sub-lethal levels of iron sulphate by African catfish affects growth and tissue lipid peroxidation. *Aquatic Toxicology* 40:51-61.

Beisel, W.R. 1982. Single nutrient and immunity. *American Journal of Clinical Nutrition* 35:417-468.

Berger, L.L. 1996. Trace minerals: Key to immunity. *Salt and Trace Minerals* 28: 1-4.

Bhaskaram, P. 1988. Immunology of iron-deficient subjects. In R.K. Chandr (Ed.), *Nutrition and Immunology* (pp. 149-168). New York: Alan R. Liss Inc.

Bullen, J.J., H.J. Rogers, and E. Griffiths. 1978. Role of iron in bacterial infection. *Current Topics on Microbiology and Immunology* 80:1-35.

Chvapil, M., A.L. Aronson, and Y.M. Peng. 1974. Relation between zinc and iron and peroxidation of lipids in liver homoginate in Cu EDTA-treated rat. *Experimental and Molecular Pathology* 20:216-227.

Desjardins, L.M. 1985. "The effect of iron supplementation on diet rancidity and the growth and physiological response of rainbow trout." Master's thesis, University of Guelph, Ontario, Canada.

Desjardins, L.M., B.D. Hicks, and J.W. Hilton. 1987. Iron catalyzed oxidation of trout diets and its effect on the growth and physiological response of rainbow trout. *Fish Physiology and Biochemistry* 3:173-182.

Gatlin, D.M. III and R.P. Wilson. 1986. Characterization of iron deficiency and the dietary iron requirement of fingerling channel catfish. *Aquaculture* 52:191-198.

Harper, H.A. 1973. *Review of Physiological Chemistry,* Fifteenth Edition. Los Altos, CA: Lange Medical Publications.

Ikeda, Y., H. Ozaki, and K. Uematsu. 1973. Effect of enriched diet with iron in culture of yellowtail. *Journal of the Tokyo University of Fisheries* 59:91-97.

Kaneko, J.J. 1980. Iron metabolism. In J.J. Keneko (Ed.), *Clinical Biochemistry of Domestic Animals* (pp. 649-669). New York: Academic Press.

Kawatsu, H. 1972. Studies of anemia in fish. V. Dietary iron deficient anemia in brook trout, *Salvelinus fontinalis. Bulletin of the Freshwater Fisheries Research Laboratory* 22:59-67.

Lall, S.P. 1989. The minerals. In J.E. Halver (Ed.), *Fish Nutrition,* Second Edition (pp. 219-257). New York: Academic Press.

Lall, S.P. and J.A. Hines. 1987. Iron and copper requirement of Atlantic salmon *(Salmo salar)* grown in sea water. Paper presented at the International Symposium on Feeding and Nutrition of Fish, Bergen, Norway, August 23-27, 1987.

Lee, Y.H., D.K. Layman, R.R. Bell, and H.W. Norton. 1981. Response of glutathione peroxidase and catalase to excess iron in rats. *Journal of Nutrition* 111:2195-2202.

Lim, C. and P.H. Klesius. 1997. Responses of channel catfish *(Ictalurus punctatus)* fed iron-deficient and replete diets to *Edwardsiella ictaluri* challenge. *Aquaculture* 157:83-93.

Lim, C., P.H. Klesius, M.H. Li, and E.H. Robinson. 2000. Interaction between dietary levels of iron and Vitamin C on growth, hematology, immune response and resistance of channel catfish *(Ictalurus punctatus)* to *Edwardsiella ictaluri* challenge. *Aquaculture* 185:313-327.

Lim, C., W.M. Sealey, and P.H. Klesius. 1996. Iron methionine and iron sulfate as sources of dietary iron for channel catfish *Ictalurus punctatus. Journal of the World Aquaculture Society* 27:290-296.

Maage, A. and H. Sveier. 1998. Addition of dietary iron (III) oxide does not increase iron status of growing Atlantic salmon. *Aquaculture International* 6:249-252.

Nakai, T., T. Kanno, E.R. Cruz, and K. Muroga. 1987. The effects of iron compounds on the virulence of *Vibrio anguillarum* in Japanese eels and ayu. *Fish Pathology* 22:185-189.

Nose, T. and S. Arai. 1976. Recent advances on studies on mineral nutrition of fish in Japan. In V.R. Pillay and W.A. Dill (Eds.), *Advances in Aquaculture* (pp. 584-590). Farnam, England: Fishing News.

NRC (National Research Council). 1993. *Nutrient Requirements of Fish.* Washington, DC: National Academy Press.

O'Dell, B.L. 1984. Bioavailability of trace elements. *Nutrition Reviews* 42:301-308.

Paripatananont, T. and R.T. Lovell. 1997. Comparative net absorption of chelated and inorganic trace minerals in channel catfish *Ictalurus punctatus* diets. *Journal of the World Aquaculture Society* 28:62-67.

Ravndal, J, T. Lovold, H.B. Bentsen, K.H. Roed, T. Gjedrem, and K.A. Rorvik. 1994. Serum iron levels in farmed Atlantic salmon: Family variation and associations with disease resistance. *Aquaculture* 125:37-45.

Roeder, M. and R. Roeder. 1966. Effect of iron on the growth rate of fishes. *Journal of Nutrition* 90:86-90.

Sakamoto, S. and Y. Yone. 1976. Requirement of red sea bream for dietary Fe - I. *Report of the Fisheries Research Laboratory,* Kyushu University 3:53-58.

Sakamoto, S. and Y. Yone. 1978a. Iron deficiency symptoms of carp. *Bulletin of the Japanese Society of Scientific Fisheries* 44:1157-1160.

Sakamoto, S. and Y. Yone. 1978b. Requirement of red sea bream for dietary iron II. *Bulletin of the Japanese Society of Scientific Fisheries* 44:223-225.

Sakamoto, S. and Y. Yone. 1979. Availabilities of three iron compounds as dietary iron sources for red sea bream. *Bulletin of the Japanese Society of Scientific Fisheries* 45:231-235.

Sealey, W.M., C. Lim, and P.H. Klesius. 1997. Influence of dietary level of iron from iron methionine and iron sulfate on immune response and resistance of channel catfish to *Edwarsiella ictaluri*. *Journal of the World Aquaculture Society* 28:142-149.

Sherman, A.R. 1992. Zinc, copper and iron nutriture and immunity. *Journal of Nutrition* 122:604-609.

Sugiura, S.H., F.M. Dong, C.K. Rathbone, and R.W. Hardy. 1998. Apparent protein digestibility and mineral availabilities in various feed ingredients for salmonid feeds. *Aquaculture* 159:177-202.

Underwood, E.J. 1977. *Trace Element in Human and Animal Nutrition,* Fourth Edition. New York: Academic Press.

Walker, R.L. and P.O. Fromm. 1976. Metabolism of iron by normal and iron deficient rainbow trout. *Comparative Biochemistry and Physiology* 55A:311-318.

Weinberg, E.D. 1974. Iron and susceptibility to infectious disease. *Science* 184: 952-958.

Chapter 10

The Role of Dietary Phosphorus, Zinc, and Selenium in Fish Health

Chhorn Lim
Phillip H. Klesius
Carl D. Webster

INTRODUCTION

Inorganic elements or minerals constitute a relatively small amount of the total body tissues. However, they are essential for normal life processes of all animals, including fish. Fish require minerals in their diets, although they can absorb several mineral elements from the surrounding water to meet part of their metabolic requirements. Seven macrominerals (calcium, chlorine, magnesium, phosphorus, potassium, sodium, and sulfur) and 16 trace minerals (aluminum, arsenic, chromium, cobalt, copper, fluorine, iodine, iron, manganese, molybdenum, nickel, selenium, silicon, tin, vanadium, and zinc) have been shown as essential in one or more animal species (Davis and Gatlin 1996). However, not all of the minerals essential for warm-blooded animals have been found to be essential in the diet for fish. The physiological functions of minerals are well defined for humans and some terrestrial animals, but this information for fish has not been well established.

Early research on mineral requirements of fish was primarily aimed at determining the optimum dietary levels necessary for good growth and prevention of deficiency signs. However, evidence from unintentional or accidental infection of fish in nutrition studies seems to indicate that most, if not all, dietary nutrients influence the immune response and/or disease resistance of fish. In the past few decades the understanding of the interrelationships between nutrition, immunity, and disease resistance in terrestrial animals has progressed rapidly. However, information on the effect of nutrition, particularly minerals, on immune function and disease resistance in fish is poorly understood. Among the minerals identified as dietary essentials, only iron, phosphorous, zinc, and selenium have been studied regarding their interactions with fish health. Iron was thoroughly discussed in the previous chapter, so it will not be included here. Thus, this chapter provides a brief overview on the effects of phosphorus, zinc, and selenium on immune responses and disease resistance of fish. Information on the requirements for, and deficiency signs of, these minerals is also included.

PHOSPHORUS

Fish require phosphorus for proper bone mineralization. Fish can absorb minerals from the water where they live, but because phosphorus is usually a limiting mineral in most natural waters, and due to its low rate of absorption from the water, fish need a supplemental dietary source of phosphorus. However, excess phosphorus increases phytoplankton growth, aquatic weed growth, and the production of algae blooms, which can cause wide fluctuations in the amount of dissolved oxygen in the water and an off-flavor in fish. Since phosphorus is an important component for the eutrophication of water, the amount of phosphorus present in the culture water is of concern, especially if discharged into surrounding waters. Thus, fish phosphorus requirements must be met without adding excess phosphorus to the water.

Requirement and Deficiency Signs

Rodehutscord and Pfeffer (1995) reported that juvenile (50 to 200 g) rainbow trout, *Oncorhynchus mykiss,* required 5 g (or 0.5 percent) available phosphorus/kg of diet. For channel catfish, *Ictalurus punctatus,* the available phosphorus requirement is 0.4 percent of the diet (NRC 1993). This requirement value was based on research using small (1 to 6 g) fish. Eya and Lovell (1998) fed small (1.8 g) channel catfish an egg white-based diet with various percentages (0.0, 0.05, 0.10, 0.25, 0.40, 0.55, and 0.85) of available phosphorus from monosodium phosphate and reported that the minimum phosphorus requirement for maximum weight gain was achieved when fishwere fed 0.42 percent available phosphorus. However, Eya and Lovell (1997a) reported that when larger (60 g) channel catfish were fed an all-plant, commercial-type diet containing 0.2 percent available phosphorus, this was sufficient for maximum weight gain when fish were grown in ponds. Robinson, Jackson, and Li (1996) reported that channel catfish fed a diet containing 0.27 percent available phosphorus showed good growth.

The net absorption of phosphorus by fish from different inorganic sources varies widely. More water-soluble sources, such as monosodium, monoammonium, and monocalcium phosphate, are highly available to channel catfish; however, dicalcium phosphate is less available than monocalcium phosphate. Pure monocalcium phosphate is 68 percent soluble in water, whereas pure dicalcium phosphate is only 38 percent soluble in water (FPC 1966). Tricalcium phosphate and coursely ground defluorinated rock phosphate are less available. Eya and Lovell (1997b) reported net absorption of phosphorus by channel catfish from such various sources as monosodium phosphate, 89 percent; monoammonium phosphate, 85 percent; finely ground defluorinated rock phosphate, 82 percent; monocalcium phosphate, 81 percent; dicalcium phosphate, 75 percent; coursely ground defluorinated rock phosphate, 55 percent; and tricalcium phosphate, 55 percent.

Phosphorus deficiency signs observed in channel catfish include reduced bone mineralization, decreased bone strength, and reduced weight gain (Lovell 1978; Wilson et al. 1982; Robinson, Jackson, and Li 1996; Eya and Lovell 1997a).

Immune Response and Disease Resistance

Although there are some reports in the literature on the effects of phosphorus on immune responses in guinea pigs, dogs, and chicks, little has been published regarding the role of phosphorus in immune function in fish. When channel catfish were challenged with a virulent strain of *Edwardsiella ictaluri,* no deaths occurred among fish fed 0.40 or 0.85 mg phosphorus/kg of diet (Eya and Lovell 1998). However, fish fed diets with 0.0, 0.05, or 0.10 mg phosphorus/kg of diet had 59 percent mortality. The break point in the mortality regression curve was determined to be 0.4 percent available dietary phosphorus. Antibody production in surviving fish indicated that a break point in the response curve was at 0.5 percent available dietary phosphorus.

The mechanisms by which dietary phosphorus affects immune responses are not well understood. Lehninger, Nelson, and Cox (1993) stated that since the synthesis of antibody protein requires the energy of two pyrophosphate bonds of adenosine triphosphate (ATP) per amide bond formed, the energy required for antibody protein production may be decreased if phosphorus is not present. Craddock and colleagues (1974) reported that an adequate supply of phosphorus was necessary for phagocytic activity of white blood cells in dogs because it provided the phosphorus used in the regeneration of ATP in the phagocytic cells.

ZINC

Zinc is required for normal growth, development, and function of all animal species. The primary functions of zinc are based on its roles as a cofactor in several enzyme systems and as a component of a large number of metalloenzymes, including carbonic anhydrase, alkaline phosphatase, carboxypeptidase, alcohol dehydrogenase, glutamic dehydrogenase, lactate dehydrogenase, ribonuclease, and DNA polymerase (NRC 1980). Fish can absorb zinc from both water and dietary sources. However, dietary zinc is more efficiently absorbed than waterborne zinc (NRC 1993).

Requirement and Deficiency Signs

The dietary zinc requirement has been determined for several fish species. The requirement values reported are 15 to 30 mg zinc/kg diet for rainbow trout and common carp, *Cyprinus carpio* (Ogino and Yang 1978, 1979), and 20 mg zinc/kg diet for channel catfish (Gatlin and Wilson 1983), blue tilapia, *Tilapia aurea* (McClain and Gatlin 1988), and red drum, *Sciaenops ocellatus* (Gatlin, O'Connell, and Scarpa 1991). In rainbow trout, zinc deficiency caused growth depression, lens cataracts, and short body dwarfism (Ogino and Yang 1978, Ketola, 1979; Satoh, Takeuchi, and Watanabe 1987). Poor growth, loss of appetite, high mortality and erosion of skin and fins were reported in zinc-deficient common carp (Ogino and Yang 1979). Zinc deficiency signs observed in red drum were slow growth, poor feed efficiency and survival, and reduced bone and scale zinc concentrations (Gatlin, O'Connell, and Scarpa 1991). Channel catfish fed zinc-deficient diets had reductions in weight gain, appetite, survival, serum zinc content, serum alkaline phosphatase activity, and bone zinc and calcium levels (Gatlin and Wilson, 1983; Scarpa and Gatlin, 1992).

The bioavailability of dietary zinc is affected by dietary levels of calcium, phosphorus and phytic acid, protein source, and form of zinc. Phytate forms a complex with transitional cations such as zinc, iron, and manganese in the gastrointestinal tract and prevents their absorption. Calcium promotes the complexing of zinc to phytates (NRC 1993). The channel catfish zinc requirement for maximum growth and maintenance of high levels of serum and bone zinc determined that soybean meal-based diets that are relatively high in phytate was 150 mg/kg as compared to 20 mg/kg when egg white-based purified diets were used (Gatlin and Wilson 1983, 1984b). McClain and Gatlin (1988) reported that blue tilapia had a zinc requirement of 20 mg Zn/kg of an egg white-based diet; however, when 1.5 percent phytate was added to the diet, there was a significant reduction in the amount of zinc measured in both scale and bone tissues. The phytate appeared to render dietary zinc unavailable, and tissue stores were being utilized. However, overt zinc deficiency signs, such as reduced growth rate, were not observed.

The availability of zinc in white fish meal-based diets, which are rich in tricalcium phosphate, to rainbow trout was very low (Satoh, Takeuchi, and Watanabe 1987). These authors suggested that supplementation of zinc at more than 40 mg/kg to the white fish meal-based diets containing 38.9 mg Zn/kg is necessary for optimum growth and prevention of dwarfism and cataracts in rainbow trout. In an earlier study, Ketola (1979) found that a white fish meal diet containing 60 mg Zn/kg was insufficient to prevent poor growth and cataracts in rainbow trout. The severity of cataracts was increased by adding extra phosphorus, calcium, sodium, and potassium to the diet. However, supplementation of Na_2EDTA at 1 percent of the diet or 150 mg Zn/kg diet overcame these problems. Gatlin, Phillips, and Torrans

(1989) reported that supplementation of high levels of zinc in practical diets for channel catfish did not adversely affect the copper status of channel catfish. Spinelli, Houle, and Wekell (1983) reported that the availability of zinc in purified diets to rainbow trout was reduced when levels of calcium and magnesium were increased.

Satoh, Takeuchi, and Watanabe (1987) showed that the availability of zinc to rainbow trout was highest in zinc sulfate, lowest in zinc chloride, and intermediate in zinc nitrate or carbonate. Paripatananont and Lovell (1995a) reported that the bioavailability of zinc from zinc methionine was greater than that of zinc sulfate for channel catfish. They reported that the bioavailability of zinc methionine for growth and for maximum bone zinc deposition was 305 percent (for bone zinc deposition) to 352 percent (for weight gain) and 482 percent (for weight gain) to 586 percent (for bone zinc deposition) of the bioavailability of zinc sulfate in egg white and in soybean diets, respectively. Feeding zinc amino acid chelate to rainbow trout resulted in greater zinc deposition in body tissue than zinc sulfate in a low calcium-phosphorus diet, but not in high calcium-phosphorus diets (Hardy and Shearer 1992). It has also been shown that the net absorption of zinc from zinc proteinate as compared to zinc sulfate was 138 percent in the purified diets and 174 percent in the practical diets (Paripatananont and Lovell 1997). Li and Robinson (1996), however, showed that the bioavailability of zinc sulfate and zinc methionine to channel catfish was similar when fish were fed a typical practical diet.

Immune Response and Disease Resistance

Studies in warm-blooded animals have suggested a relationship between zinc deficiency, poor immune response, and susceptibility to infectious disease. However, evidence on the role of zinc in fish immunity and disease resistance is not consistent. Lim, Klesius, and Duncan (1996) showed that supplementation of zinc enhanced chemotactic response of channel catfish peritoneal macrophages to *E. ictaluri* exoantigen and that zinc methionine was more effective than zinc sulfate in stimulating macrophage chemotaxis. A level of 60 mg Zn/kg diet as zinc sulfate was required to attain a chemotactic response similar to that obtained with 5 mg/kg as zinc methionine. However, the phagocytic activity of phagocytes to zymosan determined by chemiluminescence assay was suppressed by supplementation of zinc in the diets. Thus, although zinc has been found to increase macrophage chemotaxis, it may have an inhibitory effect on phagocytosis. Karl, Chvapil, and Zukoski (1973) reported that zinc significantly inhibited the phagocytic capacity of macrophages isolated from mice treated with low (0.05 percent mg $ZnCl_2$/mouse) or high (0.25 mg $ZnCl_2$/mouse) doses of zinc. They also showed that the rate at which macrophages phagocytosed bacteria was significantly slower after zinc treatment.

The serum immunoglobulin M levels of nonimmunized channel catfish was not affected by dietary zinc or calcium (Scarpa and Gatlin 1992b). Bell, Higgs, and Traxler (1984) reported no differences in the serum agglutinating antibody titer of immunized (with formalin-killed *Aeromonas salmonicida*) sockeye salmon fed diets deficient in zinc or manganese. In a more recent study, however, Paripatananont and Lovell (1995b) found that low dietary zinc significantly reduced the agglutinating antibody response of channel catfish 14 days after challenge with *E. ictaluri.* Maximum antibody titer was obtained with fish fed diets containing 15 mg Zn/kg as zinc methionine or at least 30 mg Zn/kg as zinc sulfate.

Paripatananont and Lovell (1995b) found that dietary zinc influenced the resistance of channel catfish challenged with *E. ictaluri* and that zinc methionine was three to four times more potent than zinc sulfate in protecting channel catfish against this bacterium. Lim, Klesius, and Duncan (1996), however, observed that dietary zinc did not protect channel catfish against mortality from *E. ictaluri.* The intensity of *E. ictaluri* infection, based on the number of colony-forming units/g of trunk kidney, of fish three days postchallenge was not affected by dietary zinc. Also, there was no evident trend of the effect of dietary zinc on the percentage of fish infected with *E. ictaluri,* 15 days postchallenge. Scarpa and Gatlin (1992) obtained resistance of channel catfish to *Aeromonas hydrophila* challenge in nonimmunized fish fed zinc-deficient and calcium-excessive diets, and susceptibility was observed in fish fed the replete, calcium-deficient and zinc-excessive diets. With sockeye salmon, no evidence was found that ascorbic acid and/or zinc promoted resistance against bacterial kidney disease (Bell, Higgs, and Traxler 1984).

SELENIUM

Selenium has been found to be an essential trace element for all animals studied, including fish. Selenium is a component of the enzyme glutathione peroxidase (Rotruck et al. 1973). This enzyme catalyzes reactions necessary for the conversion of hydrogen peroxide and fatty acid hydroperoxides into water and fatty acid alcohols by using reduced glutathione, thereby protecting cell membranes against oxidative damage (NRC 1993). Glutathione peroxidase acts along with vitamin E as a biological antioxidant to protect polyunsaturated phospholipids in cellular and subcellular membranes from peroxidative damage (Lovell 1989). Harper (1973) indicated that tissues or cellular components that are inherently low in glutathione peroxidase would not be affected by selenium but still would be protected by vitamin E, which acts as an antioxidant by a mechanism not involving glutathione peroxidase. Selenium also exerts the protective effects against the toxicity of heavy metals such as cadmium and mercury (Lall 1989).

Fish can absorb waterborne selenium across the gills and digestive tract (Evans 1993). The uptake of selenium as selenite across the gills is very efficient even at low waterborne concentrations. It has been reported that the level of waterborne selenium affected the dietary selenium requirement of rainbow trout (Hodson and Hilton 1983).

Requirement and Deficiency Signs

Selenium has been found to be required in the diet of Atlantic salmon, *Salmo salar;* rainbow trout; and channel catfish. The selenium requirement of fish varies with the source of selenium ingested, polyunsaturated fatty acid and vitamin E content of the diet, and the concentration of waterborne selenium (NRC 1993). The dietary selenium requirements, determined with sodium selenite, for maximum growth and glutathione peroxidase activity were 0.15 to 0.38 mg/kg diet for rainbow trout (Hilton, Hodson, and Slinger 1980) and 0.25 mg/kg for channel catfish (Gatlin and Wilson 1984a). Using the broken-line analysis, Wang and Lovell (1997) showed that minimum dietary requirements of sodium selenite, selenomethionine, and selenoyeast for weight gain of channel catfish were 0.28, 0.12, and 0.11 mg selenium/kg diet, respectively, and for liver glutathione peroxidase activity, they were 0.17, 0.12, and 0.12 mg selenium/kg diet, respectively. Data on the dietary selenium requirement of Atlantic salmon have not yet been established. However, Atlantic salmon fed selenium-deficient diets had slow growth, low glutathione peroxidase activity, reduced hematocrit, and ataxia (Poston, Combs, Leibovitz 1976; Bell et al. 1986, 1987). Rainbow trout fed a selenium-deficient diet showed reduction in growth rate, feed efficiency, and glutathione peroxidase activity (Hilton, Hodson, and Slinger 1980). In channel catfish, dietary selenium deficiency resulted in decreased growth, poor feed efficiency, and reduced glutathione peroxidase activity (Gatlin and Wilson 1984b).

Few studies have been conducted to determine the bioavailability of selenium in various selenium-containing compounds for fish. Selenium from sodium selenite or selenomethionine at levels of 1 or 2 mg/kg diet was equally effective in promoting growth and maintaining hepatic glutathione peroxidase activity of Atlantic salmon. However, liver selenium content of fish fed sodium selenite was higher than in those fed selenomethionine, whereas muscle selenium content was higher for fish fed selenomethionine than for those fed sodium selenite (Lorentzen and Julshamn 1994). The digestibility coefficients of selenium from selenomethionine, selenocystine, sodium selenite, and fish meal by Atlantic salmon have been reported to be 91.6, 52.6, 63.9, and 46.6 percent, respectively (Bell and Cowey 1989). These authors reported that the source of dietary selenium had no influence on glutathione peroxidase activities in liver and plasma, although the plasma selenium concentration was highest for fish fed selenomethionine diet. However, based on the ratio of the serum glutathione peroxidase activity to serum selenium concentration, selenocystine or sodium selenite was a better source of selenium than selenomethionine or fish meal.

The absorption of selenium in purified diets by channel catfish was 90.8 percent for selenomethionine and 62.8 percent for sodium selenite. The type of diet (purified or practical), however, had no effect on the absorption of selenium from chelated and inorganic sources (Paripatananont and Lovell 1997). Wang and Lovell (1997) compared the bioavailability of selenium from selenomethionine, selenoyeast, and sodium selenite based on weight gain, glutathione peroxidase activity, and tissue selenium content in channel catfish. They found that the relative bioavailability values of selenomethionine and selenoyeast compared to sodium selenite were 336 and 269 percent for growth, 147 and 149 percent for liver glutathione peroxidase activity, 197 and 184 percent for liver selenium, and 478 and 453 percent for muscle selenium, respectively.

Immune Response and Disease Resistance

Selenium, because of its roles in protecting cells and cell membranes against oxidative damage, plays an important role in maintaining normal immune response in terrestrial animals. However, limited information is available on the effect of dietary selenium on immune response and disease resistance in fish. Wise et al. (1993) evaluated extracellular and intracellular superoxide anion production of kidney macrophages from channel catfish fed diets containing various levels of vitamin E and selenium from sodium selenite. They reported that extracellular superoxide anion production was not affected by dietary treatments. Intracellular superoxide anion production, however, was higher for the fish fed four times the normal growth requirements of selenium and vitamin E, but not in fish fed the normal levels required for growth (0.2 mg selenium/kg and 60 mg vitamin E/kg). In a recent study with channel catfish, Wang, Lovell, and Klesius (1997) showed that serum agglutinating antibody titer to *E. ictaluri* and macrophage chemotaxis in response to *Escherichia coli* were responsive to dietary concentrations and sources of selenium. Antibody titer generally increased as dietary concentration of selenium increased, but the value was highest for fish fed selenoyeast, intermediate for fish fed selenomethionine, and lowest for fish fed sodium selenite. Macrophage chemotactic response in the presence of *E. coli* was similar for fish fed the control (0 selenium) and sodium selenite (0.4 mg selenium/kg) diets and was significantly lower than that of fish fed 0.4 mg selenium/kg diet from selenoyeast or selenomethionine.

Chinook salmon, *Oncorhynchus tshawytscha,* subclinically infected with *Renibacterium salmoninarum,* the etiological agent of bacterial kidney disease, and reared in seawater are sensitive to selenium and vitamin E deficiency. Mortality significantly increased among fish fed the selenium and vitamin E unsupplemented diet (Thorarinsson et al. 1994). Wang, Lovell, and Klesius (1997) showed that channel catfish challenged with *E. ictaluri* were also sensitive to selenium deficiency and that the source of selenium also affected the rate of mortality. Mortality significantly decreased when selenium was increased to meet the growth requirement. At this supplemental level, fish fed the selenomethionine diet exhibited significantly lower mortality than fish fed sodium selenite. Mortality value was intermediate for fish fed selenoyeast diet. These researchers suggested that dietary selenium concentration for maximum survival from *E. ictaluri* challenge was 0.20 mg/kg for fish fed selenomethionine and 0.40 mg/kg for fish fed selenoyeast and sodium selenite.

CONCLUSION

Data on the mineral requirements of fish for optimum growth and prevention of deficiency signs have accumulated rapidly, but information on the role of minerals in immune response and disease resistance is scanty, conflicting, and confined to only iron, phosphorus, zinc, and selenium. In the absence of better-defined information, however, it is reasonable to assume that adequate levels of these minerals as well as other essential dietary nutrients to meet the normal growth requirements are necessary for maintaining fish health. Deficiencies or excesses of any nutrients may have profound effects on infectious diseases and the survival of fish, largely through their effects on host defense mechanisms and the virulence of pathogens. Other factors such as composition of the diet, bioavailability of the nutrient, nutrient interactions, feeding management, duration of the experiment, environmental parameters, species, size or age, and genetic variation also influence the health of fish. Further research is needed to elucidate the roles of dietary nutrients in immune processes, identify nutrients capable of, and determine optimum levels necessary for enhancing immune response and disease resistance. It is hoped that future commercial aquaculture feeds will be formulated not only for optimum growth and feed efficiency but also for improvement of fish health.

REFERENCES

Bell, G.R., D.A. Higgs, and G.S. Traxler. 1984. The effect of dietary ascorbate, zinc, and manganese on the development of experimentally induced bacterial kidney disease in sockeye salmon *(Oncorhynchus nerka). Aquaculture* 36:293-311.

Bell, J.G. and C.B. Cowey. 1989. Digestibility and bioavailability of dietary selenium from fishmeal, selenite, selenomethionine and selenocystine in Atlantic salmon *(Salmo salar). Aquaculture* 81:61-68.

Bell, J.G., C.B. Cowey, J.W. Adron, and B.J.S. Pirie. 1987. Some effects of selenium deficiency on enzyme activities and indices of tissue peroxidation in Atlantic salmon parr *(Salmo salar). Aquaculture* 65:43-54.

Bell, J.G., B.J.S. Pirie, J.W. Adron, and C.B. Cowey. 1986. Some effects of selenium deficiency on glutathione peroxidase (EC 1.11.1.9) activity and tissue pathology in rainbow trout *(Salmo gairdneri). British Journal of Nutrition* 55:305-311.

Craddock, P.R., Y. Yawata, L. Van Saten, S. Gilberstadt, and H. Jacob. 1974. Acquired phagocytic dysfunction: A complication of hypophosphatemia of parenteral hyperalimentation. *New England Journal of Medicine* 290:1403-1407.

Davis, D.A. and D.M. Gatlin, III. 1996. Dietary mineral requirements of fish and marine crustaceans. *Reviews in Fisheries Science* 4:75-99.

Evans, D.H., 1993. *The Physiology of Fishes.* Boca Raton, FL: CRC Press.

Eya, J.C. and R.T. Lovell. 1997a. Available phosphorus requirements of food-size channel catfish *(Ictalurus punctatus)* fed practical diets in ponds. *Aquaculture* 154:283-291.

Eya, J.C. and R.T. Lovell. 1997b. Net absorption of dietary phosphorus from various inorganic sources and effect of fungal phytase on net absorption of plant phosphorus by channel catfish *Ictalurus punctatus. Journal of the World Aquaculture Society* 28:386-391.

Eya, J.C. and R.T. Lovell. 1998. Effects of dietary phosphorus on resistance of channel catfish to *Edwardsiella ictaluri* challenge. *Journal of Aquatic Animal Health* 10:28-34.

FPC (Food Protection Committee). 1966. *Food chemicals codex.* Publication 1406. Washington, DC: National Academy of Sciences, National Research Council.

Gatlin, D.M. III, J.P. O'Connell, and J. Scarpa. 1991. Dietary zinc requirement of red drum, *Sciaenops ocellatus. Aquaculture* 92:259-265.

Gatlin, D.M. III, H.F. Phillips, and E.L. Torrans. 1989. Effects of various levels of dietary copper and zinc on channel catfish. *Aquaculture* 76:127-134.

Gatlin, D.M. III. and R.P. Wilson. 1983. Dietary zinc requirement of fingerling channel catfish. *Journal of Nutrition* 113:630-635.

Gatlin, D.M. III and R.P. Wilson. 1984a. Dietary selenium requirement of fingerling channel catfish. *Journal of Nutrition* 114:627-633.

Gatlin, D.M. III and R.P. Wilson. 1984b. Zinc supplementation of practical channel catfish diets. *Aquaculture* 41:31-36.

Hardy, R.W. and K.D. Shearer. 1992. The use of zinc amino acid chelates in high calcium and phosphorus diets of rainbow trout. In H.D. Ashmead (Ed.), *The Roles of Amino Acid Chelates in Animal Nutrition* (pp. 424-439). NJ: Noyes Publication.

Harper, H.A. 1973. *Review of Physiological Chemistry,* Fifteenth Edition. Los Altos, CA: Lange Medical Publicatioms.

Hilton, W., P.V. Hodson, and S.J. Slinger. 1980. The requirement and toxicity of selenium in rainbow trout *(Salmo gairdneri). Journal of Nutrition* 110:2527-2535.

Hodson, P.V. and S.J. Hilton. 1983. The nutritional requirements and toxicity to fish of dietary and waterborne selenium. *Environmental and Biogeochemical Ecology* 35:335-340.

Karl, L., M. Chvapil, and C.F. Zukoski. 1973. Effect of zinc on the viability and phagocytic capacity of peritoneal macrophages. *Proceedings of the Society of Experimental Biological Medicine* 142:1123-1127.

Ketola, H.G. 1979. Influence of dietary zinc on cataracts in rainbow trout *(Salmo gairdneri). Journal of Nutrition* 109:965-969.

Lall, S.P. 1989. The minerals. In J.E. Halver (Ed.), *Fish Nutrition* (pp. 219-257). Second Edition. New York: Academic Press.

Lehninger, A.L., D.L. Nelson, and M.M. Cox. 1993. *Principles of Biochemistry.* Irving Place, NY: Worth Publishers.

Li, M.H. and E.H. Robinson. 1996. Comparison of chelated zinc and zinc sulfate as zinc sources for growth and bone mineralization of channel catfish *(Ictalurus punctatus)* fed practical diets. *Aquaculture* 146:237-243.

Lim, C., P.H. Klesius, and P.L. Duncan. 1996. Immune response and resistance to *Edwardsiella ictaluri* challenge when fed various dietary levels of zinc methionine and zinc sulfate. *Journal of Aquatic Animal Health* 8:302-307.

Lorentzen, A.M. and K. Julshamn. 1994. Effects of dietary selenite or seleno-methionine on tissue selenium levels of Atlantic salmon *(Salmo salar). Aquaculture* 121:359-367.

Lovell, R.T. 1978. Dietary phosphorus requirements of channel catfish. *Transactions of the American Fisheries Society* 107:617-621.

Lovell, R.T. 1989. *Nutrition and Feeding of Fish.* New York: Van Nostrand Reinhold.

McClain, W.R. and D.M. Gatlin, III. 1988. Dietary zinc requirement of *Oreochromis aureus* and effects of dietary calcium and phytate on zinc bioavailability. *Journal of the World Aquaculture Society* 19:103-108.

NRC (National Research Council). 1980. *Mineral Tolerance of Domestic Animals.* Washington, DC: National Academy Press.

NRC (National Research Council). 1993. *Nutrient Requirements of Fish.* Washington, DC: National Academy Press.

Ogino, C. and G.Y. Yang. 1978. Requirement of rainbow trout for dietary zinc. *Bulletin of the Japanese Society of Scientific Fisheries* 44:1015-1018.

Ogino, C. and G.Y. Yang. 1979. Requirement of carp for dietary zinc. *Bulletin of the Japanese Society of Scientific Fisheries* 45:967-969.

Paripatananont, T. and R.T. Lovell. 1995a. Chelated zinc reduces the dietary zinc requirement of channel catfish, *Ictalurus punctatus. Aquaculture* 133:73-82.

Paripatananont, T. and R.T. Lovell. 1995b. Responses of channel catfish fed organic and inorganic sources of zinc to *Edwardsiella ictaluri* challenge. *Journal of Aquatic Animal Health* 7:147-154.

Paripatananont, T. and R.T. Lovell. 1997. Comparative net absorption of chelated and inorganic trace minerals in channel catfish *Ictalurus punctatus* diets. *Journal of the World Aquaculture Society* 28:62-67.

Poston, H.A., G.F. Combs Jr., and L. Leibovitz. 1976. Vitamin E and selenium interrelation in the diet of Atlantic salmon *(Salmo salar):* Gross, histological and biochemical deficiency signs. *Journal of Nutrition* 106:892-904.

Robinson, E.H., L.S. Jackson, and M.H. Li. 1996. Supplemental phosphorus in practical channel catfish diets. *Journal of the World Aquaculture Society* 27:303-308.

Rodehutscord, M. and E. Pfeffer. 1995. Requirement for phosphorus in rainbow trout *(Oncorhynchus mykiss)* growing from 50 to 200 g. *Water Science and Technology* 31(10):137-141.

Rotruck, J.T., A.L. Pope, H.E. Ganther, A.B. Swanson, D.G. Hafeman, and W.G. Hoekstra. 1973. Selenium: Biochemical role as a component of glutathione peroxidase. *Science* 179:588-590.

Satoh, S., T. Takeuchi, and T. Watanabe. 1987. Availability to rainbow trout of zinc in white fish meal and of various zinc compounds. *Nippon Suisan Gakkaishi* 53: 595-599.

Scarpa, J. and D.M. Gatlin III. 1992. Effects of dietary zinc and calcium on select immune functions of channel catfish. *Journal of Aquatic Animal Health* 4:24-31.

Spinelli, J., C.R. Houle, and J.C. Wekell. 1983. The effects of phytates on the growth of rainbow trout *(Salmo gairdneri)* fed pure diets containing varying quantities of calcium and magnesium. *Aquaculture* 30:71-83.

Thorarinsson, R., M.L. Landolt, D.G. Elliott, R.J. Pascho, and R.W. Hardy. 1994. Effect of dietary vitamin E and selenium on growth, survival and the prevalence of *Renibacterium salmoninarum* infection in chinook salmon *(Oncorhynchus tshawytscha). Aquaculture* 121:343-358.

Wang, C. and R.T. Lovell. 1997. Organic selenium sources, selenomethionine and selenoyeast, have higher bioavailabililty than an inorganic selenium source, sodium selenite, in diets for channel catfish *(Ictalurus punctatus). Aquaculture* 152:223-234.

Wang, C., R.T. Lovell, and P.H. Klesius. 1997. Response to *Edwardsiella ictaluri* challenge by channel catfish fed organic and inorganic sources of selenium. *Journal of Aquatic Animal Health* 9:172-179.

Wilson, R.P., E.H. Robinson, D.M. Gatlin III, and W.E. Poe. 1982. Dietary phosphorus requirement of channel catfish. *Journal of Nutrition* 112:1197-1202.

Wise, D.J., J.R. Tomasso, D.M. Gatlin III, S.C. Bai, and V.S. Blazer. 1993. Effects of dietary selenium and vitamin E on red blood cell peroxidation, glutathione peroxidase activity and macrophage superoxide anion production in channel catfish. *Journal of Aquatic Animal Health* 5:177-182.

Chapter 11

Influence of Dietary Lipid Composition on the Immune System and Disease Resistance of Finfish

Shannon K. Balfry
David A. Higgs

INTRODUCTION

Lipids and their constituent fatty acids, together with the metabolic derivatives of some of the latter, termed eicosanoids, and other associated compounds, play essential and dynamic roles in the maintenance of optimum growth, feed efficiency, health (immunocompetence and cardiovascular function), kidney and gill function, neural and visual development, reproduction, and flesh quality (market size) of finfish species (reviewed by Higgs and Dong 2000).

In this chapter, we have restricted our focus to our present state of knowledge on the effects of dietary fatty acids on the immune response and disease resistance of various commercially important finfish species. However, before reviewing this information, we provide a brief overview of the general types of lipids and families of fatty acids that are of nutritional significance and a discussion of the types of metabolic derivatives that are elaborated from some of the key fatty acids especially important from a nutritional standpoint. Also, we briefly consider the need to partially replace marine lipid sources with alternate lipids of plant and animal origin in diets for finfish, and the possible consequences of this for fish health. Finally, we give a brief synopsis of the basic cellular and other mechanisms that comprise the immune system of finfish, so that the reader can more fully appreciate the findings of studies that have explored the interrelationship of dietary lipid composition with immune response and disease resistance. Special emphasis has been placed in this regard on how the types and levels of eicosanoid compounds can influence the immune response.

Lipids, Fatty Acids, and Eicosanoids

Lipids refer to compounds that are relatively insoluble in water but are soluble in organic solvents such as chloroform, ether, hexane, and benzene. The many types of lipids are frequently differentiated according to their polarity. In this regard, some lipids, such as triacylglycerols, wax esters, and sterol esters, are insoluble in water and are therefore called nonpolar lipids. Other lipids, such as phosphoglycerides, have varying degrees of water solubility and are termed polar lipids. These are essential components of biological membranes, and they influence their physical and functional properties.

The fatty acids within nonpolar and polar lipids generally contain a single carboxyl group and a straight unbranched carbon chain. The carbon chain may have no double bond, one double bond, or two or more double bonds, in which case the fatty acid is referred to as saturated, monounsaturated, or polyunsaturated (PUFA), respectively. The unsaturated fatty acids present in fish prey, formulated diets, and body lipids can be divided further into three major families or series, namely the oleic (n-9), the linoleic (n-6), and the linolenic (n-3). The latter two families of PUFAs have the greatest nutritional significance and are depicted in Figure 11.1. At this point, some explanation of the shorthand abbreviations for fatty acids should be provided. In Figure 11.1, for example, linolenic acid, the parent acid of the n-3 family, is abbreviated as 18:3n-3. This signifies the number of carbon atoms (18), the number of double bonds (3), and the position of the first double bond counting from the terminal methyl (CH3) group carbon to the carbon atom of the first double bond.

Finfish species are similar to other vertebrates and cannot synthesize either linoleic acid (18:n-6), the parent acid of the n-6 family, or linolenic acid (Henderson and Tocher 1987). Consequently, these fatty acids or their highly unsaturated metabolic derivatives must be of dietary origin. Also, depending upon the finfish species, the parent acid of the n-3 series or proper levels and proportions of some of the highly unsaturated members of this family alone (e.g., eicosapentaenoic acid and docosahexaenoic acid) or together with counterparts of the n-6 series of fatty acids (e.g., linoleic acid and arachidonic acid) are considered to be essential for normal growth, food utilization, health, and reproductive viability (refer to Higgs and Dong 2000 for information on the lipid and fatty acid requirements of commercially important finfish species). Further, it should be mentioned that the members of each of the families of fatty acids are created from their respective parent acids by a common enzyme system of alternating desaturases and elongases that yield a series of fatty acids of increasing unsaturation and length. It is also noteworthy that the members of one family are not interconvertible to those of another. The respective highly unsaturated fatty acids (HUFAs) of the n-9, n-6, and n-3 families of nutritional significance are eicosatrienoic acid (20:3n-9), dihomo-γ-linolenic acid (20:3n-6), arachidonic acid (20:4n-6; AA), eicosapentaenoic acid (20:5n-3; EPA), and docosahexaenoic acid

FIGURE 11.1 Probable Pathways Involved in the Desaturation and Elongation of N-6 and N-3 Series of Fatty Acids in Freshwater Fish

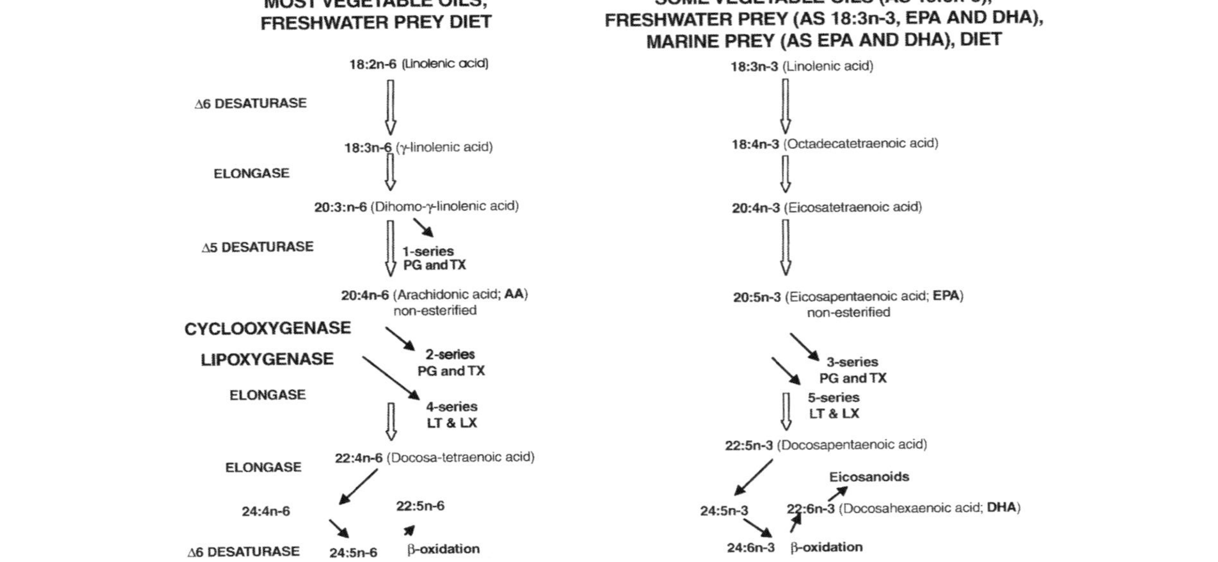

Notes: The parent acids of each family together with their respective nutritionally important highly unsaturated fatty acids and series of cyclooxygenase-derived (prostaglandins, PG, and thromboxanes, TX) and lipoxygenase-derived (leukotrienes, LT, and lipoxins, LX) compounds, collectively termed eicosanoids, are indicated (adapted from Higgs and Dong 2000). The production of 22:5n-6 and 22:6n-3 is thought to occur as illustrated rather than through Δ4 desaturation of 22:4n-6 to yield 22:5n-6 and 22:5n-3 to produce 22:6n-3 (refer to Buzzi et al. 1997; Calder 1997). There is some evidence for eicosanoid formation from DHA, for example, 14-hydroxydocosahexaenoic acid, through the action of 12-lipoxygenase (German et al. 1986).

(22:6n-3; DHA). The latter four fatty acids are progenitors of a series of compounds collectively called eicosanoids. For instance, nonesterified AA, through the action of cyclooxygenase enzymes, yields 2-series prostanoids (prostaglandins and thromboxanes) and, through the action of lipoxygenase enzymes, 4-series leukotrienes and lipoxins. Alternatively, the metabolic derivatives stemming from nonesterified EPA are 3-series prostanoids and 5-series leukotrienes and lipoxins. Collectively, these compounds play essential roles in the regulation of many physiological processes in the body and those that relate to immune response will be considered subsequently. EPA and DHA together are frequently referred to as n-3 HUFAs, but this term can also include other C20 members of the n-3 family.

Lipid Sources

Marine lipids such as herring oil, menhaden oil, and anchovy oil have been employed traditionally as sources of essential fatty acids (EFAs), nonprotein energy, and other factors of nutritional importance in finfish diets. World supplies of fish body oils, however, have varied little over the past decade (average, 1.18 million metric tons per year) even though greater demands are being placed upon these valuable commodities for expanding aquafeed markets and, to some extent, for direct use in the human diet to prevent an array of health-related problems (Higgs et al. 1995; Anonymous 1997). By the year 2010, it has been forecasted that the majority of the global supply of fish oil could be utilized in aquafeeds. Hence, the cost of the lipid fraction in finfish diets will undoubtedly rise in the future unless increased reliance is placed upon less expensive alternate lipid sources of plant and/or animal origin to furnish a significant portion of the dietary nonprotein energy needs and, in some species, the necessary dietary levels of EFAs. With respect to the latter, most studies have assessed the suitability of various alternate dietary lipid sources for finfish species in terms of growth, feed utilization, and flesh quality. By contrast, relatively little attention has been directed toward the possible adverse effects of novel lipid sources of varying lipid composition (see Table 11.1) on the immune response(s) and disease resistance of fish. The latter effects could potentially occur, especially if tissue levels of eicosanoid precursors such as AA and EPA are altered. This will be considered in more detail in the following sections.

FISH IMMUNOLOGY: GENERAL CONSIDERATIONS

Fish are susceptible to infectious diseases caused by parasites, viruses, bacteria, and fungi. There are two possible outcomes once a fish has become infected with a pathogenic organism: (1) the fish will either successfully fight the infection by preventing the growth and colonization of the patho-

TABLE 11.1. Relative Levels of N-6, N-3, and N-3 HUFAs in Selected Lipid Sources of Animal and Plant Origin

Lipid Source	n-6 Fatty Acids	n-3 Fatty Acids	n-3 HUFA
Herring oil	1	3	2 or 3
Anchovy oil	1	3	3
Menhaden oil	1	3	3
Pork lard	2	1	1
Beef tallow	1	1	ND
Chicken fat	2 or 3	1	1
Soybean oil	3	2	ND[1]
Canola oil	3	2	ND
Linseed oil	2 or 3	3	ND
Safflower oil	2 or 3[2]	1	ND
Sunflower oil	3	1	ND

Sources: Adapted from Ackman 1990; Ballestrazzi and Mion 1993; Dosanjh et al. 1988, 1998; Higgs et al. 1995; Higgs et al. unpublished data.

Notes: The numbers 1, 2, and 3 denote low (<5), medium (5-15) or high >15) percentages of the preceding families of fatty acids or n-3 HUFAs in the lipid sources.
[1] Not detected
[2] Low and high oleic acid content safflower oils have high and medium levels of n-6 fatty acids, respectively

gen and in doing so, survive, or (2) the fish may not be successful in preventing the spread of the infection, a state of physiological dysfunction (i.e., disease) will develop, and the fish will die. The different outcomes of survival or death are largely determined by the efficacy of the immune system to prevent the initial infection from occurring and/or to prevent the growth and spread of the pathogen once an infection has started.

The immune system of vertebrates, including teleost fish, can be categorized into specific acquired immunity and nonspecific innate immunity. Both use cellular and humoral (soluble) mechanisms to provide protection against infections. The specific immune system involves the recognition of a specific antigen on a pathogen, thereby providing protection against that specific pathogen. Memory of the initial pathogen exposure is a vital feature to the specific immune response. The second and subsequent immune responses to the specific antigen are more rapid and effective. Specific disease resistance is affected by factors that influence antibody production, such as the type and duration of antigen stimulation, age, temperature, and stress (Tatner 1996). In contrast, the nonspecific immune system provides an array of protective mechanisms that are inherently available to provide immediate protection against a wide variety of pathogens.

The cellular and humoral factors of both the specific and nonspecific immune systems provide fish with external and internal protection against infectious agents. The various factors comprising each of the systems can act alone or in combination to provide a range of protective mechanisms. A general review of the fish immune system follows (for detailed information see Iwama and Nakanishi 1996).

Cells of the Immune System

The leucocytes of fish, namely, granulocytes, monocytes/macrophages, thrombocytes, nonspecific cytotoxic cells, and lymphocytes, have been the subject of several reviews (Ellis 1977; Ainsworth 1992; Hine 1992). The characteristics of each cell type are briefly described here.

Fish possess polymorphonuclear cells, or granulocytes (neutrophils, eosinophils, and basophils), that contain granules, the contents of which are released upon stimulation. These granules contain substances that are capable of directly or indirectly killing pathogenic organisms. Granulocytes are highly mobile cells, traveling via the blood and lymphatic systems to sites of infection and injury, thereby playing a vital role in the inflammatory response.

Mononuclear cells in fish include the monocytes and macrophages. Monocytes are immature macrophages that are mobile throughout the circulatory system. They differentiate into macrophages in the tissues, whereupon these nonmobile cells function as phagocytes capable of a high level of killing activity. Macrophages also play important roles as antigen-presenting cells and cytokine producers.

Thrombocytes are thought to be a nucleated version of the mammalian platelet. These cells are involved in blood clotting and have recently been thought to have phagocytic properties (Secombes 1996).

Nonspecific cytotoxic cells are similar to the mammalian natural killer cells. They are capable of lysing tumor cell lines and protozoan parasites (Evans and Jaso-Friedman 1992) and are found in the blood, lymphoid tissues, and gut of fish.

Lymphocytes are the cells responsible for the specificity of the specific immune response. The two different classes of lymphocytes can be distinguished by their cell surface markers and subsequent function. B lymphocytes recognize antigen and produce specific antibodies to that antigen. T lymphocytes recognize antigen that is presented by antigen-presenting cells such as macrophages. T lymphocytes are also important sources of cytokines, which are particularly important in the inflammatory response (see Cytokines).

Physical Barriers

The mucus, skin, and scales are very effective external barriers against invading pathogens. In addition, the continual shedding of mucus prevents the attachment and subsequent colonization of pathogens, which is the first step in the infection process. Fish mucus also contains nonspecific antimicrobial factors (Alexander and Ingram 1992) and specific antibodies that help to prevent the pathogen(s) from attaching to and colonizing fish surfaces.

Nonspecific Humoral Components

Humoral components act in several ways to kill and/or prevent the growth and spread of pathogens. Many of the components use unique mechanisms to lyse the pathogen. Others act as agglutinins (aggregate cells) or precipitins (aggregate molecules). There are also opsonins that bind with the pathogen and, in doing so, facilitate its uptake and removal by phagocytic cells. Alexander and Ingram (1992) and Yano (1996) have reviewed other nonspecific humoral factors found in fish. Briefly, these factors include various lytic substances (lysozyme hemolysins, chitinase, proteinases), agglutinins/precipitins (C-reactive protein, lectins, serum amyloid P-component, α-precipitin, natural precipitins, natural antibodies, natural hemagglutinins), enzyme inhibitors (serine proteinase inhibitors, cysteine- proteinase inhibitors, metalloproteinase inhibitors, α-macroglobulin), and pathogen growth inhibitors (interferon, transferrin, caeruloplasmin, metallothionein). Antimicrobial peptides have recently been isolated from fish skin and mucosa (Cole, Weis, and Diamond 1997; Lemaître et al. 1996).

Complement refers to a series of reactions initiated by two distinct pathways—the alternative complement pathway (ACP) and the classical complement pathway (CCP). The result of complement activation by either pathway is the formation of a lytic membrane attack complex. Activation of complement by the CCP is considered to be a specific immune response because an antigen-antibody complex initiates the reaction. The ACP is considered to be part of the nonspecific immune system because it is activated by a variety of compounds and surfaces. During the sequence of events that culminates in complement-mediated lysis, many by-products (usually peptides) are formed. Some of these substances have important roles in the inflammatory immune response, such as opsonins, anaphylatoxins, neutrophil, and macrophage chemoattractants (Law and Reid 1988). The virucidal, bactericidal, parasiticidal, opsonic, and chemoattracting activity of fish complement has been reviewed by Yano (1996).

Cytokines

Cytokines are soluble factors released from cells that facilitate interactions between immune cells and thereby modulate the immune response. Several cytokines have been described in fish (Manning and Nakanishi 1996). Interleukin-1 (IL-1) is produced by macrophages and also by a variety of other cells. It has been shown to stimulate lymphocyte proliferation. Other interleukin cytokines found in fish include IL-2, IL-3, IL-4, and IL-6. Cytokines produced in fish include interferons (i.e., IFNγ), tissue necrosis factor, transforming growth factor β1, migration inhibition factor, and platelet aggregating factor. Other cytokines that are important mediators of the inflammatory response are eicosanoids.

Phagocytosis

Phagocytic cells are responsible for the clearance of foreign substances and senescent blood cells from the body. Monocytes, macrophages, neutrophils, eosinophils, and thrombocytes all appear to be phagocytic in fish. Neutrophils are the most active phagocytes, because they are highly mobile. However, macrophages have greater killer activity (Secombes and Fletcher 1992).

The process of phagocytosis can be divided into three steps (reviewed by Secombes 1996). The first is the generally passive process of attachment, which can be facilitated by opsonization of the pathogen by such substances as complement and antibody. The second stage of phagocytosis is ingestion of the foreign substance into the phagocyte. This is an active process in which pseudopodia surround the substance to create a phagosome within the cytoplasm of the phagocyte. The final stage of phagocytosis is the actual killing of the pathogen. This involves both oxygen-dependent and oxygen-independent processes (Secombes and Fletcher 1992). The oxygen-independent killing mechanism in fish phagocytes is thought to involve enzymes such as lysozyme. The oxygen-dependent killing mechanisms utilize O_2 through a respiratory burst process, which produces toxic oxygen- and nitrogen-free radicals. The reactive oxygen species produced are superoxide anions ($O^-{}_2$), hydrogen peroxide (H_2O_2), and reactive singlet oxygen (O^-). The reactive nitrogen species include nitrous oxide.

The Inflammatory Response

The inflammatory response in fish has been described in detail by Finn and Nielson (1971) and more recently by Secombes (1996). The general sequence of events is comparable to that in mammals and serves to isolate, destroy, and remove foreign substances or pathogens. Briefly, the initial step in the inflammatory response is vasodilation. This facilitates the rapid movement of neutrophils to the site of the stimulus. Monocytes and macrophages later join the neutrophils to remove and destroy the pathogen by the process of phagocytosis. The next step is the isolation of the pathogen through a healing process (Roberts 1989). If the acute inflammatory response does not remove the stimulus, a chronic inflammatory response will proceed (Roberts 1989). An aggregation of lymphocytes and macrophages at the site and the formation of granulomas with later melanization and fibrosis (Roberts 1989) characterize chronic inflammation.

Several of the aforementioned humoral immune components have very important roles in the inflammatory response in fish. For example, many of the by-products of complement activation have been shown to be vasoactive and thus promote the migration of leucocytes to the site of the inflammatory stimulus (Sakai 1992). The inflammatory response is also enhanced by leucocyte-derived eicosanoids and cytokines.

Specific Humoral (Antibody) Response

The specific humoral immune response is centered in the production of antibody in response to a specific antigen (Ag). Fish appear to respond in a similar manner to mammals, with respect to the structure of the antibody (Ab) and the effector mechanisms involved. Antibody is involved in several aspects of the immune response to infections. Binding of Ab with Ag results in neutralization of the foreign substance, thus blocking its activity and/or virulence. Ab/Ag binding also serves to enhance the phagocytosis and clearance of pathogens from the body through the processes of opsonization, agglutination, and precipitation. The formation of memory is a key feature of the specific immune response. This ensures a faster and greater response when there is subsequent exposure to the same antigen.

Cell-Mediated Cytotoxicity

Specific immune responses that are independent of Ab are collectively termed cell-mediated immunity (CMI). T lymphocytes are involved in this type of immune response as they become activated and differentiate into different functional cell types. In mammals, the cell type responsible for CMI is the cytotoxic T cell. This subpopulation of T cells recognizes Ag in a manner similar to that of all vertebrates. Fish have lymphocytes that are analogous to mammalian T cells and B cells (Clem, Miller, and Bly 1991). In mammals, T cell receptors are found on membranes of T cells, while B cells have Ig (Ab) receptors on their membranes. In fish, B cells have been identified using anti-Ig monoclonal antibody techniques. T cells have been identified in the channel catfish, *Ictalurus punctatus,* following the development of a T-lineage-specific marker (Passer et al. 1996).

Lymphocyte Proliferation Response to Mitogens

Mitogens in mammals (plant lectins such as phytohemagglutinin or concanavalin) are specific for a particular carbohydrate (CHO) moiety on the T lymphocyte. When mitogens are present in appropriate concentration, T cells will proliferate, whereas B cells will not. Accessory cells (monocytes) are required for proliferation, and they are the likely source of cytokines (Clem et al. 1985).

DIETARY FATTY ACIDS AND THE IMMUNE RESPONSE

Generally, there are three mechanisms by which dietary fatty acids may affect disease resistance and the immune system. The first is through their influence on cell membrane lipid composition. Similar to the situation in mammals, the fatty acid composition of fish diets strongly influences the fatty acid composition of cell membrane phospholipids. This, in turn, can have profound effects on disease resistance because many immune responses are based on leucocyte cell membrane interactions (e.g., phagocytosis, antigen-antibody binding, activation steps involving cytokine production). A second mechanism by which dietary fatty acids may affect the immune system involves alteration of signal transductions, possibly due to effects on protein kinase C. This mechanism was suggested following mammalian studies in which macrophages from mice ingesting fish oil had increased sensitivity to interferon-γ (Erickson and Hubbard 1993). A third mechanism whereby dietary lipid composition can affect disease resistance is through the production of immunologically active eicosanoids from nonesterified AA, EPA, DHA, and possibly another polyunsaturated fatty acid precursor, i.e. 20:3n-6 (see Figure 11.1). Baldo and Fletcher (1975) were among the first to suggest that lipid mediators may be important in modulating the immune response in fish. There has recently been great interest in studying the effects of dietary lipids on the immune response. Hence, the remainder of this chapter will focus on the current status of our knowledge of how dietary fatty acids and eicosanoids modulate the immune response and disease resistance of fish.

The majority of research to date has been conducted on teleost fish species that are commercially important for aquaculture. Hence, most of the studies have been carried out using salmonids, *Salmo* spp. and *Oncorhynchus* spp., and channel catfish. It has been suggested that there are species differences in eicosanoid production (Rowley et al. 1995). Therefore, one should not make broad generalizations with respect to the various results that have been obtained. Care should also be exercised in interpreting the results from the literature for additional reasons, which have been outlined by Blazer (1992).

Briefly, the lack of universal control diets has been cited as the main reason for caution, since the different control diets vary widely in ingredient and nutrient compositions. This makes comparisons between studies within and between species difficult to interpret, since the levels of other nutrients besides fatty acids can also influence immunocompetence.

In addition, the different environmental conditions used in the various studies may have influenced the results. For example, dissimilar fish stress caused by differences in culturing density, social factors, water quality, water temperature, and frequency of fish handling can all produce variations in circulating titers of cortisol. High cortisol titers can be immunosuppressive and may obviate any potential benefits of more favorable diet nutrient levels.

To further complicate the situation, interactions between the environment, pathogen, and fish can affect the results of an investigation. The aquatic environment is also a concern because it is difficult to ensure that all the fish actually ingest their ration of a particular diet. Further, there may be leaching of certain nutrients such as water-soluble vitamins into the aqueous environment. Genetic variation within and between fish species, strains, and individuals as well as life history stage are other variables that likely affect the immunological response to a given diet treatment. Despite the aforementioned concerns, there have been tremendous gains in knowledge regarding how dietary fatty acids affect the immune system and disease resistance in fish.

EICOSANOID SYNTHESIS AND ACTIVITY

Figure 11.1 outlines the metabolic pathways leading to the production of eicosanoids from fatty acid precursors. Macrophages appear to be the major source of eicosanoids (Pettitt, Rowley, and Barrow 1989; Pettitt et al. 1991), although granulocytes, monocytes, and thrombocytes also contribute significantly. It is uncertain whether lymphocytes are capable of synthesizing eicosanoids, but their activity can be affected profoundly by eicosanoids generated by other leucocytes. Once synthesized, eicosanoids are released, and they have a relatively short half-life in vivo (Rowley et al. 1995). Eicosanoids can be generated in a variety of different tissues in fish, including the brain, gill, liver, spleen, and heart (Knight et al. 1995), but it is the leucocytes present at the site of inflammation that are the major source of eicosanoids (Rowley et al. 1995).

The types and relative amounts of eicosanoids produced in fish appear to be different from those in mammals. Lipoxins (LX) are relatively minor eicosanoid products in mammals, whereas in fish, LX are major eicosanoid products. In support of this, Pettitt, Rowley, and Secombes (1989) found rainbow trout, *Oncorhynchus mykiss,* macrophages produce more LX than leukotrienes (LT), while mammalian macrophages tend to produce more LT than LX. In addition, Pettitt, Rowley, and Secombes (1989) found that unlike mammalian cells, trout macrophages do not require cell cooperation to produce LX. These results suggest that lipoxins are more important immunomodulators in fish than in mammals.

Leucocytes stimulated from a pathogen interaction or trauma generate eicosanoids, namely LT, LX, prostaglandins (PG), and thromboxanes (TX). Once stimulated, phospholipase enzymes located in the leucocyte cell membrane become activated and metabolize available phospholipids to generate nonesterified PUFAs such as AA, EPA, and DHA. The types and amounts of the foregoing PUFAs produced, especially AA and EPA, determine the types and amounts of the eicosanoids that are produced. Many researchers have used calcium ionophore A23187 to activate cultured leucocytes to enhance eicosanoid production. Leukotriene B production has been demonstrated following stimulation of channel catfish head kidney leucocytes (Fracalossi, Craig-Schmidt, and Lovell 1994), rainbow trout peripheral blood leucocytes (Pettitt and Rowley 1990; Pettitt, Rowley, and Barrow 1989), rainbow trout macrophages (Sharp et al. 1992), plaice, *Pleuronectes platessa,* neutrophils (Tocher and Sargent 1987), and rainbow trout thrombocytes (Lloyd-Evans et al. 1994). Stimulated trout macrophages and thrombocytes have been shown to produce increased amounts of LX (Pettitt, Rowley, and Secombes 1989; Pettitt et al 1991; Lloyd-Evans et al. 1994). Lloyd-Evans and colleagues (1994) also stimulated cultures of trout thrombocytes to produce PGE_2 and TXB_2. It is, however, important to recognize that experiments using pure cell cultures to generate eicosanoids may produce quite different results than would be found in vivo. For example, cell cooperation through cytokine production would be absent from a pure culture of a particular leucocyte.

The inflammatory response in fish is an extremely complex event that has been described previously. Eicosanoids, which are produced by fish leucocytes, act to enhance the inflammatory response (Secombes 1996). A compilation of the immunomodulatory activities of the eicosanoids is outlined in Table 11.2. In summary, the chemotactic activity of LT and LX is important in attracting more leucocytes to the site of inflammation. The vasoactive ability of PG aids in opening blood vessels to allow for the increased migration of leucocytes to the site of inflammation. Leukotrienes will also facilitate stimulation of lymphocyte proliferation and cytokine release, both of which serve to enhance antibody production. Platelet activating factor (PAF) is another lipid mediator involved in the activation and/or aggregation of platelets and leucocytes in mammals. PAF synthesis has been demonstrated in trout gill, kidney, liver, and spleen tissues, but its function in fish is yet unknown (Turner and Lumb 1989). Samples and colleagues (1997), however, found that trout leucocytes challenged with PAF show enhanced chemotactic and respiratory burst responses.

TABLE 11.2. Immunomodulatory Properties of Eicosanoids in Fish

Lipoxins
LXA_4 chemotactic/chemokinetic for rainbow trout leucocytes (Rowley et al. 1991)
LXA_4 LXB_4 inhibited rainbow trout lymphocyte proliferative response (Secombes et al. 1994)
LXA_4 stimulated the generation of plaque-forming cells in rainbow trout splenocytes (Knight and Rowley 1995)
LXA_4 enhanced the migration of rainbow trout neutrophils (Sharp et al. 1992)
Leukotrienes
LTB_4, LTB_5 stimulated lymphocyte proliferation in rainbow trout (Secombes et al. 1994)
LTB_4 enhanced the migration of dogfish, *Scyliorhinus canicula,* granulocytes (Hunt and Rowley 1986)
LTB_4 enhanced the migration of rainbow trout neutrophils (Sharp et al. 1992)
Prostaglandins
PGE_2, PGE_3 suppressed lymphocyte proliferation in rainbow trout leucocytes (Secombes et al. 1994)
PGE_2 inhibited respiratory burst activity in rainbow trout macrophages (Novoa et al. 1996)

DIETARY FATTY ACIDS, EICOSANOIDS, AND THE IMMUNE RESPONSE

Blazer and Wolke (1984) were among the first to observe the importance of diet in the immune response and disease resistance of fish. Rainbow trout showed different immune responses after they were fed either a laboratory-prepared or commercial diet. In a similar study, Blazer, Ankley, and Finco-Kent (1989) found that channel catfish also showed differences in immune responses after they were fed a control laboratory diet or commercial diets. They noted that the major differences between the diets that probably accounted for the dissimilar immune response were the total lipid content and the ratio of the n-3 to n-6 fatty acids. This suggested that the level and/or composition of dietary lipid had an important role in disease resistance.

Numerous other studies have since examined the effects that different levels of dietary fatty acids have on the fatty acid composition of tissues and cells and the subsequent effects on eicosanoid production. Bell and colleagues (1994) showed that the lipid composition of juvenile turbot, *Scophthalmus maximus,* can be modified by changes in dietary lipid composition. Atlantic salmon postsmolts fed increasing levels of dietary linoleic acid demonstrated changes in the phospholipid fatty acid composition of gills and leucocytes (Bell, Raynard, and Sargent 1991; Bell, Sargent, and Raynard 1992). In addition, the observed alterations in the phospholipid composition led to changes in eicosanoid precursors, as was reflected by increased levels of AA-derived eicosanoids (Bell, Sargent, and Raynard 1992). Therefore, by altering the dietary lipid content it is possible to alter the type and level of eicosanoids generated, which can have profound effects on fish health. Some examples of this follow.

Bell and colleagues (1993) demonstrated that Atlantic salmon postsmolts fed a diet with high linolenic acid content (from linseed oil) had increased levels of EPA in leucocyte phospholipids when compared to those fed diets with high levels of fish oil or sunflower oil. The high dietary linolenic acid content also reduced the production of AA-derived PGE_2 and TXB_2 (a stable metabolite of TXA_2). These responses resulted in increased anti-inflammatory activity and an attendant reduction in the severity of cardiac lesions. Bell and colleagues (1994) performed related experiments on juvenile turbot and obtained similar results with respect to the effect of dietary lipid composition on eicosanoid production. In this regard, fish fed diets supplemented with linseed oil were compared with those fed diets supplemented with fish oil or safflower oil. The linseed oil-fed fish had relatively higher levels of EPA and lower levels of AA in tissues, along with decreased levels of TXB_2 and pooled PGE. Further, Bell and colleagues (1995) found that increased dietary AA content resulted in significant increases in the production of PGE_2 and 6-keto$PGF1\alpha$ in tissue homogenates from juvenile turbot. Similarly, Peterson et al. (1998) found that high dietary EPA content led to decreased production of PGE_2 in rat spleen leucocytes.

It should be noted that the process by which dietary fatty acids alter cell membrane fatty acid composition and eicosanoid production is complex and does not necessarily directly reflect dietary input. For example, freshwater salmonids appear to be similar to mammals, since the substrate preference for Δ6 desaturase is 18:3n-3>18:2n-6>18:1n-9. Also, a surplus of one series of dietary fatty acids, such as n-6 fatty acids, relative to n-3 fatty acids in rainbow trout may competitively inhibit the formation of the long-chain highly unsaturated members of the series in lower concentration, for example, n-3 HUFAs in this instance. By contrast, Yu and Sinnhuber (1979) did not demonstrate the foregoing scenario in juvenile coho salmon, *Oncorhynchus kisutch.* Moreover, Bell and colleagues (1996) found that Atlantic salmon fed diets supplemented with sunflower oil (low in n-3 PUFAs) and those fed diets supplemented with linseed oil (high in n-3 PUFAs) produced approximately equal amounts of EPA. They concluded that n-3 PUFAs are better able to inhibit n-6 PUFAs from generating eicosanoids than vice versa.

In general, a diet high in n-6 PUFAs will produce relatively higher levels of the proinflammatory 2-series PGs and 4-series LTs and LXs derived from AA. Alternatively, a diet high in n-3 PUFAs will produce relatively higher levels of the anti-inflammatory 3-series PGs and 5-series LTs and LXs derived from EPA (Figure 11.1; Ashton et al. 1994). Therefore, the fatty acid composition of the diet influences the immune response because, to a large extent, it will determine which eicosanoid precursors are present in the cell membranes. It is tempting to conclude, therefore, that a diet containing high levels of n-6 PUFAs will enhance the immune response due to high levels of proinflammatory AA-derived eicosanoids, and, likewise, diets containing high levels of n-3 PUFAs will be immunosuppressive due to high levels of EPA-derived anti-inflammatory eicosanoids. This is, however, not the case, as the type of eicosanoids produced and the ultimate impact on the immune response is very complex. Factors that will affect eicosanoid production include competition for fatty acid metabolism (discussed previously), the cell type involved, and the form or source of the fatty acids in the diet. The difficulties in controlling all of the experimental conditions and other factors described previously also contribute to the variability in the results seen in the literature.

Mammalian studies have reported a direct relationship between dietary n-3 fatty acid content and immunosuppression. The immunosuppression has been linked with decreased PGE_2 production and cell-mediated immunity in dogs (Wander et al. 1997). Rodents fed diets supplemented with n-3 PUFAs have shown adverse effects on the specific and nonspecific immune systems, as seen by depressed lymphocyte function (Jeffery et al. 1996) in rats and decreased natural resistance to infection (Chang et al. 1992) in mice. Investigations into the mechanisms for this immunosuppression have reported a possible relationship between impaired cellular function and n-3 PUFAs. Costa Rosa, Safi, and Guimarães (1996) have shown that rats fed high dietary levels of n-3 fatty acids have increased immunosuppression that may be attributed to changes in glucose and glutamine metabolism, which may have led to functional changes in macrophages and lymphocytes.

The situation in fish regarding the relationship between dietary levels of n-3 PUFAs and immunosuppression is contradictory. Studies comparing the effects of high dietary levels of n-3 fatty acids versus high dietary levels of n-6 fatty acids have shown variation in the immunomodulatory properties of the different fatty acids. Nevertheless, there appear to be more reports of immunosuppression associated with high dietary n-3 fatty acid content than with high n-6 fatty acid content. Erdal and colleagues (1991) reported decreased antibody titers and likelihood of survival in Atlantic salmon fed diets with high n-3 PUFA levels. Channel catfish fed a diet containing high n-3 fatty acid content had decreased survival following challenge with *Edwardsiella ictaluri,* compared to those consuming a diet with high n-6 fatty acid content (Li et al. 1994). However, there were no differences in antibody titers between the two diet groups. Decreased disease resistance (Fracalossi and Lovell 1994) and related factors such as phagocytic capacity and killing activity (Lingenfelser, Blazer, and Gay 1995) have also been demonstrated in channel catfish fed diets high in n-3 PUFAs. Li and colleagues (1994) suggested a possible mechanism for the observed n-3 fatty acid-induced immunosuppression might be similar to that of mammals and involve decreased LTB_4 and increased LTB_5 production by macrophages and neutrophils. Fracalossi and Lovell (1994) suggested that channel catfish require a carefully balanced dietary mixture of n-3 and n-6 fatty acids to optimize the immune response in this species.

There is also evidence for immunostimulatory effects of dietary n-3 fatty acids in fish. The early work by Blazer, Ankley, and Finco-Kent (1989) and Sheldon and Blazer (1991) attributed the increased activity of head kidney macrophages to increased dietary levels of n-3 fatty acids. More recently, Thompson, Tatner, and Henderson (1996) observed that Atlantic salmon fed diets with high (n-3) to (n-6) PUFA ratios had increased B lymphocyte response and survival following experimental challenges with *Aeromonas salmonicida* and *Vibrio anguillarum.* Also, Ashton and colleagues (1994) demonstrated that head kidney supernatants derived from rainbow trout fed a diet enriched with n-3 fatty acids had greater migration stimulating ability than supernatants originating from trout fed an n-6 fatty acid-enriched diet.

Despite the conflicting effects of n-3 PUFAs on the immune response of fish, it appears that n-3-enriched diets consistently have a strengthening effect on cell membranes (Erdal et al. 1991; Klinger, Blazer, and Echevarria 1996). This is especially important for poikilotherms such as fish, where it is essential for proper membrane function to maintain biomembranes in a liquid crystalline state in relation to the prevailing water temperature. Temperature-related effects on fatty acid metabolism and immune function have been observed by Lingenfelser, Balzer, and Gay (1995). They found that channel catfish fed diets with high levels of n-3 fatty acids had enhanced immune function (especially phagocytic capacity) at low temperatures, while fish fed high dietary levels of n-6 fatty acids had enhanced disease resistance factors at higher temperatures. Fracalossi and Lovell (1994) also reported that dietary fatty acid effects on disease resistance of channel catfish were influenced by temperature, as diet-related differences were observed at high water temperature but not at low water temperature. The underlying mechanism for these temperature effects may be related to the role of n-3 fatty acids in maintaining membrane fluidity, which is particularly important for the ingestion stage of phagocytosis.

The importance of n-3 fatty acids for maintaining immune function was demonstrated by Kiron and colleagues (1995), as antibody production and macrophage killing activity were decreased in rainbow trout fed diets deficient in n-3 fatty acids. Also, Montero and colleagues (1998) found that adequate levels of n-3 HUFAs in the diet were necessary for the maintenance of alternative complement activity in gilthead seabream, *Sparus aurata.* Moreover, it is noteworthy that high dietary ratios of n-3 to n-6 fatty acids have been shown to prevent cardiomyopathy as well as the susceptibility of Atlantic salmon to transport stress (Bell et al. 1991). Therefore n-3 PUFAs are essential fatty acids in fish, not only for optimal growth and feed efficiency, but also for immunological efficacy and cardiovascular function. However, it should be emphasized that AA has also recently been found to be an essential fatty acid in turbot. In addition, other fish species such as chum salmon, *O. keta,* and tilapia, for example, *Tilapia zillii,* have been shown to have a dietary need for n-6 fatty acids, usually in the form of 18:2n-6 (Higgs and Dong 2000). In the case of juvenile turbot, Castell and colleagues (1994) showed that for optimal growth and survival, this species has a small requirement for AA in addition to that of DHA.

Hence, it is likely that future studies will reveal, as suggested by the findings of Fracalossi and Lovell (1994) for channel catfish, that the diets of finfish species will have to be tailored so that the levels and ratios of AA, EPA, and DHA in fish leucocytes are optimized for the most favorable elaboration of eicosanoid compounds and highest immunological efficacy. Further research is required to investigate the effects of using different lipid sources on the immune function and disease resistance of finfish. Long-term studies involving comprehensive examinations of the immune function and disease resistance will provide vital information on the suitability of the alternative dietary lipids for cultured finfish. This remains a challenging area for future research.

REFERENCES

Ackman, R.G. 1990. Canola Fatty Acids—An Ideal Mixture for Health, Nutrition, and Food Use. In F. Shahidi (Ed.), *Canola and Rapeseed: Production, Chemistry, Nutrition, and Processing Technology* (pp. 81-98). New York: Van Nostrand Reinhold.

Ainsworth, A.J. 1992. Fish granulocytes: Morphology, distribution, and function. *Annual Review of Fish Diseases* 2:123-148.

Alexander, J.B. and G.A. Ingram. 1992. Noncellular nonspecific defense mechanisms of fish. *Annual Review of Fish Diseases* 2:249-279.

Anonymous. 1997. Future world fish supplies should be enough for farm needs. *Fish Farming International* 24:14-16.

Ashton, I., K. Clements, S.E. Barrow, C.J. Secombes, and A.F. Rowley. 1994. Effects of dietary fatty acids on eicosanoid-generating capacity, fatty acid composition and chemotactic activity of rainbow trout *(Oncorhynchus mykiss)* leukocytes. *Biochimica et Biophysica Acta* 1214:253-262.

Baldo, B.A. and T.C. Fletcher. 1975. Inhibition of immediate hypersensitivity responses in flatfish. *Experientia* 31:495-496.

Ballestrazzi, R. and A. Mion. 1993. Lipids and teleostean fish feeding. *Rivista Italiana Acquacoltura* 28:155-173.

Bell, J.G., I. Ashton, C.J. Secombes, B.R. Weitzel, J.R. Dick, and J.R. Sargent. 1996. Dietary lipid affects phospholipid fatty acid compositions, eicosanoid production and immune function in Atlantic salmon *(Salmo salar). Prostaglandins, Leukotrienes and Essential Fatty Acids* 54:173-182.

Bell, J.G., J.D. Castell, D.R. Tocher, F.M. MacDonald, and J.R. Sargent. 1995. Effects of different dietary arachidonic acid: Docosahexaenoic acid ratios on phospholipid fatty acid compositions and prostaglandin production in juvenile turbot *(Scophthalmus maximus). Fish Physiology and Biochemistry* 14:139-151.

Bell, J.G., J.R. Dick, A.H. McVicar, J.R. Sargent, and K.D. Thompson. 1993. Dietary sunflower, linseed and fish oils affect phospholipid fatty acid composition, development of cardiac lesions, phospholipase activity and eicosanoid production in Atlantic salmon *(Salmo salar). Prostaglandins, Leukotrienes and Essential Fatty Acids* 49:665-673.

Bell, J.G., A.H. McVicar, M.T. Park, and J.R. Sargent. 1991. High dietary linoleic acid affects the fatty acid compositions of individual phospholipids from tissues of Atlantic salmon *(Salmo salar):* Association with stress susceptibility and cardiac lesion. *Journal of Nutrition* 121:1163-1172.

Bell, J.G., R.S. Raynard, and J.R. Sargent. 1991. The effect of dietary linoleic acid on the fatty acid composition of individual phospholipids and lipoxygenase products from gills and leucocytes of Atlantic salmon *(Salmo salar). Lipids* 26:445-450.

Bell, J.G., J.R. Sargent, and R.S. Raynard. 1992. Effects of increasing dietary linoleic acid on phospholipid fatty acid composition and eicosanoid production in leucocytes and gill cells in Atlantic salmon *(Salmo salar). Prostaglandins, Leukotrienes and Essential Fatty Acids* 45:197-206.

Bell, J.G., D.R. Tocher, F.M MacDonald, and J.R. Sargent. 1994. Effects of diets rich in linoleic (18:2n-6) and α-linolenic (18:3n-3) acids on the growth, lipid class and fatty acid compositions and eicosanoid production in juvenile turbot (*Scophthalmus maximus* L.). *Fish Physiology and Biochemistry* 13:105-118.

Blazer, V.S. 1992. Nutrition and disease resistance in fish. *Annual Review of Fish Diseases* 2:309-323.

Blazer, V.S., G.T. Ankley, and D. Finco-Kent. 1989. Dietary influences on disease resistance factors in channel catfish. *Developmental and Comparative Immunology* 13:43-48.

Blazer, V.S. and R.E. Wolke. 1984. Effect of diet on the immune response of rainbow trout *(Salmo gairdneri). Canadian Journal of Fisheries and Aquatic Sciences* 41:1244-1247.

Buzzi, M., R.J. Henderson and J.R. Sargent. 1997. Biosynthesis of docosahexaenoic acid in trout hepatocytes proceeds via 24-carbon intermediates. *Comparative Biochemistry and Physiology* 116B:263-267.

Calder, P.C. 1997. N-3 polyunsaturated fatty acids and immune cell function. *Advances in Enzyme Regulation* 37:197-237.

Castell, J.D., J.G. Bell, D.R. Tocher, and J.R. Sargent. 1994. Effects of purified diets containing different combinations of arachidonic and docosahexaenoic acid on survival, growth and fatty acid composition of juvenile turbot *(Scophthalmus maximus). Aquaculture* 128:315-333.

Chang, H.R., A.G. Dulloo, I.R. Vladoianu, P.F. Piguet, D. Arsenijevic, L. Girardier, and J.C. Pechère. 1992. Fish oil decreases natural resistance of mice to infection with *Salmonella typhimurium. Metabolism* 41:1-2.

Clem, L.W., N.W. Miller, and J.E. Bly. 1991. Evolution of lymphocyte populations, their interactions, and temperature sensitivities. In G.W. Warr and N. Cohen (Eds.), *Phylogenesis of the Immune System* (pp. 191-213). Boca Raton, FL: CRC Press.

Clem, L.W., R.C. Sizemore, C.F. Ellsaesser, and N.W. Miller. 1985. Monocytes as accessory cells in fish immune responses. *Developmental and Comparative Immunology* 9:803-809.

Cole, A.M., P. Weis, and G. Diamond. 1997. Isolation and characterization of pleurocidin, and antimicrobial peptide in the skin secretions of winter flounder. *Journal of Biological Chemistry* 272:12008-12013.

Costa-Rosa, L.F.B.P., D.A. Safi, and A.R.P. Guimarães. 1996. The effect of n-3 PUFA rich diet upon macrophage and lymphocyte metabolism and function. *Biochemistry and Molecular Biology International* 40:833-842.

Dosanjh, B.S., D.A. Higgs, D.J. McKenzie, D.J. Randall, J.G. Eales, N. Rowshandeli, M. Rowshandeli, and G. Deacon. 1998. Influence of dietary blends of menhaden oil and canola oil on growth, muscle lipid composition, and thyroidal status of Atlantic salmon *(Salmo salar)* in sea water. *Fish Physiology and Biochemistry* 19:123-134.

Dosanjh, B.S., D.A. Higgs, M.D. Plotnikoff, J.R. Markert, and J.T. Buckley. 1988. Preliminary evaluation of canola oil, pork lard and marine lipid singly and in combination as supplemental dietary lipid sources of juvenile fall chinook salmon *(Oncorhynchus tshawytscha). Aquaculture* 6:325-343.

Ellis, A.E. 1977. The leucocytes of fish: A review. *Journal of Fish Biology* 11:453-491.

Erdal, J.I., O. Evensen, O.K. Kaurstad, A. Lillehaug, R. Solbakken, and K.Thorud. 1991. Relationship between diet and immune response in Atlantic salmon (*Salmo salar* L.) after feeding various levels of ascorbic acid and omega-3 fatty acids. *Aquaculture* 98:363-379.

Erickson, K.L. and N.E. Hubbard. 1993. Dietary Fat and Immunology. In D.M. Klurfeld (Eds.), *Human Nutrition-A Comprehensive Treatise. Volume 8: Nutrition and Immunology* (pp. 51-78). New York: Plenum Press.

Evans, D.L. and L. Jaso-Friedman. 1992. Nonspecific cytotoxic cells as effectors of immunity in fish. *Annual Review of Fish Diseases* 2:109-121.

Finn, J.P. and N.O. Nielson. 1971. The inflammatory response of rainbow trout. *Journal of Fish Biology* 3:463-478.

Fracalossi, D.M., M.C. Craig-Schmidt, and R.T. Lovell. 1994. Effect of dietary lipid sources on production of leukotriene B by head kidney of channel catfish held at different water temperatures. *Journal of Aquatic Animal Health* 6:242-250.

Fracalossi, D.M. and R.T. Lovell. 1994. Dietary lipid sources influence responses of channel catfish *(Ictalurus punctatus)* to challenge with the pathogen *Edwardsiella ictaluri. Aquaculture* 119:287-298.

German, J.B., G.G. Bruckner, and J.E. Kinsella. 1986. Lipoxygenase in trout gill tissue acting on arachidonic, eicosapentaenoic and docosahexaenoic acids. *Biochimica et Biophysica Acta* 875:12-20.

Henderson, R.J. and D.R. Tocher. 1987. The lipid composition and biochemistry of fresh water fish. *Progress in Lipid Research* 26:281-347.

Higgs, D.A. and F.M. Dong. 2000. Lipids and fatty acids. In R.R. Stickney (Ed.), *The Encyclopedia of Aquaculture* (pp. 1-20). New York: John Wiley and Sons, Inc.

Higgs, D.A., J.S. Macdonald, C.D. Levings, and B.S. Dosanjh. 1995. Nutrition and feeding habits in relation to life history stage. In C. Groot, L. Margolis, and W.C. Clarke (Eds.), *Physiological Ecology of Pacific Salmon* (pp. 159-315). Vancouver, British Columbia, Canada: UBC Press.

Hine P.M. 1992. The granulocytes of fish. *Fish and Shellfish Immunology* 2:79-98.

Hunt, T.C. and A.F. Rowley. 1986. Leukotriene B_4 induces enhanced migration of fish leucocytes in vitro. *Immunology* 59:563-568.

Iwama, G. and T. Nakanishi. 1996. The Fish Immune System. *Fish Physiology,* Volume 15. San Diego, CA: Academic Press.

Jeffery, N.M., P.Sanderson, E.J. Sherrington, E.A. Newsholme, and P.C. Calder. 1996. The ratio of n-6 to n-3 polyunsaturated fatty acids in the rat diet alters serum lipid levels and lymphocyte functions. *Lipids* 31:737-745.

Kiron, V., H. Fukuda, T. Takeuchi, and T. Watanabe. 1995. Essential fatty acid nutrition and defense mechanisms in rainbow trout *Oncorhynchus mykiss. Comparative Biochemistry and Physiology* 111A:361-367.

Klinger, R.C., V.S. Blazer, and C. Echevarria. 1996. Effects of dietary lipid on the hematology of channel catfish, *Ictalurus punctatus. Aquaculture* 147:225-233.

Knight, J., J.W. Holland, L.A. Bowden, K. Halliday, and A.F. Rowley. 1995. Eicosanoid generating capacities of different tissues from the rainbow trout, *Oncorhynchus mykiss. Lipids* 30:451-458.

Knight, J. and A.F. Rowley. 1995. Immunoregulatory activities of eicosanoids in the rainbow trout *(Oncorhynchus mykiss). Immunology* 85:389-393.

Law, S.K.A. and K.B.M. Reid. 1988. *Complement.* IRL Press, Oxford, England.

Lemaître, C., N. Orange, P. Saglio, N. Saint, J. Gagnon, and G. Molle. 1996. Characterization and ion channel activities of novel antibacterial proteins from the skin mucosa of carp *(Cyprinus carpio). European Journal of Biochemistry* 240:143-149.

Li, M.H., D.J. Wise, M.R. Johnson, and E.H. Robinson. 1994. Dietary menhaden oil reduced resistance of channel catfish *(Ictalurus punctatus)* to *Edwardsiella ictaluri. Aquaculture* 128:335-344.

Lingenfelser, J.T., V.S. Blazer, and J. Gay. 1995. Influence of fish oils in production catfish feeds on selected disease resistance factors. *Journal of Applied Aquaculture Issue* 5:37-48.

Lloyd-Evans, P., S.E. Barrow, D.J. Hill, L.A. Bowden, G.E. Rainger, J. Knight, and A.F. Rowley. 1994. Eicosanoid generation and effects on the aggregation of thrombocytes from the rainbow trout, *Oncorhynchus mykiss. Biochimica et Biophysica Acta* 1215:291-299.

Manning, M.J. and T. Nakanishi. 1996. The specific immune system: Cellular defenses. In G.K. Iwama and T. Nakanishi (Eds.), *The Fish Immune System. Fish Physiology* Volume 15 (pp. 159-205). San Diego, CA: Academic Press.

Montero, D., L. Tort, M.S. Izquierdo, L. Robaina, and J.M. Vergara. 1998. Depletion of serum alternative complement pathway activity in gilthead seabream caused by α-tocopherol and n-3 HUFA dietary deficiencies. *Fish Physiology and Biochemistry* 18:399-407.

Novoa, B., A. Figueras, I. Ashton, and C.J. Secombes. 1996. In vitro studies on the regulation of rainbow trout *(Oncorhynchus mykiss)* macrophage respiratory burst activity. *Developmental and Comparative Immunology* 20:207-216.

Passer, B.J., C.-I.H. Chen, N.W. Miller, and M.D. Cooper. 1996. Identification of a T lineage antigen in the catfish. *Developmental and Comparative Immunology* 20:441-450.

Peterson, L.D., N.M. Jeffery, F. Thies, P. Sanderson, E.A. Newsholme, and P.C. Calder. 1998. Eicosapentaenoic and docosahexaenoic acids alter rat spleen leukocyte fatty acid composition and prostaglandin E_2 production but have different effects on lymphocyte functions and cell-mediated immunity. *Lipids* 33:171-180.

Pettitt, T.R. and A.F. Rowley. 1990. Uptake, incorporation and calcium-ionophore-stimulated mobilization of arachidonic, eicosapentaenoic and docosahexaenoic acids by leucocytes of the rainbow trout, *Salmo gairdneri. Biochimica et Biophysica Acta* 1042:62-69.

Pettitt, T.R., A.F. Rowley, and S.E. Barrow. 1989. Synthesis of leukotriene B and other conjugated triene lipoxygenase products by blood cells of the rainbow trout, *Salmo gairdneri. Biochimica et Biophysica Acta* 1003:1-8.

Pettitt, T.R., A.F. Rowley, S.E. Barrow, A.I. Mallet, and C.J. Secombes. 1991. Synthesis of lipoxins and other lipoxygenase products by macrophages from the rainbow trout, *Oncorhynchus mykiss. Journal of Biological Chemistry* 266:8720-8726.

Pettitt, T.R., A.F. Rowley, and C.J. Secombes. 1989. Lipoxins are major lipoxygenase products of rainbow trout macrophages. *Federation of European Biochemical Societies Letters* 259:168-170.

Roberts, R.J. 1989. The pathophysiology and systematic pathology of teleosts. In R.J. Roberts (Ed.), *Fish Pathology* (pp. 55-91). London: Baillier Tindall.

Rowley, A.F., J. Knight, P. Lloyd-Evans, J.W. Holland, and P.J. Vickers. 1995. Eicosanoids and their role in immune modulation in fish—A brief overview. *Fish and Shellfish Immunology* 5:549-567.

Rowley, A.F., T.R. Pettitt, C.J. Secombes, G.J.E. Sharp, S.E. Barrow, and A.I. Mallet. 1991. Generation and biological activities of lipoxins in the rainbow trout—An overview. *Advances in Prostaglandin, Thromboxane and Leukocyte Research* 218:557-560.

Sakai, D.K. 1992. Repertoire of complement in immunological defense mechanisms of fish. *Annual Review of Fish Diseases* 2:223-247.

Samples, B.L., G.L. Pool, G.I. Pritchard, and R.H. Lumb. 1997. Lipid mediator mechanisms in fish. *The Progressive Fish-Culturist* 59:106-117.

Secombes, C.J. 1996. The nonspecific immune system: Cellular defenses. In G.K. Iwama and T. Nakanishi (Eds.), *The Fish Immune System. Fish Physiology,* Volume 15 (pp. 63-105). San Diego, CA: Academic Press.

Secombes, C.J., K. Clements, I. Ashton, and A.F. Rowley. 1994. The effect of eicosanoids on rainbow trout, *Oncorhynchus mykiss,* leucocyte proliferation. *Veterinary Immunology and Immunopathology* 42:367-378.

Secombes, C.J. and T.C. Fletcher. 1992. The role of phagocytes in the protective mechanisms of fish. *Annual Review of Fish Diseases* 2:53-71.

Sharp, G.J.E., T.R. Pettitt, A.F. Rowley, and C.J. Secombes. 1992. Lipoxin-induced migration of fish leukocytes. *Journal of Leukocyte Biology* 51:140-145.

Sheldon, Jr. W.M. and V.S. Blazer. 1991. Influence of dietary lipid and temperature on bactericidal activity of channel catfish macrophages. *Journal of Aquatic Animal Health* 3:87-93.

Tatner, M.F. 1996. Natural changes in the immune system of fish. In G.K. Iwama and T. Nakanishi (Eds.), *The Fish Immune System. Fish Physiology,* Volume 15 (pp. 255-287). San Diego, CA: Academic Press.

Thompson, K.D., M.F. Tatner, and R.J. Henderson. 1996. Effects of dietary (n-3) and (n-6) polyunsaturated fatty acid ratio on the immune response of Atlantic salmon, *Salmo salar* L. *Aquaculture Nutrition* 2:21-31.

Tocher, D.R. and J.R. Sargent. 1987. The effect of calcium ionophore A23187 on the metabolism of arachidonic and eicosapentaenoic acid in neutrophils from a marine teleost fish rich in (n-3) polyunsaturated fatty acids. *Comparative Biochemistry and Physiology* 87B:733-739.

Turner, M.R. and R.H. Lumb. 1989. Synthesis of platelet activating factor by tissues from the rainbow trout, *Salmo gairdneri. Biochimica et Biophysica Acta* 1004: 49-52.

Wander, R.C., J.A. Hall, J.L. Gradin, S.-H. Du, and D.E. Jewell. 1997. The ratio of dietary (n-6) to (n-3) fatty acids influences immune system function, eicosanoid metabolism, lipid peroxidation and vitamin E status in aged dogs. *Journal of Nutrition* 127:1198-1205.

Yano, T. 1996. The nonspecific immune system: Humoral defense. In G.K. Iwama and T. Nakanishi (Eds.), *The Fish Immune System. Fish Physiology,* Volume 15. San Diego, CA: Academic Press.

Yu, T.C. and R.O. Sinnhuber. 1979. Effect of dietary ω3 and ω6 fatty acids on growth and feed conversion efficiency of coho salmon *(Oncorhyncus kisutch). Aquaculture* 16:31-38.

Chapter 12

Immunostimulants in Fish Diets

Ann L. Gannam
Robin M. Schrock

INTRODUCTION

The use of immunostimulants in fish diets has accelerated in recent years as more production-grade diets are fortified with a variety of natural substances that promise to heighten innate immunity. Promotions on the use of such diets cite increased health and survival of fish fed enhanced diets, compared to fish fed other production diets, but do not provide documentation of the claims. Laboratory research continues to focus on injection, immersion, intragastric or intestinal administration of immunostimulants, methods that are not practical for large-scale production applications (Raa 1996; Sakai 1999). Investigations of the effects of orally administered immunostimulants have occurred only in the past decade, and caution must be used in drawing conclusions from the studies. Production level studies have not been conducted, and beyond anecdotal evidence, few published reports support claims of uniform benefits from immunostimulants in aquaculture applications. Fish farmers have experienced variable results, although laboratory trials have produced positive immunostimulatory effects at the molecular and cellular level. The effects measured in vitro or in vivo in controlled laboratory experiments cannot be assured under less uniform production conditions.

The use of immunostimulants as diet additives has met with uncertainty and skepticism because the actions of these compounds are poorly understood and the results of feeding trials have not shown consistently positive effects. Some products are of unclear nature and origin. A convincing explanation of their mode of action or documentation of efficacy under practical conditions is not available. The exact mechanism of immunostimulant absorption is not known. However, immunostimulants are promoted in aqua-

---

The authors wish to thank the U.S. Fish and Wildlife Service and the Bonneville Power Administration, Project 87-401, for research support.

culture as a means to overcome the immunosuppressive effects that occur in normal aquaculture operations, due to stressors (Thompson et al. 1993; Jeney et al. 1997) or unavoidable consequences of high-density culture (Vadstein 1997). Immunostimulants might be used as a prophylactic treatment in anticipation of expected seasonal outbreaks of known endemic diseases (Nikl, Evelyn, and Albright 1993) or a suppressive treatment for latent or sublethal pathogens. Culture conditions promoting chronic, subacute, or acute disease outbreaks include crowding, handling, accumulation of wastes, ambient flora and fauna, oxygen levels, exposure to sunlight, and water temperature. Unlike vaccines, immunostimulants simultaneously elevate the overall resistance of animals to many infectious agents by stimulating the nonspecific immune response.

Loosely defined, an immunostimulant is a substance that enhances the immune system by interacting directly with cells of the system activating them. Some responses that are routinely reported are macrophage activation, increased phagocytosis by neutrophils and monocytes, increased lymphocyte numbers, increased serum immunoglobulins, and increased lysozyme (Raa 1996; Sakai 1999). Immunostimulants that are effective in fish diets in a laboratory setting act within the nonspecific immune system at several levels. The first immune system defenses are substances found in mucus secreted by endothelial cells, macrophages attacking pathogens directly, and include many lytic and agglutinating factors. Proteins and enzymes act directly with molecules on the microbe's surface to inhibit bacterial growth or facilitate phagocytosis. The most common immunostimulants are nonvirulent microorganisms or their by-products. The compounds are recognized by the cellular components of the nonspecific immune system and initiate the same humoral and cellular response as pathogenic organisms. The evolutionary history of each aquatic species determines the individual immune factors that are present, and the magnitude and success of their response against immunomodulatory agents or pathogens.

Biological rationale for immunostimulants in the fish's diet is based on the evolutionary history of immune system development in aquatic organisms (Manning, Grace, and Secombes 1982; Chevassus and Dorson 1990; Wiegertjes et al. 1996). Survival in the aquatic environment requires an immune system that can combat the constant challenge of waterborne pathogens. Immunomodulators present in the diet stimulate the nonspecific immune system, while antigenic substances such as bacterins or vaccines initiate the more prolonged process of antibody production and acquired immunity (specific immune response) (Anderson 1993). Aquatic organisms evolved immediate, generalized responses to compensate for the continual exposure and delayed, specific responses that require time for acquired immunity to develop. Fish immunologists have concentrated their investigations of immunostimulation on laboratory research designed to explain the actions of individual immune response components to immunomodulation. Responses have generally been associated with the nonspecific immune system, although antigen-antibody-based enhancement has been reported (Raa

1996; Sakai 1999). Prophylactic and therapeutic administration of immunomodulators will need to be adapted to each cultured species in anticipation of recognized pathogens, under known environmental conditions.

The types of immunostimulants available to aquaculture and their different sources and mechanisms of action will be explored, and a discussion of why their performance may vary in production applications will be included to advance the discussion of immunostimulants in fish diets beyond reviews of molecularly based experimental results so that we may examine practical considerations that may affect immunostimulant efficacy on a production level.

THE GENETIC BASIS OF THE IMMUNE RESPONSE

The genetic basis of differences in innate disease resistance in aquatic organisms is not well understood. Investigations of differences among species are increasing (Chevassus and Dorson 1990; Wiegertjes et al. 1996; Balfry, Heath, and Iwama 1997). An understanding of nonspecific immunity (Dalmo, Ingebrigtsen, and Bøgwald 1997) and specific or acquired immunity in fishes (Anderson 1993) is needed to determine the contributions of individual immune factors to disease resistance (Manning, Grace, and Secombes 1982; Chevassus and Dorson 1990; Wiegertjes et al. 1996; Dalmo, Ingebrigtsen, and Bøgwald 1997). It is accepted that the specific immune response of the host may impart lasting disease resistance (Anderson 1993; Wiegertjes et al. 1996), but the mounting of an antigen-initiated immune response requires prolonged response time. In aquatic organisms, the reticuloendothelial system (RES) provides early, nonspecific protection against the microbiological challenges that surround fishes at every life stage (Tatner and Manning 1985; Dalmo, Ingebrigtsen, and Bøgwald 1997). In the plaice, *Pleuronectes platessa* L., passive immunity transferred to young fish from the mother has been attributed to lectins (Bly, Grimm, and Norris 1986). Antimicrobial factors found in fish eggs, including lysozyme and lectin, have been shown to enhance disease resistance against *Aeromonas* spp. (Yousif, Albright, and Evelyn 1994; 1995). Further studies of individual species will be needed to determine the timing of immune maturation for establishing immunostimulation feeding protocols.

Nonspecific humoral defense substances, including lysozyme and lectin, are present throughout the life history in blood, mucus, gills, skin, and the intestine (Dalmo, Ingebrigtsen, and Bøgwald 1997). In selective breeding programs to increase disease resistance, a significant positive correlation has been observed between growth rate and survival (Fjalestad, Gjedrem, and Gjerde 1993). The increase in growth rate seen with glucan immunostimulation, although not genetic, may be acting on a general fish condition and on the immune system. A more recent study of five Atlantic salmon, *Salmo salar,* strains with differential susceptibility to *Aeromonas salmonicida* (furunculosis) showed that complement and antiprotease activities were positively correlated with resistance, although heritability has yet to be determined (Marsden et al. 1996). Related secretory products of the nonspecific immune system produced by granulocytes and monocytes in the blood and endothelial cells and macrophages in the tissues are active in defending the host against infection in the skin, gills, and intestines. The kidney, spleen, and liver are primary sites of the endothelial cells and macrophages that engulf invading pathogens which reach the blood. Immunostimulants exploit the system of humoral and cellular response triggered by the organism's innate recognition of certain biochemical molecules.

NUTRITIONAL CONTRIBUTIONS TO THE IMMUNE RESPONSE

Nutritional aspects of immune function are very important because diet can have a great impact on the immune response in fish (Landolt 1989; Waagbø 1994). Blazer (1992) presents a good overview of the role of nutrition in disease resistance in fish. Vitamins C, B_6, E, and A and the minerals iron and fluoride were identified as micronutrients that could affect disease resistance. Vitamins and minerals are not immunostimulants in the strict sense because they enhance the immune system by providing substrates and serving as cofactors necessary for the immune system to work properly. Vitamin C at 1,000 mg/kg diet enhanced the nonspecific immune response by increasing oxidative burst, pinocytosis, and lysozyme activity in 146 g rainbow trout, *Oncorhynchus mykiss* (Verlhac et al. 1998). Vitamin C increases hemolytic activity when given in high doses, and complement activity is enhanced because vitamin C is involved in formation of a subcomponent of complement C1. However, dietary vitamin C was not found to compensate for the downregulation of the immune system in 15 g Atlantic salmon stressed by confinement (Thompson et al. 1993). Vitamin A fed to 20 g Atlantic salmon at a normal diet concentration (1.95 mg/kg diet) and at a high (15 mg/kg diet) level increased humoral antiprotease and bactericidal activity significantly (Thompson et al. 1994). Vitamin A increases the nonspecific cytotoxic cells that can kill a variety of transformed cells. Supplemental vitamin E (60 and 2,500 mg/kg diet) has caused an increase in phagocytic response of macrophages of small (6 g) channel catfish, *Ictalurus punctatus* (Wise et al. 1993). Three-gram channel catfish fed 4.0 mg folic acid/kg diet had maximal ratios of leucocytes and lymphocytes to erythrocytes, indicating that immunocompetence may increase as the dietary dose exceeds the requirement (1.01 ± 0.27 mg/kg diet) for weight gain (Duncan et al. 1993).

Polyunsaturated fatty acids of the linolenic series (n-3) found in fish diets have been shown to increase disease resistance in fish (Blazer 1992; Rowley et al. 1995). In addition, cellular membranes become stronger and more resistant to lysis. Fluidity of the cell membrane depends on the fatty acid composition. Therefore, leukocyte functions may also be affected. However, high levels of n-3 polyunsaturated fatty acids can cause oxidative stress unless counteracted by antioxidants such as vitamins E and C or minerals (Erdal et al. 1991). Selenium, zinc, and copper are minerals that are components of the antioxidant enzymes glutathione peroxidase and zinc-copper superoxide dismutase (Shils and Young 1988). The ratio of fatty acids is also important for disease resistance. Results of the diet trial with Atlantic salmon (18 to 20 g) suggest that fish fed the low ratio of n-3 to n-6 polyunsaturated fatty acids may be less resistant to infection (Thompson, Tatner, and Henderson 1996). In contrast, channel catfish fed diets supplemented with menhaden oil (high in n-3 fatty acids) had higher mortalities when challenged with *Edwardsiella ictaluri* (Li et al. 1994). Fracalossi and Lovell (1994) also found that channel catfish fed oils high in n-3 fatty acids had higher mortalities when challenged with *E. ictaluri.* The adverse effect of feeding high dietary levels of n-3 fatty acids observed in channel catfish but not in Atlantic salmon is probably due to the differences in n-3 fatty acid requirements of these species.

The effects of diet composition, feeding rates, and food efficiencies at different temperatures will need to be considered when evaluating diet-borne immunostimulants. Sakai (1999) reviewed the sparse literature on growth effects of immunostimulants and found inconsistent results among species. Comparisons of different commercial products should include a knowledge of all components of the diet used. A single component of a commercial diet, for example, vitamin C, may affect the efficacy of immunostimulants. The nutritional attributes of substances added to diets for their immunostimulatory properties should be investigated for interactions with other nutritional components. Using 146 g rainbow trout, Verlhac and colleagues (1998) found a significant interaction ($P < 0.01$) between the glucan (MacroGard, KS Biotec-Mackzymal, Tromso, Norway*) and vitamin C at 150 or 1,000 mg/kg in the diet, enhancing complement interaction through the classical pathway. Substances administered as immunostimulants in the diet may have a separate, positive effect on fish condition by increasing growth rates, as was demonstrated in sea bream, *Sparus aurata* (Mulero et al. 1998). Landolt (1989) reviewed findings for enhancement of general disease resistance with diets containing various levels of minerals and other nutritional substances. The importance of these observations is that similar effects are seen with immunostimulants among a variety of species.

IMMUNOSTIMULANT SOURCES AND MOLECULAR STRUCTURE

Sources of immunostimulants for aquaculture use may be chemically or biologically produced (Nikl, Albright, and Evelyn 1991; Raa 1996; Sakai 1999). However, the latter predominate in both research and the few feeding studies that exist (Robertsen, Engstad, and Jørgensen 1994; Raa 1996; Sakai

---

*Mention of manufacturer or trade name does not mean endorsement.

1999). The biological sources of most immunostimulants added to commercial fish diets include mycelial fungus and yeasts and a few bacterial preparations. Nonspecific immunity in higher animals, such as fish, has developed over evolutionary time in an environment with large numbers of these cells and their by-products. Fish and other aquatic organisms are constantly exposed to the waterborne pathogens and substances that provide the trigger to mount an immune response. Through the process of selection, substances that promote immune stimulation, including pathogen invaders, have produced different levels of immune response in fishes occupying vastly different environments.

Most of the immunostimulants in fish diets are polysaccharides derived from bacteria, fungi or yeasts, and plants. The substances may be the cells themselves or preparations from the cell walls containing the ß-1,3- and ß-1,6-glucan molecules that initiate the nonspecific immune response. An emphasis in fisheries research has been placed on ß-glucans (Robertsen, Engstad, and Jørgensen 1994), preferred in aquaculture because they occur naturally and are less likely to cause concerns about residue in food fish or water quality. Structure of the glucans is very important for activity; ß-glucans without any or with few side chains are of limited activity in fish. Figure 12.1 shows part of a generic glucan. Note the ß-(1,6)- and ß-(1,3)-glucan branches. Yeast glucans are the most commonly used immunomodulators in aquaculture. In processing, a layer of protein and lipids has to be removed to expose the active ß-glucans found on the inner wall of yeast cells. The proportion of ß-1,6 side chains to the number of glucose molecules in the ß-1,3 backbone are important because fewer and shorter side chains may reduce the activity of the glucan in fish.

Physical treatment and processing of the glucans can affect length of the side chains and, thus, the potency of glucan products (Saito, Yoshioka, and Uehara 1991). The ratio of ß-1,6 side chains to the number of glucose molecules in the ß-1,3 backbone may also be affected. Several yeast species are used consistently in the production of many commercial glucan products. The most commonly used sources of glucan products include *Saccharomyces cerevisiae* (baker's yeast), the source of MacroGard and M-glucan, and the fungal preparations *Schizophyllum commune* and *Sclerotium glucanicum* (Robertsen et al. 1990; Nikl, Albright, and Evelyn 1991; Chen and Ainsworth 1992; Engstad, Robertsen, and Frivold 1992; Matsuyama, Mangindaan, and Yano 1992; Jeney and Anderson 1993a; Jørgensen, Lunde, and Robertsen 1993; Jørgensen et al. 1993; Jørgensen and Robertsen 1995; Siwicki, Anderson, and Rumsey 1994; Sung, Kou, and Song 1994; Santarém, Novoa, and Figueras 1997). The molecular structure of MacroGard has been examined in some detail. It is a branched molecule with 83 percent of the glucose units bound in ß-1,3 linkages, 5 percent ß-1,6 branching points, 6 percent ß-1,6 linkages, and 5 percent of the glucose molecules in the nonreducing terminal positions. The complete structure of the mol-

FIGURE 12.1. ß-(1,3)-Glucan Backbone of a Generic Glucan Structure, Including a ß-(1,3)-Glucan Branch and a ß-(1,6)-Glucan Branch

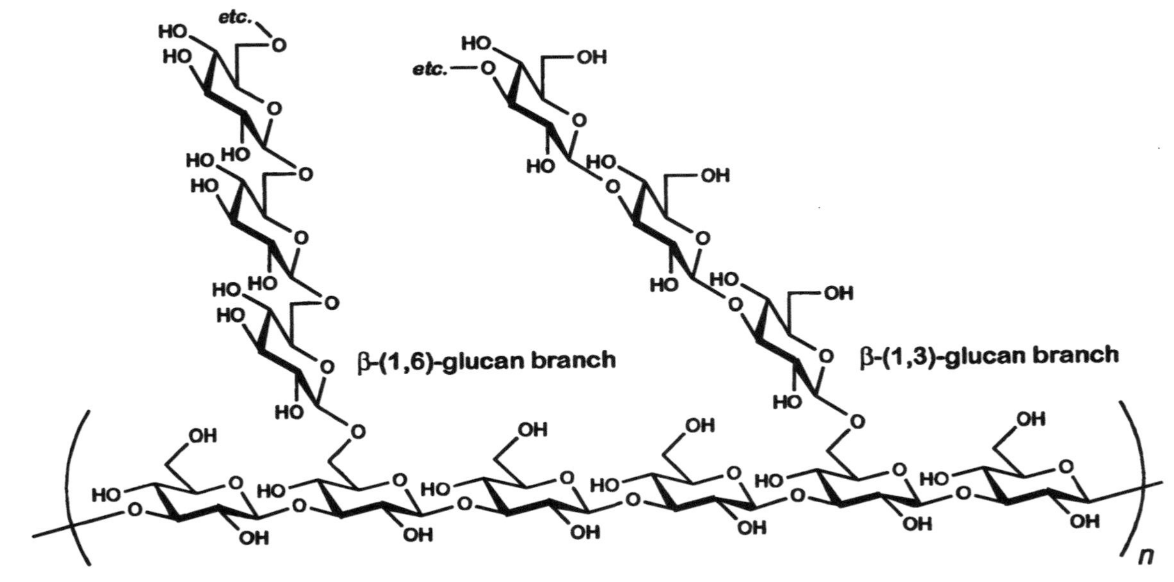

Source: Courtesy of Dr. Peter Livant.

ecule is not known. MacroGard would have greater efficacy if the ß-1,6 polyglucose side chains were cleaved, leaving only the ß-1,3 side chains on the ß-1,3-glucose backbone structure (Robertsen, Engstad, and Jørgensen 1994). The apparent contradiction in which the structure of MacroGard is most active may be due to the possibility that different forms are more active depending on the mode of administration and which indicators of immune function are examined. Robertsen (1999) suggested that oral administration of high molecular weight ß-glucans may induce both mucosal and systemic responses in vertebrates. Lentinan from *Lentinus edodes* (shiitake mushrooms) has two glucose branches for every five ß-1,3 glucose units. Schizophyllan from *Schizophyllum commune* contains one glucose branch for every third glucose in the ß-1,3 backbone. VitaStim-Taito (VST) is produced from an *S. commune* preparation. Scleroglucan, an excretion product of *S. glucanum,* is similar in form to schizophyllan. Several studies have compared the immunostimulatory effect of other types of substances to ß-glucans (Nikl, Albright, and Evelyn 1991; Siwicki, Anderson, and Rumsey 1994), but in oral administration trials, the yeast ß-glucans predominate. An earlier study investigated ten polysaccharides to compare the effect of chemical structure on the ability to activate the alternative complement pathway and macrophages (Yano, Matsuyama, and Mangindaan 1991). The ß-1,6-glucosidic side chains of the ß-1, 3-glucan branched side chains found in lentinan, schizophyllan, and scleroglucan preparations were identified as potentiators of the alternative complement pathway in fish.

Saito, Yoshioka, and Uehara (1991) found in vitro that high molecular weight (1,3)-ß-D-glucans in the single helix form were the most potent in activating coagulation factor G from limulus amebocyte lysate. In a later study, Aketagawa et al. (1993) also found the single helix was more active. Using curdlan, ß-1,3 glucan from the culture medium of the bacterium *Alcaligenes faecalis,* purified paramylon from *Euglena gracilis,* and schizophyllan from *S. commune,* the authors determined that the glucan with the single helical conformation had a greater ability to activate limulus coagulation factor G. However, in fish, the triple helical conformation as found in VitaStim-Taito (Nikl, Evelyn, and Albright 1993) appears to have greater activity than the single helix.

Other compounds that may have an immunostimulatory effect include polysaccharides containing sugars other than glucose (glycans) (Yano, Matsuyama, and Mangindaan 1991). Five-gram Atlantic salmon have been immunized with an injection of a polysaccharide from the broth culture supernatant of *Aeromonas salmonicida,* having mannose as the major constituent (Bricknell et al. 1997). Peptides in fish by-products can also enhance the nonspecific defense and immune response in fish. Bøgwald et al. (1996) used Atlantic cod, *Gadus morhua,* muscle protein hydrolysate in an intraperitoneal injection in Atlantic salmon. Superoxide anion production was enhanced in the head kidney leucocytes when assayed at two and four days postinjection. It was suggested that fish protein hydrolysate could be used in fish diets as a prophylactic treatment. Muramyl dipeptides from mycobacterial cell wall preparations found in Freund's complete adjuvant can also cause an immunostimulatory effect (Olivier, Evelyn, and Lallier 1985). Gildberg and Mikkelsen (1998) used acid peptides extracted from cod stomach hydrolysate as an immunostimulant in diets for Atlantic cod fry. Results showed that some stimulation of the nonspecific immune system occurred in the cod fry. The algae, *Spirulina,* fed to 18 g channel catfish at 2.7 percent of the diet, enhanced antibody response to a thymus-dependent antigen but not a thymus-independent antigen (Duncan and Klesius 1996b).

Different immunostimulants may prove effective for different life stages, based on solubility. An algal extract, laminaran, more soluble than the fungal and yeast glucans, has also proven to activate macrophages (Dalmo and Seljelid 1995) and to increase respiratory burst activity in anterior kidney leucocytes of Atlantic salmon (Dalmo, Bøgwald, et al. 1996). The authors promoted laminaran for its superior solubility compared with other ß-1,3-glucans, a factor that should be considered in candidate substances for diet applications. Absorption of laminaran from water by yolk-sac larvae of Atlantic halibut, *Hippoglossus hippoglosus,* through the skin and intestinal epithelium (Strand and Dalmo 1997) suggests that it may have the potential to enhance immunity in early life stages before the development of acquired immunity.

Products of bacteria may also stimulate nonspecific immune response in fish. However, research has predominantly focused on the effects of individual components of immune response at the cellular level, rather than considering the whole aquatic organism. Bacterins (antigenic compounds prepared from bacteria) are prepared in a variety of ways for immunizations to provide protection against diseases (Jeney and Anderson 1993a), even when antibodies are not detected following immunization (Anderson 1993). The bacterin may act as an immunostimulant on a nonspecific level, providing the broad-spectrum responses offered by other bacterial products (Jeney and Anderson 1993a; Raa 1996). Chen and colleagues (1998) used extracellular products of *Mycobacterium* spp. mixed with Freund's adjuvant for intraperitoneal injection to stimulate nonspecific response in Nile tilapia, *Oreochromis nilotica.* In that study, neutrophil activity, oxygen radical production, and serum lysozyme activity increased. The phasing of the peaks of activities of these immune factors emphasized the importance of the time of sampling on the analysis of effects. Glucans have been promoted as adjuvants for bacterins and vaccines, but contradictory evidence of enhanced immunization effects and nonspecific immune stimulation have been found with bacterins of different origins. Glucans alone may promote a similar nonspecific response to a *Yersinia ruckeri* O-antigen in rainbow trout (Jeney and Anderson 1993a). They may also increase nonspecific stimulation, compared with a bacterin alone, to *Aeromonas salmonicida* in Atlantic salmon (Aakre et al. 1994). In turbot, *Scophthalmus maximus,* ß-glucan did not act as an adjuvant with a vibriosis vaccine (de Baulny et al. 1996). In rainbow trout, extracellular products of *Mycobacterium* spp. acted on the specific immune system, inducing high antibody titers in vaccinated fish (Chen et al. 1996). The combined effects of several immunomodulatory substances and enhancement of both the nonspecific and specific immune systems should be studied based on the particular immune factors they stimulate.

Synthetic substances are often chemical copies of peptides found in natural cell walls, which activate macrophages and cells of the specific immune response, including T and B lymphocytes. The mode of action includes inhibition of peptidases in the macrophage and lymphocyte cell walls and stimulation of macrophage secretion of cytokines that promote monocyte and granulocyte production. Originally developed to treat helminths in ruminants, synthetic preparations such as levamisole continue to be investigated as dietary immune enhancers (Mulero et al. 1998). The compound also acts as an immunostimulant, causing a nonspecific immune response in fish. Mulero, Estaban, and Meseguer (1998) reported that levamisole targets lymphocytes to increase lymphokine production, which in turn activates macrophages. This indirect effect may explain inconsistencies in results among different experimenters. Increased disease resistance and growth in 100 g gilthead bream (*Sparus aurata* L.) against *Vibrio anguillarum* were demonstrated in a dietary study with levamisole. Phagocytosis and respiratory burst activity increased, and a delayed rise in serum complement occurred. The levamisole-treated group showed increased phagocytosis of *Vibrio anguillarum* by head kidney leucocytes and lower mortality after injection challenge with *Vibrio.* However, levamisole has been found to have an immunosuppressive effect in higher doses in sea bream. Results of in vitro trials of the same immunostimulant may not be predictive of potential immune enhancement in diet trials (Mulero et al. 1998). This phenomenon has been demonstrated with natural immunostimulatory substances as well. Yeast glucans imparted enhanced disease resistance in channel catfish against *E. ictaluri* when administered intraperitoneally (Chen and Ainsworth 1992), but not when administered orally (Duncan and Klesius 1996a). The discovery that levamisole also stimulated phagocytic activity and increased disease resistance in trout and carp should prompt future examination of other aquaculture therapeutics and diet additives as potential immune enhancers or inhibitors.

MODES OF ACTION OF IMMUNOSTIMULANTS

Excellent reviews of the literature concerning immunostimulants for fish have been compiled by Robertsen, Engstad, and Jørgensen (1994), Raa (1996), and Sakai (1999). The emphases of these reviews have been laboratory studies that investigate intraorganismal responses in a few commercial fish species after treatment primarily with ß-1,3-glucans. Investigations of the effects of different types of orally administered immunostimulants during the past decade have focused on individual measurements of immune stimulation, with very few investigations on the absorption of immunostimulants (Ingebrigtsen et al. 1993; Sveinbjørnsson et al. 1995). Measured nonspecific responses include macrophage and phagocytic activity, killing activity, reactive oxygen species, and chemiluminescent response. Humoral responses include increases in serum complement and lysozyme activities, two immune substances associated with both nonspecific and specific immune response. Intraorganismal changes are associated with differences in disease resistance after challenge with pathogens known to cause mortality in cultured fish. The present chapter concentrates on the limited number of studies in which immunostimulation has been originated in the diet, with emphasis on areas of concern for production-level applications. Immunostimulants in fish diets act by inducing or suppressing different nonspecific defense mechanisms in fish.

Investigations of immunostimulants in fish diets have focused primarily on the humoral and cellular components of nonspecific response at the stage after absorption into the blood. Measurement of immune factors in the serum show that humoral changes occur when immunostimulants are administered, including a selective increase in lysozyme and an increase in complement in 65 to 220 g Atlantic salmon (Engstad, Robertsen, and Frivold 1992). Increases in lysozyme have been measured in trout (Jørgensen et al. 1993), Atlantic salmon (Engstad, Robertsen, and Frivold 1992), and yellowtail, *Seriola quinqueradiata* (Matsuyama, Mangindaan, and Yano 1992), after intraperitoneal injection or oral administration of yeast glucans. Glucans administration in Atlantic salmon increased serum protease and lysozyme activity after four days, and there was a peak in activity at ten days during an *A. salmonicida* infection (Møyner et al. 1993). Lysozyme was found to increase after two weeks in turbot (30 g) fed yeast glucan (de Baulny et al. 1996), an effect that persisted in a group fed the immunostimulant and administered a vaccine. Further investigations should include tests for immediate increases of humoral substances found in the external mucus where pathogens are first encountered. If the immunostimulant is water soluble, direct incidental exposure to immunostimulants in the diet would be expected externally, in addition to the response initiated after the diet was first ingested. The consequences of preexisting high lysozyme levels before an infection or the presence of increased lysozyme due to latent infection before immunostimulation must be considered when determining the timing, dosage, and measurement of effects of enhanced diets. High blood leucocyte numbers were accompanied by increases in oxidative radical production and phagocytic activity in yellowtail (Matsuyama, Mangindaan, and Yano 1992) and rainbow trout, respectively (Jeney and Anderson 1993b; Siwicki, Anderson, and Rumsey 1994).

Studies of bacterial infections reveal that differences in how bacteria infect the organism will require different prophylactic immunostimulants. *Vibrio anguillarum* is transported across the epithelium of the anterior intestine to the lamina propria to the liver, blood, and other tissues (Grisez et al. 1996). Yeast glucans are effective oral immunostimulants against *V. anguil-*

larum in salmon and as an adjuvant in a vaccine for *Vibrio*. The complete mechanism of action of most immunostimulants must still be defined based on experiments using intraperitoneal administration because the type of immune factors examined in oral trials is limited (Sakai 1999).

Intraperitoneal injections of glucans have both systemic and local effects on the nonspecific defense system of the fish. Injections of fungal preparations, schizophyllan, scleroglucan, and lentinan in common carp, *Cyprinus carpio,* at 2 to 10 mg/kg fish enhanced the phagocytic activity of the pronephros phagocytic cells (Yano, Mangindaan, and Matsuyama 1989). Survival of the fish treated with the glucans was improved. If yeast glucans are injected in 20 to 30 g Atlantic salmon at a high dose, 100 mg/kg fish, there is no protection for one week, but maximum protection occurs after three to four weeks. If a low dose (2 to 10 mg/kg) is used, protection occurs at day seven, then declines (Robertsen et al. 1990). The prophylactic use of ß-1,3-glucan for channel catfish culture has been proposed, based on laboratory trials of injected glucan (Chen and Ainsworth 1992). Each immunostimulant acts differently, and the mode of action is unclear. Several factors affect the efficacy of glucans. Mode of delivery should be examined. When fed, protection is lower than when administered via injection. One reason for the lower protection could be that if the glucans are being fed at a certain percent body weight, all of the fish may not be getting a full dose, and a positive effect may not be observed. In addition, the high molecular weight of some glucans raises the question of whether they are directly absorbed by the fish or need further digestion.

According to Sveinbjørnsson and colleagues (1995), the aminated polysaccharide ß-1,3-D-polyglucose was taken up through epithelial cells in the intestine. The glucan, labeled with ^{125}I-tyramine cellobiose, was traced after intragastric and rectal administration. Thirty minutes after the dose was given to Atlantic salmon (50 to 80 g), radioactive material was found in the hindgut epithelial cell vacuoles, liver, spleen, heart, and kidney. Dalmo, Bøgwald, and colleagues (1996), using large Atlantic salmon, also determined that ß-1,3-D-glucan (laminaran) was absorbed into vesicles in the intestinal epithelial cells and was found in anterior kidney macrophages. The tissue distribution of radioactive-labeled, aminated ß-1,3-glucan (immediately and up to 24 hours after intravenous, intraperitoneal, and peroral administration in Atlantic salmon) suggests that regular daily administration of glucans, even orally, would maintain the presence of glucans in the identified tissues (Ingebrigtsen et al. 1993). Jørgensen and Robertsen (1995) found evidence of glucan receptors in Atlantic salmon macrophages. The secretion of complement components and lysozyme activity was associated with increased phagocytotsis in mammals and has been demonstrated in yellowtail after stimulation with ß-1,3-glucan (Matsuyama, Mangindaan, and Yano 1992). A membrane receptor protein was found for ß-1,3-D-glucan in crustacean blood (Cerenius et al. 1994) and, when activated with glucan or other lipopolysaccharides, initiated phagocytosis and encapsulation of microorganisms by blood cells. Rombout and van den Berg (1989) and Rombout, Bot, and Taverne-Thiele (1989) showed that both soluble and particulate antigens are taken up by epithelial cells of the second gut segment of the hindgut and transported to intraepithelial macrophages. In addition, immunological defense cells, such as lymphocytes, granulocytes, macrophages, and monocyte-like cells, were identified in the intestinal mucosa of the common carp. Pathogens that can pass the physical barriers of skin and mucus and the secreted nonspecific humoral factors encounter the cells of the mucosal lining. The mucus-producing cells are found in the mucosal lining with the epidermal phagocytic cells that produce the nonspecific humoral factors. Macrophages, phagocytes, and granulocytes are the effector cells of the immune systems, which are enhanced by immunostimulants in fish diets. Through absorption, the immune function and disease resistance of the entire organism is stimulated. After introduction to the fish, glucan cleared more quickly from the blood than from the kidney and liver. An important finding was that glucan absorbed in the stomach persisted in the blood much longer than that administered rectally. Therefore, the possibility of immediate stomach absorption, followed by continued, slower absorption in the intestine, should be considered in designing diet studies. Daily feeding should maintain measurable levels of glucan in the organs investigated in these two studies, but adequate dosages need to be determined. The possibility of bioaccumulation and immune system inhibition should be evaluated. The glucans in the Sveinbjørnsson et al. (1995) and Dalmo et al. (1996) studies were given to the fish by intubation, perorally and peranally. The experimental method of administration raises the question of how well the glucan would be absorbed in a diet matrix.

The use of glucans in a bath treatment for fish has not been examined in depth. In one study, young 20 to 30 g rainbow trout were immersed in a glucan preparation from barley (100 μg/ml). Exposure to the glucan caused the nonspecific disease resistance mechanisms to be activated. Blood samples taken at 240 hours showed the total number of neutrophils was still high compared with control fish. Phagocytic cell activity increased 24 hours after exposure and reached a maximum at 72 to 96 hours (Jeney and Anderson 1993b). Nikl, Evelyn, and Albright (1993) used the glucan preparation VitaStim-Taito (VST) in a bath for 6.5 g chinook salmon, *Oncorhynchus tshawytscha.* Two forms of VST were used: one was highly purified, with a molecular weight (MW) of 4.75×10^5 and the other was a crude form with an MW of 1 to 2×10^6. Neither form provided protection to the fish as a bath. Ingebrigtsen and colleagues (1993) showed the persistence of high concentrations of glucans in the gastrointestinal contents after peroral administration. Therefore, in intensive fish culture where high levels of immunostimulatory substances are used in diets, there is the possibility that sufficient amounts of immunostimulants could be excreted to promote some level of nonspecific immune response.

How does immune system stimulation on the molecular or cellular level translate into disease resistance and improved survival and health of individuals at aquaculture facilities or hatcheries? This question remains unanswered. Phagocytosis activated by an algal ß-1,3-glucan, laminarin, has been localized in the skin and intestinal epithelial layer of Atlantic halibut yolk-sac larvae (Strand and Dalmo 1997). This study documented direct immunostimulation of secondary nonspecific immune responses by localization of absorbed immunostimulants in the epithelium of larval fishes during the period before development of specific immune response. Manning, Grace, and Secombes (1982) reviewed earlier studies that reported phagocytosis by free macrophages in the skin, gut, gills, kidney, and connective tissue of trout at four days and in common carp at two weeks posthatch. The ability of fish to respond to immunostimulation may be exploited very early in their life history to reduce mortality in yolk-sac larvae. Strand and Dalmo (1997) administered the laminaran in water; therefore, the possibility of immediate stimulation of phagocytic epidermal cells by direct contact with waterborne diet particles should be explored. Other reports of poor absorption of radio-labeled aminated ß-1,3-polyglucose in Atlantic salmon after peroral (Ingebrigtsen et al. 1993) and intragastric administration in Atlantic salmon (Sveinbjørnsson et al. 1995) suggest that dosages to predict uptakes need further study before results can be applied at production levels.

The nonspecific immune system in fish is often called more primitive than the specific response system. The nonspecific system responds to general classes of nonself substances, allowing quick mobilization of humoral and cellular components of the immune system, with key tissues or organs participating in the defense. Contradictory findings among studies may be attributed to different immunostimulant doses, molecular forms, modes of administration, and the measured immune factors. Ultimately, enhanced disease resistance, as demonstrated by increased survival after pathogen challenge is the desired outcome. Prophylactic immunomodulators should increase both cell numbers and secretion of antimicrobial agents in anticipation of different pathogens. Immunostimulants may also be selected for their efficacy features if known pathogen outbreaks occur routinely.

PATHOGENS AND IMMUNOSTIMULANTS

Gram-negative pathogens used in disease challenge experiments following immunostimulation are pathogens common in the culture of salmonids, as are Gram-positive pathogens found in warm-water, tropical, marine, or shellfish culture. Disease challenge experiments to test for enhanced disease resistance after immunostimulation have focused on pathogens common in the high culture densities used in aquaculture, such as *V. anguillarum, V. ordalii, V. vulnificus, A. salmonicida, A. hydrophila, Yersinia ruckerii, Edwardsiella tarda, E. ictaluri, Pasteurella piscicida, Enterococcus seriolicida, Streptococcus* sp., and yellow-head baculovirus. Several of these pathogens are considered primary pathogens. Outbreaks of *V. anguillarum* and *A. salmonicida* occur even under the best of conditions. Secondary pathogens such as *A. hydrophila* and some *Pseudomonas* species require that the fish be compromised in some way. Aquaculture provides many compromises from crowding, poor water quality, and handling stress. Other opportunistic diseases such as those caused by fungi abound. Potential target pathogens for immunostimulant treatment are other common aquaculture pathogens, for example, *Piscirickettsia salmonis.* Outbreaks typically occur weeks after transfer to seawater, the assumed reservoir of infection. Because the history of detection and identification of the disease in the natural environment is recent compared to other diseases of cultured fish (Fryer et al. 1990), the mode of infection is less well understood. The compromise of kidney, spleen, and liver suggests that immunostimulant therapy may be used to increase macrophage activity in those tissues. Diseases that invade the skin or muscle cause focal dermomyonecrosis, and ulcerative and necrotic lesions would be logical candidates for immunostimulant prophylactic measures against *A. salmonicida* and *V. anguillarum.* These are primary pathogens that cause periodic outbreaks even under the best culture conditions and are responsible for chronic infections that reduce fish quality by affecting skin and muscle. Chronic, proliferative diseases that cause lesions in the haematopoietic tissue, such as *Renibacterium salmoninarum* and *Mycobacterium* spp., can survive within macrophages because they are not affected by heightened phagocytic activity or respiratory burst activity. These bacteria have not proven susceptible to heightened immunostimulation in research trials.

Glucan administration has proven to increase disease resistance against *E. ictaluri* in channel catfish, against *A. salmonicida* and various *Vibrio* species in salmonids, and against pathogens common to farmed marine species such as *P. piscicida* (Matsuyama, Mangindaan, and Yano 1992). The effectiveness of the treatment apparently depends not only on dose but also on the innate resistance of the fish species to the pathogen and on environmental influences such as water temperature and dissolved oxygen. Consideration must also be given to virulence of the bacterial strain. It is known that recurring outbreaks of the same disease at an aquaculture facility may not be caused by the same strain. Documentation of both typical and atypical *A. salmonicida* outbreaks on fish farms shows that bacterial strains must be considered in the evaluation of the efficacy of particular immunostimulants against a disease. Differences in effectiveness of immunostimulant treatment, as measured by survival after disease challenge, may be the result of difference in virulence. The development of resistant strains of bacteria because of antibiotic treatments must also be considered. The defense mechanism developed by the fish species may also be changed by the antibiotic treatments.

RELATIONSHIP TO OTHER REGULATORY SYSTEMS

The interesting link among different regulatory systems in fish is found in the head kidney. The head kidney combines in one organ the site of haematopoesis, antibody, and cortisol production, which are all important features of both the immune and endocrine systems (Weyts et al. 1999). Protein and carbohydrate metabolisms are influenced by corticosteroids, a link between nutrition and endocrine function found in a tissue where glucans have been localized after absorption (Ingebrigtsen et al. 1993; Sveinbjørnsson et al. 1995). Many nonspecific humoral response factors act between the regulatory systems during times of stress. A review of the reticuloendothelial system in fish (Dalmo, Ingebrigtsen, and Bøgwald 1997) describes much of the current understanding of immune components of both the specific and nonspecific systems and of their levels of interaction.

DOSAGE AND ABSORPTION

A major task remaining for the promoters of immunomodulant diets is the determination of prophylactic and therapeutic regimens for aquaculture. Tissue distribution after absorption by intestinal uptake has been determined for organs important in both nonspecific and specific immunity (Sveinbjørnsson et al. 1995). Limited research shows that peroral administration permits more than trace amounts of immunomodulator to reach lymphoid tissues (Ingrebrigtsen et al. 1993). In larval halibut, the algal preparation laminarin was absorbed from the posterior intestine (Strand and Dalmo 1997). After intravenous administration, laminaran accumulated in the spleen and kidney in Atlantic salmon (Dalmo et al. 1995) and in the heart of Atlantic cod (Dalmo, Ingebrigtsen, et al. 1996). Different glucan preparations were used; therefore, levels of uptake may have been influenced by molecular structure. Intravenous injection trials show that many forms of immunostimulants leave the blood stream to be engulfed by macrophages and endothelial cells in the spleen, kidney, heart, and liver (Dalmo et al. 1995; Dalmo, Ingebrigtsen, et al. 1996). It can be presumed that once absorbed into the blood after ingestion from the diet, the immunostimulants make their way to the lymphoid tissues. This assumption disallows any structural molecular changes in the immunostimulant during residence in the gastrointestinal tract, a presumption that needs to be tested before specific immunostimulants are promoted for use against specific pathogens.

Numerous questions remain about the dosage, length of time to feed immunostimulants, and persistence of effects for every immunostimulant and every fish species. Dosage and duration cannot be separated because the duration of feeding affects not only the efficacy but also the possible accumulation and, therefore, the persistence in the fish. Results from intraperitoneal injection of channel catfish with a baker's yeast-derived glucan demonstrated that phagocytic activity increased and mortality decreased in fish treated with glucan and challenged with *E. ictaluri* (Chen and Ainsworth 1992). The researchers concluded that glucan has a systemic, although short-lived effect, and that continual administration might be necessary to sustain effective levels. During injection trials in the same study, the timing of glucan injection in relation to antigen administration greatly affected antibody titer levels. It should be considered that all effects might be lessened or delayed in feeding trials where absorption is decreased (Dalmo, Bøgwald et al. 1996). Also, there are cost considerations of long-term administration via diet and the high dosages needed to compensate for low absorption when administered in diets.

Intraperitoneal injections of glucans have both systemic and local effects on the nonspecific defense system of the fish. Yeast glucans injected at a high dose (100 mg/kg of fish) provided no protection for week one, but maximum protection occurred after three to four weeks. A low dose (2 to 10 mg/kg) provided protection on day seven, then declined (Robertsen et al. 1990). Peroral administration of immunostimulants allows better control of dosage than feeding trials, providing results that can be expected when a particular substance has passed through or been absorbed in the gastrointestinal tract. Dosages would be a more critical variable in diet trials. Laminaran, a ß-1,3-D-glucan, elicited increased superoxide anion production and increased lysosomal acid phosphatase activity in anterior head kidney leucocytes in Atlantic salmon, in samples collected two days after peroral treatment (Dalmo, Bøgwald, et al. 1996). Production-level feeding trials should now be developed and coordinated with absorption studies. The question of an effective dosage will be further complicated by different feeding strategies applied in culture operations.

The effect a dose of a specific glucan may have against different pathogens should also be considered in immunostimulant diet applications. M-glucan conferred a higher relative percentage protection against *V. anguillarum* than against *V. salmonicida* or *Yersinia ruckeri,* and the timing of the challenge after glucan treatment was also important (Robertsen et al. 1990). This research demonstrated that as a general prophylactic measure, some level of disease resistance might be afforded, but dosage, timing of administration, and target pathogens would all have to be considered separately in developing feeding protocols. Significant differences have been demonstrated repeatedly in mortality between controls and fish administered immunostimulants of varying sources (Robertsen, Engstad, and Jørgensen 1994; Raa 1996; and Sakai 1999); therefore, progress from the laboratory to practical diet studies should be attempted. In the same study, a higher dosage resulted in higher mortality in the treated group than in controls. The authors speculated that the higher glucan amount may have loaded phagocytic cells, rendering them less available to phagocytose the bacteria. This example reflects the complexity of dosage and duration interactions. Elucidation for every step of nonspecific response and interactions with the specific immune system during immunostimulation and disease challenge will be needed to explain the interrelated mechanisms of interaction.

The type or form of immunostimulant used and the immune factors that are chosen for measurement both determine when a response will occur.

Limited diet-based studies are available to distinguish between either the differences in the time of the response created by these factors or the differences attributed to holding conditions that may contribute to stress. Results of the sequence and response time of the individual nonspecific immune factors should be studied to determine what frequency of feeding of immunostimulatory substances is necessary to sustain increased nonspecific immune activity. It should be determined if the physical presence of the immune substance may inhibit macrophages and phagocytes by loading them. Detailed results of laboratory studies must be repeated on the production level to determine the dosages, how the different immunostimulants are absorbed, where they are retained and for how long, and when maximum protection occurs.

BASAL IMMUNOSTIMULATORY STATUS

Before the expense and haphazard application of immune enhancers occur, baseline levels and the timing and magnitude of innate nonspecific immune response to specific pathogens by a particular fish species should be better understood. Factors common to all experiments that may confound experimental results will be considered first. The simplest case would be pathogen-free fish, raised in pathogen-free water with no immunostimulatory substances present in the diet. Table 12.1 presents estimated effects of the contribution that each factor would make to immunostimulation under common conditions in aquaculture. The table begins by assuming no effect of latent innate resistance to the pathogen, no effects of other diet additives on basal immunity, and no effect of treated water (UV radiation, ozonation). The table does assume different levels of background immunostimulation determined by the disease status of the fish at the time they were fed an immunostimulant in the diet. Immunostimulant diets and natural waters containing natural immunostimulatory substances are assigned a single-level effect. The table is designed to present the concept of preexisting immune stimulation. Pathogen-free fish, raised in pathogen-free water with no immunostimulants in the diet, would not experience an effect, assuming that the immune system is in a resting state for the life stage being considered.

This simple depiction of possible compounded effects of varied sources of immunostimulation serves to show the complexity of interactions that may affect experimental results; they should be considered for practical applications. The table does not take into account additional confounding factors that have not been quantified, in order to allow speculation about interactions of more complex aspects of the fish, environment, or diet in immune stimulation. Also, the presentation does not allow for differences in innate immunity of fish species or stocks against specific pathogens. Innate immunity would modulate an individual organism's response to pathogens, naturally occurring immunostimulants in the water, types of immuno-

TABLE 12.1. Determining Background Immune Status of Aquatic Organisms

Water Source	Pathogen	Dietary Immunostimulant	Level of Immune Stimulation
0	0	0	0
+	0	0	+
0	0	+	+
+	0	+	++
0	+	0	+
++	+	0	+++
0	+	+	++
+	+	+	+++
0	++	0	++
+	++	0	+++
0	++	+	+++
++	++	+	+++++
0	+++	0	+++
+	+++	0	++++
0	+++	+	++++
+	+++	+	+++++

Confounding Variables

Cultured aquatic species	**Environment**
Innate disease resistance	Natural immunostimulant background
Life history stage	Temperature
Acquired immunity	Season
Pathogen	**Dietary immunostimulant**
Strain virulence	Molecular structure
Mode of infection	Mode of action
Presence of antibodies	Dose and absorption efficiency

Note: The magnitude of the immune response to the absence (0) or presence (+) of exogenous sources of immune stimulation is determined by quantitative exposure (0, +, ++, +++) and many confounding variables. Different sources of immune stimulation may elicit varying degrees of response: water—treated (0), spring or well (+), or river (++); pathogen-carrier state (+), subacute (++), or acute (+++) infections; and dietary immunostimulants in varying forms and doses (0, +, ++, +++).

stimulants used, and even dosage. The table assumes no immunostimulation by spring and well water, an assumption that would need to be tested at each facility. Furthermore, seasonal changes in immune competence, as well as those that occur throughout the fish's life history, will also affect results.

PATHOGEN-HOST INTERACTIONS AND INNATE DISEASE RESISTANCE

Many sites of entry are available to pathogens, and opportunities increase in cultured fish due to crowding and stress. Pathogen invasions occur at all external sites and through gastrointestinal routes. Knowledge of the localization of macrophages in absorptive tissues and their mobilization after immune stimulation by immunomodulating substances or specific pathogens is crucial for the assessment of enhanced diets. An opportunistic pathogen such as *A. salmonicida* occurs commonly in cultured fish, due to abrasions caused by crowding, and is inhibited by immunostimulants of several biological origins in Atlantic salmon; rainbow trout (Raa et al. 1992); coho, *Oncorhynchus kisutch;* and chinook salmon, *O. tshawytscha* (Nikl, Albright, and Evelyn 1991; Nikl, Evelyn, and Albright 1993; Siwicki, Anderson, and Rumsey 1994). When a pathogen first encounters the fish's external surface, mucus forms the first physical barrier, and the epithelium is the second. Macrophages, the effector cells of immunostimulants, are first encountered in the gills and gut, then in the internal organs such as the kidney and spleen. Macrophage stimulation is cited as the most frequently observed change in fish fed immunostimulants. Pathogens that are effectively phagocytized by macrophages would be expected to be those most effectively combated by prophylactic feeding. Conversely, pathogens that have to be stopped by mucus or skin would be poor candidates for attack with immunostimulants.

Studies are needed to determine the proper time to administer immunostimulants to fish. *Vibrio anguillarum,* for example, can invade the host within 30 minutes of contact; therefore, immunostimulants would need to be fed in anticipation of a potential challenge. *Vibrio anguillarum* is the normal microflora of brackish or saltwater, and there are often outbreaks when fish are transferred to netpens. Prophylactic treatment would need to occur before transfer (Galeotti 1998). The correct dosage and feeding period would allow the immunostimulant to reach the target organ. The macrophages are then primed to resist the pathogen. Studies also need to be conducted to determine if immunostimulants are useful when fed to fish that face a long emigration before reaching the estuary or ocean. Does the immunostimulatory effect persist long enough to warrant the use of immunostimulants as a method to enhance seawater transfer, for example, in hatchery fish that need to migrate hundreds of miles to the ocean? Based on the current literature in which persistence of effects was no longer than several weeks, it is questionable to assume that immunostimulators in aquaculture diets can be applied universally. Coastal populations with very short emigration times to the ocean would be better candidates for immunostimulant diets than fish with longer migration distances.

The mode of infection and spread of fish pathogens within individual fish have not been well documented. The spread of *Vibrio* administered in live feed to larval turbot was detailed by Grisez et al. (1996), and it appears that the bacteria are endocytosed in the anterior intestine. This information may explain a potential problem with immunostimulant therapy. From the studies locating glucan after administration in diet, absorption is in the posterior intestine, and transport to the liver, spleen, and kidney (Sveinbjørnsson et al. 1995). Timing of immunostimulant administration would need to be early enough to ensure adequate absorption to tissues other than the posterior intestine. Because the two locations of absorption are not the same, it is suggested that the selected immunostimulant will need to address this difference. There are some factors in *Vibrio* spp. that may explain the variation in disease resistance achieved by different immunostimulants. The bacteria, *V. anguillarum,* infect the blood, kidney, spleen, gills, and intestinal tracts, which are sites of increased nonspecific response to immunostimulants. On the other hand, *Vibrio ordalii* forms colonies in tissues with low response to immunostimulants and therefore might not be exposed to engulfment and processing by macrophages.

Aeromonas salmonicida is very common in cultured salmonids. Unlike *Vibrio,* virulence is associated with the A-layer, and a macromolecular refractive protein barrier of *A. salmonicida* protects the bacteria from bacteriophages. Phage receptors may be blocked by the protein macromolecules. Resistance to serum complement has been reported. However, another serum component, hemolysin, has been found to protect salmonid erythrocytes from *A. salmonicida* (Rockey et al. 1989). *Aeromonas hydrophila* infects warm-water species such as catfish. Latent, chronic, and acute infections that occur affect different tissues (Grizzle and Kiryu 1993). Therefore, the selected immunostimulant would need to act on all of these sites to afford disease resistance. If immunostimulants are to be considered as a prophylactic measure, feeding would need to be continual to combat the continual exposure to this disease that is ubiquitous in natural waters and transmitted by resistant carriers. Outbreaks are generally associated with stress or compromised hosts. Caution should be taken when using immunostimulants because immune inhibition has been documented when compounds are administered at high levels or for extended periods of time (Robertsen et al. 1990; de Baulny et al. 1996; Sakai 1999).

Varied susceptibility among different salmonid species to *Ceratomyxa shasta,* a parasite encountered during seaward emigration, suggests a further possible application of immunostimulants in hatchery fish. In regions where infection is anticipated, prophylactic administration of the immunostimulant might prove effective before hatchery release. *Cerotomyxa shasta* invades the host through the intestine in the mucosal epithelial cells. After the skin, these cells are the second site of nonspecific response and should be investigated for the mechanism of defense to the parasite in fish resistant to the pathogen.

It is commonly accepted that specific immunostimulating substances can increase different components of the nonspecific and specific immune systems and that disease resistance against some common fish pathogens is conferred via control of dosage, duration of feeding, and timing of pathogen challenge (Robertsen, Engstad, and Jørgensen 1994; Raa 1996; and Sakai

1999). The degree of disease resistance, as measured by survival after challenge, differs considerably among studies. The innate resistance of fish species to common bacteria explains, in part, the differences—just as basal diet formulations need to be tailored to fish and their culture environment. It is, however, apparent that considerable effort will need to be expended to evaluate immunostimulant-enhanced diets for production applications.

LIFE STAGE DIFFERENCES

Larval fish or fry begin to feed when their immune systems are immature. Defense against pathogens at this stage is restricted to the nonspecific immune system (Manning, Grace, and Secombes 1982) or limited passive immunity transferred from the mother (Bly, Grimm, and Morris 1986). The primitive status of the teleost immune system is similar to that of commercially important crustaceans, such as crayfish, *Pacifastacus leniusculus,* and may be similarly based on the presence of membrane receptor proteins for glucan-binding proteins. In crayfish, fungal 1,3-ß-D-glucans convert the prophenloxidase activating system to its active form, resulting in stimulation of phagocytosis of microbial cells (Cerenius et al. 1994).

In fish, a recognition mechanism on the macrophages has developed for ß-glucans (Jørgensen and Robertsen 1995). Phagocytosis, the primary macrophage function promoted by stimulation of the nonspecific immune system in fish, is known to develop within the first few weeks posthatch in a number of species. Manning, Grace, and Secombes (1982) summarized studies describing the delayed development of the lymphoid system in young fish, compared to nonspecific free macrophage activity. It is during this critical period when nonspecific immunity offers the sole protection. Thus, the use of immunostimulants in production diets should be studied. Free macrophages phagocytosed carbon particles in fish as young as four days posthatch, and the particles were found to accumulate under the skin, in the gut and gills. It is clear that even before feeding begins, a primitive mechanism exists for young fish to process some types of immunostimulants. Months after phagocytosis, particles were found to localize in other tissues such as the spleen, kidney, heart, and blood. These tissues had developed sites of macrophage activity. Strand and Dalmo (1997) investigated the absorption of ß-(1,3)-glucan in yolk-sac larvae of Atlantic halibut. The soluble ß-(1,3)-glucan was administered in water, and absorption in skin and intestinal epithelial cells was verified by location of the radio- and fluorescein-labeled laminaran. Low molecular weight substances were absorbed through the skin epithelium, whereas macromolecules were absorbed by the intestine. Only high molecular weight substances were retained in the larvae, and some excreted, degraded glucan was noted. In a hatchery setting, residual diet pellets may partially dissolve while in the raceway or pond, and some added immune stimulation may be imparted by this route as a "bath."

The sites of pathogen entry in fish are external surfaces and the gastrointestinal tract after ingestion. Early phagocytic activity in young fish was demonstrated in free macrophage activity, and the effects of immunostimulants were achieved by oral or bath presentation. Differences need to be determined in the timing of the onset of the stages in immunity development among species. It is clear that, until the maturation of lymphoid tissue allows for stimulation of the specific immune system, immunostimulatory substances in the diet would offer some protection against opportunistic diseases. The different lymphoid organs responsible for mounting an immune response mature at different times. The sequence of these events should be understood for each fish species to determine the timing of the types of immunostimulants fed to stimulate either nonspecific or specific immunity. Once the specific immune system has developed, a better understanding of the interactions of nonspecific and specific immunity is critical to determine sources of immune enhancement and suppression. Macrophages are closely involved in specific immunity. The simultaneous effects of immunostimulants on both the nonspecific and specific immune systems may explain some contradictions in immunostimulant research. Developmental changes add another component to the ability of the organisms to respond to immunostimulation. Thus, experiments should be conducted over extended periods of time to define different nutritional effects, and to identify immunological demands that are the result of developmental changes in the fishes.

SPECIES DIFFERENCES

The effectiveness of immunostimulants, in particular ß-(1,3)-glucans, in effecting the increase of a number of immune functions in macrophages is no longer in dispute. Discrepancies of studies using different fish species and strains of pathogens are hypothesized to be based on differences in levels of oxygen radicals (O^{-2}) produced. The bacteria may resist or neutralize certain levels of O^{-2}, or the bacterial extracellular products may be cytotoxic and affect the fish's ability to fight the pathogen (Santarém, Novoa, and Figueras 1997). Increases in serum lysozyme activity have been attributed to increases in the number of phagocytes excreting lysozyme, as well as increases in lysozyme production in these cells (Engstad, Robertsen, and Frivold 1992; Robertsen, Engstad, and Jørgensen 1994). However, bacterial strain virulence also affected results (Santarém, Novoa, and Figueras 1997).

The difference in efficacy among cold-water species is complex. Injection of Atlantic salmon with M-glucan resulted in an increased activity of lysozyme and complement, coinciding with resistance to some pathogenic bacteria, in particular *V. salmonicida*, *V. anguillarum*, and *Y. ruckeri* (Robertsen et al. 1990). Subsequent experiments by the same research group determined that the glucan derived from the yeast *Saccharomyces cerevisiae* increased the enzyme lysozyme. Levels of lysozyme, an immune substance used to demonstrate immune response, were found to be distinctive in families of fish selected for producing different cortisol (hormone involved in stress response) levels (Fevolden, Refstie, and Røed 1992). Lysozyme demonstrated a stronger genetic component than cortisol. Therefore, the effect of a selective increase in lysozyme as a measurement and/or mode of immune enhancement in different fish species needs to be investigated. Resting or basal levels of immune factors in individual fish species should be explored to predict possible effects of immunostimulators. The relationship between immune factors and stress events marked by increases in cortisol and eventually to disease resistance has been studied for the purpose of developing disease-resistant fish strains. Therefore, inheritance and situational or environmental factors may influence an immune factor that is closely related to the bactericidal activity of macrophages.

STRESS AND DISEASE RESISTANCE

In addition to conferring disease resistance, immunostimulants may have a positive effect in an aquaculture situation, to counter the immunosuppressive effects of stress, especially those associated with an increase in corticosteroids, including reduced respiratory burst activity of phagocytes and leucocyte bactericidal activity (Barton and Iwama 1991). Stresses have been associated with reduced immunocompetence that affected lymphocyte numbers and antibody production; both are components of the specific immune response. It is the elements of the nonspecific response, such as complement, that are of interest when considering immunostimulatory diets. A combination of nutrition and immunostimulants may work synergistically to reduce the effects of stress experienced by fish in an aquaculture setting. Jeney and colleagues (1997) found that feeding the glucan preparation, VitaStim-Taito at 0.1 percent in the diet to rainbow trout (200 to 300 g) several weeks before transportation helped to reduce the negative effects of stress on the nonspecific immune response. Mobilization of monocytes during periods of stress prepares the fish to combat disease challenges at a time when downregulation of the immune system compromises their ability to counter infection. Kanazawa (1997) determined that the polyunsaturated fatty acid DHA [22:6(n-3)] at 2 percent in the diet increased the tolerance of red sea bream, *Pagrus major*, larvae exposed to air and low oxygen. Red sea bream had a higher tolerance for increased water temperature when soybean lecithin (phospholipids) was included in the diet. In a study of crowding stress, gilthead sea bream, 12 g fish were stocked into tanks at a high density (over 40 kg/m^3). Another group of fish were held at low density. Fish were fed either a control diet or diets supplemented with 250 mg/kg of either vitamin C or vitamin E. The fish at the high density fed the vitamin E-supplemented diet had the same level of alternative complement activity as the low-density fish, which was significantly higher (P<0.05) than that of the high-density fish fed the control or vitamin C-supplemented diets. Therefore, additional vitamin E seems to protect the complement system against stress-related reduction of activity (Montero et al. 1999).

The relationship between disease resistance and stress response has been well explored in salmon. It has been shown that fish bred for differential response to stress exhibit differential susceptibility to a number of diseases (Fevolden, Refstie, and Røed 1992; Fevolden et al. 1993). An important factor linking stress and disease susceptibility is lysozyme, found to have a genetic heritability greater than cortisol. Lysozyme levels change during stress episodes, as well as in immunostimulant diet trials (Engstad, Robertsen, and Frivold 1992; Jørgensen, Lunde, and Robertsen 1993; Sung, Kou, and Song 1994; Thompson et al. 1993; de Baulny et al. 1996). The acknowledged stress of aquaculture is an important consideration when using research results to predict production-level performance.

ENVIRONMENTAL INFLUENCES

Recent studies that investigate the effects of environmental factors on macrophage activity are important to our understanding of immune stimulation (Le Morvan et al. 1997) and the effect that different holding conditions may have on its enhancement. In common carp, phagocytic activity of head kidney cells was enhanced at lower temperatures, as measured by phagocytosis of a yeast immunostimulator, *S. cerevisiae* (Le Morvan et al. 1997). A similar effect had been seen in rainbow trout (Hardie, Fletcher, and Secombes 1994), accompanied by increased production of macrophage activating factor in leucocytes. Observed in both warm- and cold-water species, this temperature effect on macrophages raises the question of the effectiveness of a single immunostimulant in diets to counter pathogens common at various temperatures. The possibility of seasonal and environmental changes should also be considered. Efficiency of the host's natural defense mechanisms may change at different temperatures that present different immunostimulatory environments. An observation that should also be applied to quantification of immune enhancement is to avoid a restrictive use of indicator assays and systems. In testing immunostimulatory substances to improve the health and condition of cultured fish, it would be of great value to develop large-scale studies and to examine survival under production conditions and disease challenges in environments representative of fish culture stocking densities.

Investigations using important commercial species continue to use injections of glucans to demonstrate the activation of particular components of the nonspecific immune response. An important influence on the efficacy of the glucan is the virulence of the particular pathogen. This promotes the case not only for the necessity of testing each individual fish species for effects based on their innate, inherited disease resistance but also for testing different strains of pathogens. Laboratory strains may not possess the same killing power as a more virulent strain that fish may encounter in a production facility.

SUMMARY AND CONCLUSIONS

It is not surprising that uncertainty remains in the use of immunostimulants in aquaculture. Diverse levels of disease resistance to the same pathogen exist among related species. The complex sequence of nonspecific stimulation, interacting with the specific immune system when activated by pathogen invasion, is not yet understood. The evolutionary genetics of the development of the nonspecific immune system in fishes in an environment crowded with a variety of bacteria, yeasts, and fungi helps explain the complexity of the interactions that must be considered. Interpretation of results reported for individual immune factors has been complicated by different types of immunostimulants and modes of administration used. Few studies have been repeated in the laboratory or applied in production facilities. Predictions of levels of response as measured by disease resistance will need to be based on much more complex investigations than provided by current research. Future research should attempt to separate nonspecific immune factor stimulation from preexisting, conferred disease resistance against specific pathogens. Genetics of the species, life history stage, and the culture environment all interact with the type and dosage of immunostimulant to contribute to the efficacy of the immunostimulatory substance.

REFERENCES

Aakre, R., H.I. Wergeland, P.M. Aasjord, and C. Endresen. 1994. Enhanced antibody response in Atlantic salmon (*Salmo salar* L.) to *Aeromonas salmonicida* cell wall antigens using a bacterin containing ß-1,3-M-glucan as adjuvant. *Fish & Shellfish Immunology* 4:47-61.

Aketagawa, J., S. Tanaka, H. Tamura, Y. Shibata, and H. Saito. 1993. Activation of limulus coagulation factor G by several (1, 3)-ß-D-glucans: Comparison of the potency of glucans with identical degree of polymerization but different conformations. *Journal of Biochemistry* 113:683-686.

Anderson, D.P. 1993. Specific Immune Response in Fish. *Fish Diseases Diagnosis and Prevention Methods, An International Workshop* (pp. 17-20). Olsztyn, Poland: FAO Project GCP/INT/526/JNP.

Balfry, S.K., D.D. Heath, and G.K. Iwama. 1997. Genetic analysis of lysozyme activity and resistance to vibriosis in farmed chinook salmon, *Oncorhynchus tshawytscha* (Walbaum). *Aquaculture Research* 28:893-899.

Barton, B.A. and G.K. Iwama. 1991. Physiological changes in fish from stress in aquaculture with emphasis on the response and effects of corticosteroids. *Annual Review of Fish Diseases* 1:3-26.

Blazer, V.S. 1992. Nutrition and disease resistance in fish. *Annual Review of Fish Diseases* 2:309-323.

Bly, J.E., A.S. Grimm, and I.G. Morris. 1986. Transfer of passive immunity from mother to young in a teleost fish: Haemaglutinating activity in the serum and eggs of plaice, *Pleuronectes platessa* L. *Comparative Biochemistry and Physiology* 84A(2):309-313.

Bøgwald, J., R.A. Dalmo, R.M. Leifson, E. Stenberg, and A. Gildberg. 1996. The stimulatory effect of a muscle protein hydrolysate from Atlantic cod, *Gadus morhua* L., on Atlantic salmon, *Salmo salar* L., head kidney leucocytes. *Fish & Shellfish Immunology* 6:3-16.

Bricknell, I.R., T.J. Bowden, J. Lomax, and A.E. Ellis. 1997. Antibody response and protection of Atlantic salmon *(Salmo salar)* immunized with an extracellular polysaccharide of *Aeromonas salmonicida. Fish & Shellfish Immunology* 7:1-16.

Cerenius, L., Z. Liang, B. Duvic, P. Keyser, U. Hellman, E.T. Palva, S. Iwanaga, and K. Söderhall. 1994. Structure and biological activity of a 1,3-ß-glucan-binding protein in crustacean blood. *The Journal of Biological Chemistry* 269(47): 29462-29467.

Chen, D. and A.J. Ainsworth. 1992. Glucan administration potentiates immune defense mechanisms of channel catfish, *Ictalurus punctatus* Rafinesque. *Journal of Fish Diseases* 15:295-304.

Chen, S.-C., T. Yoshida, A. Adams, K.D. Thompson, and R.H. Richards. 1996. Immune response of rainbow trout to the extracellular products of *Mycobacterium* spp. *Journal of Aquatic Animal Health* 8:216-222.

Chen, S.-C., T. Yoshida, A. Adams, K.D. Thompson, and R.H. Richards. 1998. Nonspecific immune response of Nile tilapia, *Oreochromis nilotica,* to the extracellular products of *Mycobacterium* spp. and to various adjuvants. *Journal of Fish Diseases* 21:39-46.

Chevassus, B. and M. Dorson. 1990. Genetics of resistance to disease in fish. *Aquaculture* 85:83-107.

Dalmo, R.A., J. Bøgwald, K. Ingebrigtsen, and R. Seljelid. 1996. The immunomodulatory effect of laminaran [ß(1,3)-D-glucan] on Atlantic salmon, *Salmo salar* L., anterior kidney leucocytes after intraperitoneal, peroral and peranal administration. *Journal of Fish Diseases* 19:449-457.

Dalmo, R.A. , K. Ingebrigtsen, and J. Bøgwald. 1997. Nonspecific defense mechanism in fish, with particular reference to the reticuloendothelial system (RES). *Journal of Fish Diseases* 20:241-273.

Dalmo, R.A., K. Ingebrigtsen, J. Bøgwald, T.E. Horsberg, and R. Seljelid. 1995. Accumulation of immunomodulatory laminaran [ß(1,3)-D-glucan] in the spleen and kidney of Atlantic salmon, *Salmo salar* L. *Journal of Fish Diseases* 18: 545-553.

Dalmo, R.A., K. Ingebrigtsen, B. Sveinbjørnsson, and R. Seljelid. 1996. Accumulation of immunomodulatory laminaran [ß(1,3)-D-glucan] in the heart, spleen, and kidney of Atlantic cod, *Gadus morhua* L. *Journal of Fish Diseases* 19:129-136.

Dalmo, R.A. and R. Seljelid. 1995. The immunomodulatory effect of LPS, laminaran and sulphated laminaran [ß(1,3)-D-glucan] on Atlantic salmon, *Salmo salar* L., macrophages in vitro. *Journal of Fish Diseases* 18:175-185.

de Baulny, M.O., C. Quentel, V. Fournier, F. Lamour, and R. Le Gouvello. 1996. Effect of long-term oral administration of ß-glucan as an immunostimulant or an adjuvant on some nonspecific parameters of the immune response of turbot *Scophthalmus maximus. Diseases of Aquatic Organisms* 26:139-147.

Duncan, P.L. and P.H. Klesius. 1996a. Dietary immunostimulants enhance nonspecific immune responses in channel catfish but not resistance to *Edwardsiella ictaluri. Journal of Aquatic Animal Health* 8:241-248.

Duncan P.L. and P.H. Klesius. 1996b. Effects of feeding *Spirulina* on specific and nonspecific immune responses of channel catfish. *Journal of Aquatic Animal Health* 8:308-313.

Duncan, P.L., R.T. Lovell, C.E. Butterworth, Jr., L.E. Freeberg, and T. Tamura. 1993. Dietary folate requirements determined for channel catfish, *Ictalurus punctatus. Journal of Nutrition* 123:1888-1897.

Engstad, R.E., B. Robertsen, and E. Frivold. 1992. Yeast glucan induces increase in lyozyme and complement-mediated haemolytic activity in Atlantic salmon blood. *Fish & Shellfish Immunology* 2:287-297.

Erdal, J.I., O. Evensen, O.K. Kaurstad, A. Lillehaug, R. Solbakken, and K. Thorud. 1991. Relationship between diet and immune response in Atlantic salmon (*Salmo salar* L.) after feeding various levels of ascorbic acid and omega-3 fatty acids. *Aquaculture* 98:363-379.

Fevolden, S.E., R. Nordmo, T. Refstie, and K.H. Røed. 1993. Disease resistance in Atlantic salmon *(Salmo salar)* selected for high or low responses to stress. *Aquaculture* 109:215- 224.

Fevolden, S.E., T. Refstie, and K.H. Røed. 1992. Disease resistance in rainbow trout *(Onchorhynchus mykiss)* selected for stress response. *Aquaculture* 104:19-29.

Fjalestad, K.T., T. Gjedrem, and B. Gjerde. 1993. Genetic improvement of disease resistance in fish: An overview. *Aquaculture* 111:65-74.

Fracalossi, D.M. and R.T. Lovell. 1994. Dietary lipid sources influence responses of channel catfish *(Ictalurus punctatus)* to challenge with the pathogen *Edwardsiella ictaluri. Aquaculture* 119: 287-298.

Fryer, J.L., C.N. Lannan, L.H. Garcés, J.J. Larenas, and P.A. Smith. 1990. Isolation of a rickettsiales-like organism from diseased coho salmon *(Oncorhynchus kisutch)* in Chile. *Fish Pathology* 25(2):107-114.

Galeotti, M. 1998. Some aspects of the application of immunostimulants and a critical review of methods for their evaluation. *Journal of Applied Ichthyology* 14:189-199.

Gildberg, A. and H. Mikkelsen. 1998. Effects of supplementing the feed to Atlantic cod *(Gadus morhua)* fry with lactic acid bacteria and immuno-stimulating peptides during a challenge trial with *Vibrio anguillarum. Aquaculture* 167:103-113.

Grisez, L., M. Chair, P. Sorgeloos, and F. Ollevier. 1996. Mode of infection and spread of *Vibrio anguillarum* in turbot *Scophthalmus maximus* larvae after oral challenge through live feed. *Disease of Aquatic Organisms* 26:181-187.

Grizzle, J.M. and Y. Kiryu. 1993. Histopathology of gill, liver, and pancreas, and serum enzyme levels of channel catfish infected with *Aeromonas hydrophila* complex. *Journal of Aquatic Animal Health* 5:36-50.

Hardie, L.J., T.C. Fletcher, and C.J. Secombes. 1994. Effect of temperature on macrophage activation and the production of macrophage activating factor by rainbow trout *(Oncorhynchus mykiss)* leucocytes. *Developmental and Comparative Immunology* 18:57- 66.

Ingebrigtsen, K., T.E. Horsberg, R. Dalmo, and R. Seljelid. 1993. Tissue distribution of the immunostimulator animated ß1,3-polyglucose in Atlantic salmon *(Salmo salar)* after intravenous, intraperitoneal and peroral administration. *Aquaculture* 117:29-35.

Jeney, G. and D.P. Anderson. 1993a. Enhanced immune response and protection in rainbow trout to *Aeromonas salmonicida* bacterin following prior immersion in immunostimulant. *Fish & Shellfish Immunology* 3:51-58.

Jeney, G. and D.P. Anderson. 1993b. Glucan injection or bath exposure given alone or in combination with a bacterin enhance the nonspecific defense mechanisms in rainbow trout *(Oncorhynchus mykiss). Aquaculture* 116:315-329.

Jeney, G., M. Galeotti, D. Volpatti, Z. Jeney, and D.P. Anderson. 1997. Prevention of stress in rainbow trout *(Oncorhynchus mykiss)* fed diets containing different doses of glucan. *Aquaculture* 154:1-15.

Jørgensen, J.B., H. Lunde, and B. Robertsen. 1993. Peritoneal and head kidney cell response to intraperitoneally injected yeast glucan in Atlantic salmon, *Salmo salar* L. *Journal of Fish Diseases* 16:313-325.

Jørgensen, J.B. and B. Robertsen. 1995. Yeast ß-glucan stimulates respiratory burst activity of Atlantic salmon (*Salmo salar* L.) macrophages. *Developmental and Comparative Immunology* 19:43-57.

Jørgensen, J.B., G.J.E. Sharp, C.J. Secombes, and B. Robertsen. 1993. Effect of yeast-cell glucan in the bactericidal activity of rainbow trout macrophages. *Fish & Shellfish Immunology* 3:267-277.

Kanazawa, A. 1997. Effects of docosahexaenoic acid and phospholipids on stress tolerance of fish. *Aquaculture* 155:129-134.

Landolt, M.L. 1989. The relationship between diet and the immune response of fish. *Aquaculture* 79:193-206.

Le Morvan, C., P. Clerton, P. Deschaux, and D. Troutaud. 1997. Effects of environmental temperature on macrophage activities in carp. *Fish & Shellfish Immunology* 7:209-212.

Li, M.H., D.J. Wise, M.R. Johnson, and E.H. Robinson. 1994. Dietary menhaden oil reduced resistance of channel catfish *(Ictalurus punctatus)* to *Edwardsiella ictaluri. Aquaculture* 128:335-344.

Manning, M.J., M.F. Grace, and C.J. Secombes. 1982. Developmental aspects of immunity and tolerance in fish. In R.J. Roberts (Ed.), *Microbial Disease of Fish* (pp. 31-46). London: Academic Press.

Marsden, M.J., L.C. Freeman, D. Cox, and C.J. Secombes. 1996. Nonspecific immune responses in families of Atlantic salmon, *Salmo salar,* exhibiting differential resistance to furunculosis. *Aquaculture* 146:1-16.

Matsuyama, H., R.E.P. Mangindaan, and T. Yano. 1992. Protective effect of schizophyllan and scleroglucan against *Streptococcus* sp. infection in yellowtail, *Seriola quinqueradiata. Aquaculture* 101:197-203.

Montero, D., M. Marrero, M.S. Izquierdo, L. Robaina, J.M. Vergara, and L. Tort. 1999. Effect of vitamin E and C dietary supplementation on some immune parameters of gilthead seabream *(Sparas aurata)* juveniles subjected to crowding stress. *Aquaculture* 171:269- 278.

Møyner K., K.H. Røed, S. Sevatdal, and M. Heum. 1993. Changes in nonspecific immune parameters in Atlantic salmon, *Salmo salar* L., induced by *Aeromonas salmonicida* infection. *Fish & Shellfish Immunology* 3:253-265.

Mulero, V., M.A. Estaban, and J. Meseguer. 1998. In vitro levamisole fails to increase seabream (*Sparus aurata* L.) phagocyte functions. *Fish & Shellfish Immunology* 8:315- 318.

Mulero, V., M.A. Estaban, J. Muñoz, and J. Meseguer. 1998. Dietary intake of levamisole enhances the immune response and disease resistance of the marine teleost gilthead seabream (*Sparus aurata* L.). *Fish & Shellfish Immunology* 8:49-62.

Nikl, L., L.J. Albright, and T.P.T. Evelyn. 1991. Influence of seven immunostimulants on the immune response of coho salmon to *Aeromonas salmonicida. Diseases of Aquatic Organisms* 12:7-12.

Nikl, L., T.P.T. Evelyn, and L.J. Albright. 1993. Trials with an orally and immersion-administered ß-1,3 glucan as an immunoprophylactic against *Aeromonas salmonicida* in juvenile chinook salmon *Oncorhynchus tshawytscha. Diseases of Aquatic Organisms* 17:191-196.

Olivier, G., T.P.T. Evelyn, and R. Lallier. 1985. Immunity to *Aeromonas salmonicida* in coho salmon *(Oncorhynchus kisutch)* induced by modified Freund's complete adjuvant: Its nonspecific nature and the probable role of macrophages in the phenomenon. *Developmental and Comparative Immunology* 9:419-432.

Raa, J. 1996. The use of immunostimulatory substances in fish and shellfish farming. *Reviews in Fisheries Science* 4(3):229-288.

Raa, J., G. Roerstad, R. Engstad, and B. Robertsen. 1992. The use of immunostimulants to increase resistance of aquatic organisms to microbial infections. In M. Shariff, R.P. Subasinghe, and J.R. Arthur (Eds.), *Diseases in Asian Aquaculture I* (pp. 39-50). Proceedings of the First Symposium in Asian Aquaculture. Fish Health Section. Bali, Indonesia: Asian Fisheries Society.

Robertsen, B. 1999. Modulation of the nonspecific defense of fish by structurally conserved microbial polymers. *Fish & Shellfish Immunology* 9:269-290.

Robertsen, B., R.E. Engstad, and J.B. Jørgensen. 1994. ß-glucans as immunostimulants in fish. Modulators of fish immune responses. In J.S. Stolen and T.C. Fletcher (Eds.), Volume 1. *Models for Environmental Toxicology, Biomarkers, Immunostimulators* (pp. 83-99). Fair Haven, NJ: SOS Publications.

Robertsen, B., G. Rørstad, R. Engstad, and J. Raa. 1990. Enhancement of nonspecific disease resistance in Atlantic salmon, *Salmo salar* L., by a glucan from *Saccharomyces cerevisiae* cell walls. *Journal of Fish Diseases* 13:391-400.

Rockey, D.D., L.A. Shook, J.L. Fryer, and J.S. Rohovec. 1989. Salmonid serum inhibits hemolytic activity of the secreted hemolysin of *Aeromonas salmonicida. Journal of Aquatic Animal Health* 1:263-268.

Rombout, J.H.W.M., H.E. Bot, and J.J. Taverne-Thiele. 1989. Immunological importance of the second gut segment of carp. II. Characterization of mucosal leucocytes. *Journal of Fish Biololgy* 35:167-178.

Rombout, J.H.W.M. and A.A. van den Berg. 1989. Immunological importance of the second gut segment of carp. I. Uptake and processing of antigens by epithelial cells and macrophages. *Journal of Fish Biology* 35:13-22.

Rowley, A.F., J. Knight, P. Lloyd-Evans, J.W. Holland, and P.J. Vickers. 1995. Eicosanoids and their role in immune modulation in fish—A brief overview. *Fish & Shellfish Immunology* 5:549-567.

Saito, H., Y. Yoshioka, and N. Uehara. 1991. Relationship between conformation and biological response for (1,3)-ß-D-glucans in the activation of coagulation factor G from limulus amebocyte lysate and host-mediated antitumor activity: Demonstration of single-helix conformation as a stimulant. *Carbohydrate Research* 217:181-190.

Sakai, M. 1999. Current research status of fish immunostimulants. *Aquaculture* 172:63-92.

Santarém, M., B. Novoa, and A. Figueras. 1997. Effects of ß-glucans on the nonspecific immune responses of turbot (*Scophthalmus maximus* L.). *Fish & Shellfish Immunology* 7:429-437.

Shils, M.E. and V.R.Young (Eds.). 1988. *Modern Nutrition in Health and Disease.* Seventh Edition. Philadelphia, PA: Lea and Febiger.

Siwicki, A.K., D.P. Anderson, and G.L. Rumsey. 1994. Dietary intake of immunostimulants by rainbow trout affects nonspecific immunity and protection against furunculosis. *Veterinary Immunology and Immunopathology* 41:125-139.

Strand, H.K. and R.A. Dalmo. 1997. Absorption of immunomodulating ß-(1,3)-glucan in yolk-sac larvae of Atlantic halibut, *Hippoglossus hippoglosus* (L.). *Journal of Fish Diseases* 20:41-49.

Sung, H.H., G.H. Kou, and Y.L. Song. 1994. Vibriosis resistance induced by glucan treatment in tiger shrimp *(Penaeus monodon). Fish Pathology* 29(1):11-17.

Sveinbjørnsson, B., B. Smedsrød, T. Berg, and R. Seljelid. 1995. Intestinal uptake and organ distribution of immunomodulatory aminated ß-1,3-D-polyglucose in Atlantic salmon (*Salmo salar* L.). *Fish & Shellfish Immunology* 5:39-50.

Tatner, M.F. and M.J. Manning. 1985. The ontogenic development of the reticulo-endothelial system in the rainbow trout, *Salmo gairdneri* Richardson. *Journal of Fish Diseases* 8:35- 41.

Thompson, I., T.C. Fletcher, D.F. Houlihan, and C.J. Secombes. 1994. The effect of dietary vitamin A on the immunocompetence of Atlantic salmon (*Salmo salar* L.). *Fish Physiology and Biochemistry* 12: 513-523.

Thompson, I., A. White, T.C. Fletcher, D.F. Houlihan, and C.J. Secombes. 1993. The effect of stress on the immune response of Atlantic salmon (*Salmo salar* L.) fed diets containing different amounts of vitamin C. *Aquaculture* 14:1-17

Thompson, K.D., M.F. Tatner, and R.J. Henderson. 1996. Effects of dietary (n-3) and (n-6) polyunsaturated fatty acid ratio on the immune response of Atlantic salmon, *Salmo salar* L. *Aquaculture Nutrition* 2: 21-31.

Vadstein, O. 1997. The use of immunostimulants in marine larviculture: Possibilities and challenges. *Aquaculture* 155:401-417.

Verlhac, V., A. Obach, J. Gabaudan, W. Schüep, and R. Hole. 1998. Immunomodulation by dietary vitamin C and glucan in rainbow trout *(Onchorhynchus mykiss). Fish & Shellfish Immunology* 8:409-424.

Waagbø, R. 1994. The impact of nutritional factors on the immune system in Atlantic salmon, *Salmo salar* L.: A review. *Aquaculture and Fisheries Management* 25:175-197.

Weyts, F.A.A., N. Cohen, G. Flik, and B.M.L. Verburg-van Kemenade. 1999. Interactions between the immune systems and the hypothalamo-pituitary-interrenal axis in fish. *Fish & Shellfish Immunology* 9:1-20.

Wiegertjes, G.F., R.J.M. Stet, H.K. Parmentier, and W.B. Van Muiswinkel. 1996. Immunogenetics of disease resistance in fish: A comparative approach. *Developmental and Comparative Immunology* 20(6):365-381.

Wise, D.J., J.R. Tomasso, T.E. Schwedler, V.S. Blazer, and D.M. Gatlin III. 1993. Effect of vitamin E on the immune response of channel catfish to *Edwardsiella ictaluri*. *Journal of Aquatic Animal Health* 5:183-188.

Yano, T., R.E.P. Mangindaan, and H. Matsuyama. 1989. Enhancement of resistance of carp *Cyprinus carpio* to experimental *Edwardsiella tarda* infection, by some ß-1,3-glucans. *Nippon Suisan Gakkaishi* 55(10):1815-1819.

Yano, T., H. Matsuyama, and R.E.P. Mangindaan. 1991. Polysaccharide-induced protection of carp, *Cyprinus carpio* L., against bacterial infection. *Journal of Fish Diseases* 14:577-582.

Yousif, A.N., L.J. Albright, and T.P.T. Evelyn. 1994. In vitro evidence for the antibacterial role of lysozyme in salmonid eggs. *Diseases of Aquatic Organisms* 19:15-19.

Yousif, A.N., L.J. Albright, and T.P.T. Evelyn. 1995. Interaction of coho salmon *Oncorhynchus kisutch* egg lectin with the fish pathogen *Aeromonas salmonicida*. *Diseases of Aquatic Organisms* 21:193-199.

Chapter 13

Mycotoxins in Fish Feeds

Bruce B. Manning

INTRODUCTION

Mycotoxins are secondary metabolites produced by fungal organisms that are commonly referred to as molds. Mycotoxins can be very simple compounds such as moniliformin or those which are extremely chemically complex such as the trichothecene type of toxins. Mold toxins vary in their toxicity toward different species of animals, but generally they are more toxic to the juveniles of any species. Mycotoxins are produced under conditions of adequate free water, ideal relative humidity, optimum temperature, and suitable substrate such as grain starch in the presence of oxygen that allow for development of the mold organism. Preharvest conditions will influence the production of mycotoxins in grains and other agricultural crops. The ideal conditions for production of the different mycotoxins vary considerably; aflatoxin production in corn usually occurs during periods of high temperatures with prolonged periods of severe drought associated with extensive insect activity. The production of the *Fusarium* mycotoxins, on the other hand, usually occurs under field conditions of cool temperatures and high moisture at harvest (CAST 1989).

Molds are ubiquitous organisms that are found on most grains and oil seeds. Although mold organisms may contaminate the grains used to manufacture feed for agricultural animals, in most cases their numbers are relatively low, and they are kept from multiplying by harvesting grains under proper conditions and storing them in well-managed facilities. Proper harvesting conditions include harvesting at the proper stage of maturity and at the ideal moisture content.

At moisture levels above 12 percent it is usually necessary to lower the moisture content of harvested grains such as corn by using gas-fired dryers to obtain the ideal storage moisture. Adequate storage facilities are important for keeping mold development to a minimum. These include sound storage structures that exclude environmental moisture from feed grains.

Most of the mold organisms that infect grains and plant protein products used to manufacture fish diets belong to genera *Aspergillus, Penicillium,* and *Fusarium.* Toxigenic species of these genera produce toxins that have the potential to reduce growth and health status of fish and other agriculturally important animals consuming contaminated feeds. Examples of some toxigenic mold species and the mycotoxins they produce are shown in Table 13.1. These molds have been determined to produce under natural conditions the mycotoxin attributed to them in significant quantities to be harmful to at least one species of farm animals. In some cases, a particular species of mold may produce more than one mycotoxin.

Concerns about mycotoxin contamination of human foods and feeds for animals gained prominence after outbreaks of a condition of turkey poults in England in 1960 referred to as Turkey "X" disease, which was, at the time, a disease of unknown etiology (Blount 1961). The condition killed at least 100,000 young turkeys during that period and also occurred in other species of young poultry (i.e., ducklings and pheasants). Subsequent scientific investigations determined that outbreaks of this condition had a common factor, the inclusion of peanut meal infected with the mold *Aspergillus flavus.* The causative agent was later isolated and purified by chromatographic procedures and given the name aflatoxin (Sargeant, O'Kelly, et al. 1961; Sargeant, Sheridan, et al. 1961). Aflatoxin was determined to be composed of four structurally similar compounds, which were designated aflatoxins B_1, B_2, G_1, and G_2 (Hartley, Nesbitt, and O'Kelly 1963).

EFFECT OF AFLATOXIN IN FEEDS ON FISH AND OTHER ANIMALS

Examples of the harmful effects of mycotoxins on fish include the devastating effect that the consumption of aflatoxin-contaminated diet had on rainbow trout, *Onchorhyncus mykiss.* The first recognized incidences of mycotoxicosis of fish occurred more than 35 years ago in trout hatcheries in the United States when rainbow trout, which had been fed pelleted diets that were prepared with cottonseed meal contaminated with aflatoxin, developed liver tumors. The tumors, referred to as hepatomas, were later diagnosed as hepatocellular carcinomas (HCC) (Halver 1969). Later evaluation of these events indicated that morbidity as high as 85 percent occurred during these epizootics (Post 1987). Experimentally, aflatoxin B_1 (AFB_1) was shown to be a potent liver carcinogen of fish. Rainbow trout receiving diet containing 0.4 parts per billion (ppb) AFB_1, which is the most toxic of the aflatoxins, for 15 months had a 14 percent incidence of HCC (Lee et al. 1968). In another study, feeding a diet containing 20 ppb AFB_1 for 8 months resulted in a 58 percent incidence of HCC in trout, and feeding continued to 12 months resulted in an 83 percent incidence (Schoenhard et al. 1981). Aflatoxins B_2, G_1, and G_2, the other common aflatoxins, have low carcinogenicity in trout compared to AFB_1 (Lee, Hendricks, and Bailey 1991).

TABLE 13.1. List of Selected Mycotoxins and the Fungal Organisms That Biosynthesize the Mycotoxins

Mold Toxin	Fungal Organism[1]	Primary Biological Response of Susceptible Animals[2,3]
Aflatoxin B_1, B_2, G_1, G_2	*Aspergillus flavus, A. parasiticus*	Liver carcinomas (HCC)
Cyclopiazonic acid	*Aspergillus flavus, A. versicolor*	Neurotoxin
Ochratoxin A	*Aspergillus alutaceus (ohraceus)*	Kidney toxin, also liver
Sterigmatocystin and versicolorin A	*Penicillium luteum, Aspergillus versicolor*	Hepatotoxin, carcinogenic
Ochratoxin A	*Penicillium verrucosum*	Kidney toxin, also liver
Fumonisin B_1, B_2, B_3, B_4	*Fusarium moniliforme*	Hepatotoxin, disrupts sphingolipid metabolism. ELEM in equines, PPE of swine
Moniliforme	*Fusarium proliferatum*	Tricarboxylic cycle disruption, cardiac lesion
Trichothecenes T-2 toxin	*Fusarium sporotrichioides*	Gastrointestinal disturbances, dermal irritation, mouth lesions, and anemia
Deoxynivalenol	*Fusarium graminearum*	Feed refusal syndrome, vomiting
Zearalenone	*Fusarium graminearum*	Estrogenic syndrome of swine
Ergot alkaloids Ergotamine Ergocystine	*Claviceps purpurea*	Ergot toxicity, ergotism, loss of extremities

Notes:
1 Some species of fungal organisms produce more than one mycotoxin.
2 Additional biological and clinical responses may occur with each mycotoxicosis.
3 HCC = hepatocellular carcinoma; ELEM = equine leukoencephalomalacia; PPE = porcine pulmonary edema.

Among the salmonid species, it has been demonstrated that rainbow trout have greater sensitivity to aflatoxin than coho salmon, *O. kisutch*. Feeding a diet containing 20 ppb AFB_1 resulted in a 96 percent incidence of HCC in rainbow trout and 0 percent in the coho salmon. The species difference was attributed to the greater ability of trout hepatocytes to convert AFB_1 to the 8,9-epoxide, the proposed carcinogen, which forms adducts with DNA. Also, these DNA adducts may persist longer in trout than in coho salmon (Coulombe, Bailey, and Nixon 1984; Coulombe 1993). Species differences have been observed in other animals. For instance, AFB_1 is more toxic to the rat than the mouse. The rat readily converts AFB_1 in the liver to its 8,9-epoxide, forming mutagenic adducts with DNA more efficiently than the mouse, and the rat has lower activity of the reduced glutathione (GSH-S-transferase) system to detoxify AFB_1 through the formation of conjugates of glutathione. Other conjugates of AFB_1 are formed with sulfate and glucuronic acid in liver (Coulombe 1993). Degen and Neumann (1981) demonstrated with in vitro systems that mouse liver was two to four times more active in conjugating the 8,9-epoxide of AFB_1 than rat liver, thus having greater capacity to detoxify AFB_1.

Among fishes, the channel catfish, *Ictalurus punctatus,* appears to be one of the most resistant when exposed to AFB_1. There have been no reports of HCC in channel catfish exposed to AFB_1 (Halver 1969). Jantrarotai and Lovell (1990b) found that feeding of 10,000 μg purified AFB_1/kg diet for ten weeks resulted in decreased growth rate and moderate internal lesions, but no HCC, whereas feeding 2,154 μg AFB_1/kg diet did not suppress growth or produce internal lesions. In another feeding study, conducted by Chavez-Sanchez, Martinez, and Osorio Moreno (1994), Nile tilapia, *Oreochromis nilotica,* showed significant reductions in growth rate when fed diets containing at least 1,880 μg AFB_1/kg, but not when fed 940 μg AFB_1/kg diet. Precancerous liver lesions were observed in tilapia receiving levels of AFB_1 of 1,880 μg/kg diet and above, but none developed into carcinomas during the 75-day experiment. These studies indicate that certain species of warm-water fish are less sensitive to aflatoxin than trout and other species of cold-water fish. Jantrarotai, Lovell, and Grizzle (1990) postulated, based on previous reports (Strength et al. 1982), that channel catfish are able to more effectively detoxify and excrete aflatoxin than rainbow trout because of the activity of the glucuronyltransferase and sulfotransferase detoxification systems. Rainbow trout have only the glucuronyltransferase system for detoxification of aflatoxin (Loveland, Nixon, and Bailey 1984).

OTHER MYCOTOXINS PRODUCED BY ASPERGILLUS *MOLDS*

Other mycotoxins are elaborated by fungi that belong to the genus *Aspergillus.* Mycotoxins that may possibly appear in fish diets include ochratoxin A (OA), cyclopiazonic acid (CPA), sterigmatocystin, and versicolorin A. The latter two are biosynthetic precursors to AFB_1 but lack the same level of carcinogenicity to rainbow trout, so their possible toxic effects on fish have been overshadowed by the much greater toxicity of AFB_1. By comparison, when 14-day trout embryos were exposed by immersion in a 0.5 parts per million (ppm) aqueous suspension of sterigmatocystin for one hour there was a 13 percent incidence of liver tumors after one year, but a 53 percent incidence of tumors with comparable concentrations of aflatoxin (Hendricks et al. 1980). The toxicity of purified CPA to channel catfish was evaluated by Jantrarotai and Lovell (1990a) and found to be more toxic than purified AFB_1. When catfish fingerlings were fed diets containing five levels of CPA for ten weeks, the subchronic toxic level was determined to be 100 μg/kg of diet. Significant reductions in growth rate were observed for these fish, but there were no other toxicity symptoms at this inclusion level. Hematological values of catfish were not affected in any of the treatments, including feeding 10,000 μg CPA/kg diet, although fish fed this level had histological lesions of the trunk kidney and stomach. These findings are important because CPA is produced by *A. flavus* and, with aflatoxin, may concurrently contaminate grains used to manufacture fish diets (Jantrarotai and Lovell 1990a). The effect of concurrent contamination by these two mycotoxins on fish has not been reported. Ochratoxin A, which is produced by toxigenic strains of *Aspergillus alutaceus* (formerly *A. ochraceus*), has been named as the causative agent in the fatal human disease Balkan endemic nephropathy (Marquardt and Frohlich 1992). In most animals studied, including fish, the primary target organ of OA intoxication is the kidney. Exposure of rainbow trout to OA by intraperitoneal injection of an LD50 dose of 4.67 mg/kg resulted in necrosis of kidney tubular cells. There was also necrosis of the liver (Doster, Sinnhuber, and Wales 1972).

MYCOTOXINS PRODUCED BY PENICILLIUM *MOLDS*

Some of the mycotoxins that are biosynthesized by molds of the *Aspergillus* species are also produced by *Penicillium* molds. Included in this group are sterigmatocystin and versicolorin A. Both of these mycotoxins, as mentioned previously, are precursors to AFB_1, but *Penicillium* molds do not have the biosynthetic pathways necessary to complete the synthesis of AFB_1. Ochratoxin A is produced by *P. verrucosum* as well as *A. alutaceus,* but these two mold organisms, which are classified as storage fungi, prefer different substrates and climatic conditions for optimum production of this mycotoxin. Production of OA is optimized when *P. verrucosum* uses grains such as corn or wheat for substrates, and *A. alutaceus* has better mycotoxin production on oil seeds such as soybeans and peanuts. Optimum temperature range for the two fungi is reflective of the substrates they prefer; *P. verrucosum* has higher mycotoxin production at a lower temperature range (4 to 31°C), which is a typical temperature range for upper Midwest grain production areas, while higher mycotoxin production by *A. alutaceus* occurs at a higher temperature range (12 to 37°C), the typical range for oil seeds grown in warmer climates (Marquardt and Frohlich 1992). Cyclopiazonic acid is also produced by fungi in the *Penicillium* genus.

The possibility of cross contamination of grains and oil seeds with two mycotoxins exists, especially where a single isolate of the fungal species is capable of producing both mycotoxins, as has been found with a considerable number of *A. flavus* isolates that produce both CPA and AFB_1. This feature may be limited by observations that some toxigenic fungal organisms have lower production of known mycotoxins in the presence of competitive microorganisms. This was demonstrated by Chelack et al. (1991), where OA production by *A. alutaceus* on barley was limited in the presence of competing organisms when compared to the greater production of OA on sterilized barley. Cross contamination may possibly occur at receiving facilities in the process of mixing lots of grain from various sources that are contaminated with different mycotoxins.

FUSARIUM *MYCOTOXINS AND THEIR EFFECT ON FISH AND OTHER ANIMALS*

Molds that belong to the *Fusarium* genus produce a diverse group of mycotoxins. The effect of these mycotoxins on warm-water fish has received much attention in the past several years because they can contaminate the feed grains that are so heavily relied on in diets. One of the most studied mycotoxins has been fumonisin B_1 (FB_1), which is produced by *Fusarium moniliforme,* a fungal organism that infects developing corn. Fumonisin B_1 is one of the several fumonisins that contaminate corn, rice, and other small feed grains, but it is usually the predominant chemical variant of the fumonisins and is considered to be the most toxic. Its toxicity varies dramatically by species of animal exposed to it. All equines are extremely sensitive to FB_1; consumption of diet containing as little as 5 mg/kg over several days causes the fatal condition known as equine leukoencephalomalacia (ELEM), which results in liquifaction of the brain. Swine are sensitive to FB_1 at low dietary levels; consumption of diet containing 15 to 20 mg FB_1/kg causes the development of porcine pulmonary edema (PPE), or filling of the pleural cavity with fluid. Consumption of corn contaminated with FB_1 has been linked to human esophageal cancer in areas of the world where corn that has not been maintained under satisfactory storage conditions is a dietary staple (Norred 1993).

Channel catfish appear to be fairly tolerant of FB_1, when compared to equines or swine. Lumlertdacha et al. (1995) found that fingerlings weighing 1.5 g showed reduced growth when fed a practical type of diet containing 20 mg/kg for 10 weeks, while 30 g catfish required 80 mg/kg FB_1 to cause a significant reduction in growth after 12 weeks (Lumlertdacha et al. 1995). Li, Raverty, and Robinson (1994) showed similar findings in that catfish fingerlings with an initial weight of 5 g fed 40 mg FB_1/kg had significant reduction in growth after 12 weeks. In a feeding study that compared the toxicity of FB_1 supplied as purified FB_1 with that supplied in crude form from *F. moniliforme* culture material, it was observed that only 50 mg FB_1/kg from the crude source was required to significantly reduce growth rate of 5 g catfish compared to 200 mg/kg of purified FB_1 (Manning 1998). Tilapia, *Oreochromis niloticus,* with initial weight of 22 g experienced significantly reduced weight gain when fed a semipurified diet containing 75 mg FB_1/kg from culture material (Jain, Auburn University, unpublished data).

A consistent response of all animals exposed to FB_1 is alteration of tissue concentrations of the sphingolipids, sphingosine and sphinganine. This occurs as a result of FB_1 competitive inhibition of the enzyme, ceramide synthetase, which converts sphinganine to ceramide, thus causing tissue accumulation of sphinganine. Since the tissue concentrations of sphinganine increase substantially, while those of sphingosine are changed only slightly, an elevation of the ratio of sphinganine to sphingosine concentrations occurs. Elevation of the sphinganine-to-sphingosine ratio has been proposed to be a biomarker for exposure to FB_1 (Riley, Wang, and Merrill 1994). Channel catfish and tilapia in the feeding studies cited previously exhibited elevated sphinganine-to-sphingosine ratios as a result of consuming FB_1. There were marked increases in the sphingolipid ratios of liver, kidney, and serum of catfish from the study conducted by Lumlertdacha and colleagues (1995), which closely followed the reduction in growth rate in an FB_1 dose-related manner (Goel et al. 1994).

Manning (1998) found that catfish fingerlings fed FB_1 from the crude culture material source had higher elevated sphinganine-to-sphingosine ratios than fingerlings fed purified FB_1 at the same levels, indicating that some other unidentified compound present in the crude culture material enhanced the toxicity of FB_1. Tilapia in the study described previously also had sphingolipid ratios elevated in a dose-related response to higher dietary levels of FB_1 (Jain, Auburn University, unpublished data). The toxicity of FB_1 appears to be greater in younger, smaller fish (Lumlertdacha et al. 1995). The dietary levels of fumonisin contamination described in the foregoing studies are obtainable under practical culture conditions, since samples of feed-grade corn may contain FB_1 levels of 100 to 400 mg/kg.

The effects of another *Fusarium* mycotoxin, moniliformin produced by *F. proliferatum,* was evaluated in a ten-week feeding study conducted with 1.5 g channel catfish fingerlings by Yildirim et al. (2000). Two dietary levels of fumonisin B_1 were included in this study to evaluate the effects of two combinations of the mycotoxins. The addition of five levels of moniliformin supplied from *F. proliferatum* culture material to a semipurified diet resulted in significantly lower weight gains for catfish fed the lowest level of moniliformin, 20 mg/kg diet. Feeding FB_1 at the lower level, 20 mg/kg, resulted in significantly lower weight gain compared to the control fish. Hematocrit was significantly lowered for fish fed 60 mg moniliformin/kg or 40 mg FB_1/kg diet. Because the mode of action of moniliformin is to disrupt intermediary metabolism at the conversion of pyruvate to the tricarboxylic acid cycle starting intermediate, acetyl-CoA, consumption of this mycotoxin results in increased serum pyruvate levels. Serum pyruvate for the catfish fed 60 mg moniliformin/kg diet was significantly higher than control fish. Combination of the two mycotoxins further enhanced this response because feeding each mycotoxin at 40 mg/kg resulted in an increased serum pyruvate level above that observed when moniliformin was fed at 60 mg/kg. Combinations of the two mycotoxins, however, did not enhance the fumonisin effect, since there were no additional increases in sphinganine-to-sphingosine ratios above the ratio observed for fish fed only FB_1. Corn can commonly be found contaminated with moniliformin at levels that could provide dietary amounts which are toxic to channel catfish and possibly to tilapia.

Other *Fusarium* mycotoxins that have significant toxic effects on agriculturally important animals other than fish include T-2 toxin, deoxynivalenol (DON) or vomitoxin, diacetoxyscirpenol (DAS), and zearalenone (ZEN). The effects of these mycotoxins has been extensively studied in farm animals. Deoxynivalenol, DAS, and T-2 toxin belong to a group of mycotoxins referred to as trichothecenes, all of which have structural similarities characterized as by having the tetracyclic 12,13-epoxy-trichothecane moiety of the sesquiterpenoids. Exposure of cattle, poultry, and swine to the trichothecenes resulted in oral lesions and gastrointestinal disturbances such as vomiting, diarrhea, and intestinal inflammation and hemorrhaging. Hematological alterations such as anemia and leukopenia have been observed for animals exposed to trichothecenes (CAST 1989).

Feed refusal syndrome resulting from loss of appetite, which is sometimes associated with vomiting, has been observed for animals, especially swine and poultry, consuming grains contaminated with trichothecenes. These and other behavioral changes may be the result of a neurotransmitter imbalance in the brain. The mechanism of trichothecene toxicity is due, in part, to the ability of these mycotoxins to inhibit protein synthesis in liver and other tissues resulting in hyperaminoacidemia. The increased levels of circulating amino acids include tryptophan, which is a precursor of the neurotransmitter serotonin (Smith 1992). Thus, consumption of trichothecenes may increase brain concentrations of tryptophan and cause increased synthesis of serotonin. Serotonin is known to play a role in the control of appetite (Smith and Seddon 1998). Feeding fingerling rainbow trout (initial weight 1.0 g) graded levels of T-2 toxin (0 to15 mg/kg feed) in a 16-week trial resulted in depressed growth, feed efficiency, hematocrit, blood hemoglobin, and food acceptance for fish fed dietary T-2 concentrations greater than 2.5 mg/kg. Additionally, force-feeding 15 mg T-2 toxin/kg diet to adult trout resulted in intestinal hemorrhaging and regurgitation of diet (Poston, Coffin, and Combs 1982). In a study with juvenile rainbow trout, progressively lower weight gains were observed when they were fed various levels of 1.0 to 13.0 mg DON/kg feed for four weeks. Results from another segment of the study showed that feed refusal syndrome was encountered when the trout received 20 mg DON/kg diet. No vomiting behavior was observed with either group of fish, and normal feeding behavior resumed when the trout were offered an uncontaminated diet. The results of this study indicate that sensitivity of juvenile rainbow trout to dietary DON, with respect to feeding behavior (refusal), is similar to that of the pig (Woodward, Young, and Lun 1983).

Zearalenone is a nonsteroidal metabolite with estrogen-like effects that is produced by *Fusarium graminearum* and other *Fusarium* molds. Zearalenone is a common contaminant of corn and also wheat, wheat by-products, barley, oats, and sorghum. The production of this mycotoxin is promoted by high humidity and low temperatures. Although not chemically similar to steroidal estrogenic compounds, ZEN is capable of producing estrogen-like responses in a variety of animals, including swine, cattle, and sheep. Swine exhibit the most dramatic responses to ingestion of grains contaminated with ZEN, even at low doses (1 mg/kg). These include swollen vulva and mammary glands and enlargement of the uterus (Coulombe 1993). Young male pigs are also affected by exposure to ZEN, resulting in testicular atrophy and enlarged nipples (Coulombe 1993). Zearalenone appears to affect sensitive animals by competing for cytoplasmic estrogenic receptors, thus preventing endogenous estrogenic compounds from achieving normal physiological function. Reports of the effects of ZEN on fish are limited.

Fusaric acid, 5-butylpicolinic acid, is a phytotoxin that is abundantly produced by many species of *Fusarium* fungi, including *F. moniliforme.* Bacon et al. (1996) found that 75 strains of *Fusarium* fungi produce fusaric acid, many of which also produce other mycotoxins of concern to producers of livestock and fish. Evaluation of a group of feed grains for fusaric acid content revealed that samples of corn, wheat, and barley contained an average of 12 mg/g (Smith and Sousadias 1993). Fusaric acid appears to have low toxicity to most animals in which it has been studied, usually requiring high dietary levels to elicit any toxic response. At toxic levels, fusaric acid produces behavioral responses that are similar to deoxynivalenol and other trichothecene mycotoxins (e.g., loss of appetite and vomiting). Fusaric acid elevates brain concentrations of tryptophan and serotonin (Chaouloff et al. 1986), as does DON, but by a different mechanism. Fusaric acid, which is derived from tryptophan and bears a structural similarity to it, competes for the normal tryptophan binding sites on blood albumin, causing elevated free- blood tryptophan (Chaouloff et al. 1986). As a result there is increased uptake of free tryptophan in the brain and increased serotonin synthesis.

While fusaric acid has low toxicity to most animals, its importance as a mycotoxin may be in its ability to synergize other *Fusarium* mycotoxins such as DON and FB_1. Studies to evaluate the synergistic effect of fusaric acid with *Fusarium* mycotoxins have shown that toxicity of these mycotoxins is enhanced in the presence of nontoxic levels of fusaric acid. Swine fed practical diets consisting of blends of *Fusarium*-contaminated grains to provide various combined levels of DON and fusaric acid had reduced growth and diet intake when fed only 1.1 μg DON and 15 μg fusaric acid/g diet (Smith, MacMillan, and Castillo 1997). Bacon, Porter, and Norred (1995) observed that fertile chicken eggs injected with combinations of fusaric acid at a nonlethal concentration and graded concentrations of FB_1 exhibited a synergistic toxic response. Injection of 5 μg/egg of FB_1 in combination with 5 μg/egg of fusaric acid caused an embryonic death rate of approximately 50 percent, compared to no embryonic death for eggs injected with 5 μg/egg of FB_1 without fusaric acid. As previously mentioned, a study comparing the toxicity of FB_1 from culture material with purified FB_1 fed to channel catfish showed that the crude form was about four times more toxic than purified FB_1 derived from the same culture material, using reduction in growth and elevation of sphingolipid ratio as criteria (Manning 1998). The component of culture material that made it more toxic to catfish was not identified, but in a subsequent feeding study, an aqueous chromatographic fraction of the culture material that had no toxicity to catfish fingerlings enhanced the toxicity of a purified fraction containing FB_1 (Manning 1998).

OTHER MYCOTOXINS: THE ERGOT ALKALOIDS

Another group of mycotoxins that have not received the attention given to AFB_1 and FB_1 is the ergot alkaloids. These toxins are produced by members of the genus *Claviceps,* which usually infect grasses that are cultivated for forage or production of small grains such as wheat or rye. During the course of development of the infective fungal organism in the head of the immature grain, some of the developing kernels are replaced by reproductive components referred to as conidia. These grain replacements darken and harden into the sclerotia of ergot. Grain samples containing ergot are easily identified by the black sclerotinia bodies, which usually have the same size and shape of the grain that they replaced. It is the sclerotia that contain the potentially harmful ergot alkaloids. The many ergot alkaloids may evoke different physiological responses. One of the most well known of these being diminished blood circulation that occurs in ergotism of humans resulting from consumption of ergot-contaminated grain (Lacey 1991). Diminished circulation to the extremities often causes dry gangrene and loss of the affected extremity. Ergot alkaloids are physiologically potent compounds, and their effects on fish that consume diets containing small grains, such as wheat, or by-products of these grains are not understood and should be examined.

EFFECT OF MYCOTOXINS ON THE IMMUNE RESPONSE

Evidence suggests that consumption of diets contaminated with mycotoxins suppresses the immune systems of animals. Mycotoxins that apparently compromise the immune system include aflatoxin, T-2 toxin, deoxynivalenol, ochratoxin A, and fumonisin B_1. The effect of mycotoxins on the immunological function of terrestrial animals and poultry has been examined extensively, while their effect on that of fish has received comparatively little attention.

Trichothecene toxins adversely affect the immune systems of poultry, swine, and other animals. Consumption of trichothecene mycotoxins causes suppression of immune response by reducing the numbers of circulating phagocytes, resulting in reduction of both phagocytic activity and chemotaxis by macrophages. These responses reduce the uptake and presentation of antigenic substances to antibody-producing B cells, resulting in lowered production of IgA and IgM. Another mechanism for reduced production of immunoglobulins by the trichothecene mycotoxins is through their ability to inhibit protein synthesis. Decreased weights of the bursa of Fabricius, the organ in avian species that is the site of B cell production, were observed after broiler chickens were fed diets containing 4 or 8 mg T-2 toxin/kg (Wyatt, Hamilton, and Burmeister 1973).

Aflatoxin causes suppression of the immune system when the 8,9-epoxide of AFB_1 forms adducts of DNA and RNA that inhibit protein synthesis, resulting in reduced production of certain cell components such as C4 complement and lymphokines (e.g., interleukines, by the liver and T lymphocytes). Aflatoxin suppresses phagocytosis by macrophages, which alters subsequent processing and presentation of antigen to B lymphocytes and causes lowered disease resistance through decreased antibody production (Pier 1991). Chickens with aflatoxicosis were found to have (1) a reduction in the phagocytotic activity of heterophils; (2) a reduction in the percentage of heterophils engaged in phagocytosis; (3) a reduction in spontaneous and chemotactic locomotion by macrophages; (4) a reduction in destruction of phagocytized bacteria; and (5) impaired cellular and serum factors needed for optimal phagocytosis (Chang and Hamilton 1979). Decreased immunoglobulin concentrations were observed in broiler chicks after administration of aflatoxin (Giambrone et al. 1978).

Ochratoxin A impairs the immune system through inhibition of protein synthesis by blocking the important translational enzyme phenylalanine tRNA synthetase. Thus, this toxin reduces levels of the circulating antibodies IgG and IgM. Ochratoxin A decreases phagocytic activity of the macrophages, diminishing antigen clearance and presentation to B cells, which further decreases the production of antibodies.

Fumonisin B_1 may diminish the immune response of animals by inhibiting protein synthesis through interference with the incorporation of valine into protein of rat primary hepatocytes (Norred et al. 1990, 1992). As mentioned previously, FB_1 has been shown to disrupt sphingolipid metabolism. Sphingolipids and their breakdown products are involved in cell regulatory functions associated with cell surface receptors and critical functions of other cellular components (Hunnan and Bell 1989). Thus, alteration of sphingolipid metabolism as a result of FB_1 toxicosis may be a factor in diminishing immune system response of animals consuming diets contaminated with FB_1 (Lumlertdacha and Lovell 1995). In vitro treatment of chicken macrophages with various doses of FB_1 (0.5, 5.0, and 10 μg FB_1/ml) resulted in significant cytotoxicity, morphological alterations, and depression of phagocytic potential, suggesting that consumption of diets contaminated with FB_1 may increase susceptibility of chickens exposed to FB_1 to bacterial infections (Qureshi and Hagler 1992). Channel catfish fed a practical diet containing 80 mg FB_1/kg prepared with *Fusarium moniliforme* culture material with a known level of FB_1 experienced significantly greater mortality after challenge with *Edwardsiella ictaluri* infection than control fish. Antibody production of catfish fed diets containing either 20 or

80 mg FB_1/kg was significantly lower than that of control fish (Lumlertdacha and Lovell 1995).

SAMPLING TECHNIQUE FOR MYCOTOXIN ANALYSIS OF FEED COMMODITIES

Sampling technique is very important to accurately determine the concentration and type of mycotoxin contamination in a lot or shipment of feed grains. Usually, it is desirable to sample lots of grains or diet ingredients for mycotoxin analysis before they are unloaded, or at least before they become part of the grain inventory of the animal diet production facility. Sampling of incoming loads of grain should be accomplished by a method that will provide representative samples that accurately reflect the true mycotoxin content of the grain lot. If sampling is being done on a load of grain contained in a hopper of a truck-and-trailer combination, a standard grain sampling probe provides a suitable means to collect samples that are combined, thoroughly blended, and subsampled for laboratory analysis. When collecting samples with a probe, it is important to develop a standard, routine collection pattern that will yield a final representative sample. Size of the sample is also important to reducing sample variance. The recommended sample proportion is 0.5 kg per 1,000 kg grain in a shipment. Following this approach, the sample size for a truck-and-trailer combination would be approximately 11.4 kg (25 lbs), since the typical shipment in this type of truck contains 22,727 kg (50,000 lbs). Railcars might be sampled in a different manner, such as sampling the grain from the conveying device as it is being unloaded. Additional discussion of various types of equipment to be used in grain sampling from railcars or trucks is provided by Whitaker, Dickens, and Giesbrecht (1991).

After the sample has been collected, the current recommended practice, especially when analyzing feed grains for aflatoxin, is to coarse grind the entire sample, thoroughly blend it in a suitable mixer, and take a subsample equivalent to about $^1/_{20}$ (ca. 500 g) of the total sample for laboratory analysis. The subsample should be reground to pass through a no. 20 sieve. The rationale for grinding the complete sample before subsampling when analyzing for aflatoxin is that contaminated corn, cottonseed, or peanuts usually contain individual kernels, seeds, or nuts that have very high aflatoxin concentrations, and unrepresentative inclusion or exclusion of these in a small sample may give misleading analytical results for the aflatoxin concentration of the grain shipment. Samples of grain are ground efficiently with a disk mill.

SCREENING, ANALYSIS, AND DETECTION OF MYCOTOXINS

Advances in methods for screening, analysis, and detection of mycotoxins during the past ten years have allowed grain receiving and shipping elevators and terminals to identify and evaluate the mycotoxin content of grains and diet ingredients in a timely manner. These methods usually use a monoclonal antibody that has been developed specifically for the mycotoxin. Monoclonal antibodies allow for quick isolation of particular mycotoxins after preliminary sample extraction and cleanup. Extraction of the mycotoxin from the grain or diet sample is usually accomplished with a 70 to 80 percent aqueous solution of an organic solvent, such as methanol. Once the mycotoxin has been extracted, a cleanup procedure is employed to separate the solvent-mycotoxin mixture from the grain residue. One or more filtration steps result in a mycotoxin preparation that is suitable for further analysis.

One analysis method that has decreased the time required for accurate determination of mycotoxin content is the direct competitive enzyme-linked immunosorbent assay, or ELISA. The clarified, extracted toxin is mixed with an enzyme-toxin conjugate, and a small quantity of the mixture is placed in an antibody-coated microwell. Wells with standards of known concentrations of aflatoxin are prepared at the same time. During a short incubation period, free toxin from the grain sample and bound toxin in the conjugate compete for binding sites on the antibody. The unbound conjugate is rinsed from the well and a substrate, which changes to a blue color in the presence of the conjugate enzyme, is added to the well. A red stopping solution is added to each well, and the color of the solution in the sample wells is evaluated in comparison to the color in the standard wells. If more free mycotoxin from the sample is bound to the antibody in the microwell, the color of the solution will appear more red, but if there is less free toxin, more enzyme-linked toxin of the conjugate will bind to the antibody-coated well, resulting in the development of a blue color. More intense blue color indicates less mycotoxin in the sample. Visual comparison of the color changes in the sample wells with those of the standards provides very good estimates of mycotoxin concentration. Increased sensitivity can be achieved by using additional standards and measuring the subtle changes in blue color with a photometric microwell reader.

Another use of monoclonal antibody technology is found in mycotoxin isolation with the immunoaffinity column. This approach uses a small column that is similar to a 1 cc tuberculin syringe, into which a small quantity of antibody-coated resin has been placed. The clarified extract containing the mycotoxin is passed through the column and becomes bound to the antibody in a typical antigen-antibody reaction. After cleanup with water, the mycotoxin is eluted from the column with 100 percent methanol. At this point, the eluate can be analyzed by liquid chromatography or directly evaluated for mycotoxin content with a fluorometer.

High-performance liquid chromatography (HPLC) can also be used to determine mycotoxin concentration in contaminated grains. Various chromatographic procedures have been employed to isolate and cleanup the mycotoxin of interest from the grain extract prior to determination by HPLC. Many of these are presented in *Official Methods of Analysis of*

AOAC International (AOAC 1995), but use of the immunoaffinity columns described previously provide convenient, accurate means of preparing grain samples for HPLC analysis. Reversed-phase chromatography is the most commonly used HPLC method for the separation of mycotoxins. Columns packed with silica that has been modified with a nonpolar C-18 ligand are preferred. These columns are generally used with moderately polar solvent systems employing aqueous solutions of methanol or acetonitrile, or combinations of these two solvents. The usual methods of detection, once the chromatographic separation has been completed, are ultraviolet (UV) or fluorescence detection. Each of these methods of detection has its own advantages or disadvantages. Ultraviolet detection does not require a sample to be derivatized with a chromatogen compound and therefore can be used efficiently without additional processing, but it sometimes does not have the degree of sensitivity required. Fluorescence detection, on the other hand, usually has greater sensitivity but may require derivatization to detect weakly fluorescent compounds. The aflatoxins offer an example of this situation. Aflatoxins B_2 and G_2 have strong fluorescent characteristics, but aflatoxins B_1 and G_1 have weak fluorescent properties and require derivatization with halogen compounds of either bromine or iodine, or trifluoroacetic acid, prior to detection (Steyn, Thiel, and Trinder 1991). Other methods of detection may be used with HPLC analysis; these include refractometry and multiple photodiode array detector.

Thin-layer chromatography was the original chromatographic technique used to detect and quantify mycotoxins, beginning with aflatoxin. This technique continues to be used because of its comparative simplicity and ability to detect low concentrations of mycotoxins. Gas chromatography (GC) plays a limited role in the analysis of mycotoxins, being used mainly for analysis of the trichothecenes. Newer liquid chromatographic (LC) techniques in combination with mass spectrometry (MS), often referred to as LC-MS, are used to confirm the identification of mycotoxins with a confidence unobtainable with the sole use of LC techniques. Extensive information on mycotoxin analysis is found in *Official Methods of Analysis of AOAC International* (AOAC 1995).

PREVENTION OF MYCOTOXIN CONTAMINATION IN GRAINS AND FISH FEEDS

Proper storage of grains and feedstuffs to prevent contamination with mycotoxins at levels that may cause toxicoses in cultured fish should take into account that mold growth usually occurs under conditions of high moisture content (>12 percent), high relative humidity, and in the presence of adequate oxygen. Assuming that high levels of mycotoxin contamination have not occurred in the field prior to harvest and again during harvest before storage of the grains begins, every precaution should be taken to protect these products from conditions that promote the development of mycotoxins. Additional conditions, other than high moisture content, that promote development of mold growth and contamination with mycotoxins include storage of grains containing large amounts of extraneous materials such as weeds, which usually have a higher moisture content than the grain and can form pockets of moldy grain, and storage of grains or complete fish feeds in facilities that allow exposure of these products to water intrusion from precipitation or water condensation as a result of fluctuations in air temperature. Contamination of feed grains and finished fish diets with mycotoxins can be prevented by accepting grains for diet manufacture that have moisture levels of 12 percent or less, removing debris from grains with a scalping processs prior to storage, and storing grains in clean, structurally sound bins or buildings. Most complete fish diets are processed into physical forms, by steam pelleting or extrusion, that are suitable for application to water when feeding fish. After the pelleting process has been completed, pellets must be adequately dried and cooled to prevent mold growth. Moisture levels of less than 10 percent are recommended, and in some instances a moisture level of 9 percent is desirable for fish diets that will be bagged. Additionally, some manufacturers of bagged channel catfish diets incorporate surface application of liquid mold inhibitors containing propionic acid to extruded pellets. Mold inhibitors of propionic acid have been used safely for human foods and animal diets for many years. Often overlooked, rodent and insect infestation of feed grains can promote the development of molds with subsequent production of mycotoxins by damaging the outer protective layers of stored grain, allowing the invasion and growth of fungal organisms. Control of these pests is important to maintaining nutrient quality and the mycotoxin-free condition of stored grains.

TREATMENTS TO REDUCE MYCOTOXIN CONCENTRATIONS IN FISH FEEDS

Ammoniation of grains and cottonseed meal that were contaminated with aflatoxin was developed many years ago in response to the need to find a means to make these feedstuffs safe for use in animal diets. The ammoniation detoxification process works by chemically altering the structure of aflatoxin to form products that are not toxic to animals. Because the process leaves corn and other diet ingredients dark and discolored, its main application has been for grain and cottonseed designated for cattle diets.

A more promising approach has been the use of adsorbents, which, when added to complete animal diet, bind aflatoxin and render the toxin incapable of being absorbed in the digestive tract. These substances fall into the general classification of clays, with many different chemical and structural characteristics. Commonly, these clays are referred to as hydrated sodium calcium aluminosilicates (HSCAS) and contain various aluminosilicate compounds in combination with sodium, calcium, and other cations such as iron, magnesium, and potassium. These HSCAS compounds exist in internally hydrated forms containing up to 23 percent moisture and appear to efficiently bind only aflatoxin, rather than other mycotoxins, because of the unique chemical (1,3-diketone moiety) and physical (rigid coplanarity) characteristics of aflatoxin (Taylor 1999). Broiler chicks fed a diet containing 4 mg AFB_1/kg had a normal growth rate and prevention of typical AFB_1 toxicity signs (i.e., changes in organ weights and altered serum chemistry), when the diet contained 1 percent of a commercially available HSCAS (Ledoux et al. 1999). Other studies with poultry also have shown that HSCAS preparations effectively bind aflatoxin and reduce its toxic effects, but these have a limited effect on the trichothecene mycotoxins T-2 toxin and diacetoxyscirpenol (Kubena et al. 1990, 1993). After in vitro evaluation of clay adsorbents indicated that clay compounds could sequester CPA, Dwyer and colleagues (1997) found that CPA was not effectively bound in diets fed to broiler chickens that contained 1 percent additions of three binding clays. Claims that these and other adsorbent compounds bind mycotoxins other than aflatoxin may be based on in vitro testing rather than on in vivo evaluation. Ledoux and Rottinghaus (1999) stated that in vivo studies with poultry conducted in their laboratory indicated that none of the inorganic adsorbents tested provided protection against fumonisin B_1, moniliformin, ochratoxin A, or vomitoxin.

These HSCAS compounds are incorporated into diets at inclusion rates below 2 percent during the diet manufacturing process. They have proved effective in binding aflatoxin and in preventing aflatoxicosis in poultry, swine, and other animals (CAST 1989). Currently, the HSCAS compounds are classified as generally recognized as safe (GRAS) for animal diets at inclusion levels of 2 percent or less by the U.S. Food and Drug Administration (FDA). These compounds should be evaluated in fish diets for efficacy in preventing aflatoxicosis of fish prior to general use.

REGULATORY CONTROL OF MYCOTOXIN CONTAMINATION

After identification of aflatoxin as the causative agent of the liver carcinomas (HCC) observed in trout, and the demonstration that similar pathological lesions observed in other animals and humans were associated with exposure to aflatoxin, the FDA in 1965 placed an upper-level concentration of 30 ppb aflatoxin that could be present in commodities used in manufacturing diets for animals. Subsequently, after improvements in the analytical methods for detection and quantification of aflatoxin were achieved, the FDA lowered the limit to 20 ppb, where it has remained since 1969. The 20 ppb action level for aflatoxin applies to commodities and food (except milk: AFM_1, 0.5 ppb) intended for human use, dairy diets, or diets for immature animals. Other action levels of aflatoxin contamination that are currently enforced include (1) 100 ppb for corn designated for breeding cattle, breeding swine, or mature poultry; (2) 200 ppb intended for finishing swine; and (3) 300 ppb in corn intended for finishing beef cattle. Additionally, under FDA regulations, the maximum aflatoxin level of cottonseed meal is 300 ppb when designated for beef cattle, swine, and poultry. Corn not having a designated use must not contain more than 20 ppb (Meronuck and Xie 1999). The FDA action level for aflatoxin is the level above which corn, cottonseed meal, and feeds can be condemned and are subject to seizure. Exceptions to this regulation have been implemented by the FDA when aflatoxin levels of corn have been generally higher than usual due to unfavorable regional growing conditions during certain years. Blending aflatoxin-contaminated corn with noncontaminated corn is an unacceptable practice and is subject to legal action by the FDA.

The FDA places advisory levels on deoxynivalenol (DON) to prevent the effects of this mycotoxin on animals that are sensitive to it. The FDA advisory levels for DON include (1) 1 ppm for finished wheat products intended for human consumption; (2) 10 ppm for grain and grain by-products intended for beef and feedlot cattle older than four months and chickens, provided these ingredients do not exceed 50 percent of the diet; (3) 5 ppm DON in grain and grain by-products intended for swine, with the recommendation that these ingredients not exceed 20 percent of the diet; and (4) 5 ppm on grains and grain by-products intended for all other animals, with the recommendation that these ingredients not exceed 40 percent of the diet (Meronuck and Xie 1999). Advisory levels by the FDA for mycotoxins do not carry penalties of condemnation and seizure by the FDA, but they are intended to provide guidelines for marketing and utilization of grains and grain by-products containing DON. The FDA has considered implementing an advisory level for fumonisin in grains intended for human consumption.

CONCLUSION

Mycotoxins can contaminate feed grains and finished diets intended for the production of agriculturally important animals, including fish. After initial concern about the toxic effects of mycotoxins on various species of salmonid fish, there has been increasing interest during recent years about their toxicity to cultured warm-water fish. Diets fed to warm-water fish, such as channel catfish and tilapia, are being formulated to rely more on plant sources and less on animal sources for dietary protein and energy. Because of the heavy reliance on grains and other plant products in the production of fish, it is important to understand how contamination of these fish diet components by molds and the mycotoxins they produce can affect fish production and health. It has been demonstrated that many of these mycotoxins can lower productive efficiency of cultured fish by reducing growth rate, impairing immunity, and in some cases causing increased mortality. In the past several years there has been a greater awareness of mycotoxins, resulting in the development of management strategies to curtail their production in feed grains and fish diets. Also, development and continued improvement of rapid tests for mycotoxins, such as ELISA and immunoaffinity tests, in grains and fish diets will greatly aid in the control of these toxins. Additional research should be conducted to further evaluate the effects of mycotoxins where limited information is available for fish. The synergistic effects on fish production of the various combinations of mycotoxins that may be present in fish diets need to be better understood. Finally, more emphasis should be directed toward improving the understanding of the effect of mycotoxins on the immunological responses of fish.

REFERENCES

AOAC. (Association of Analytical Chemists). 1995. *Official Methods of Analysis of AOAC International,* Sixteenth Edition. Arlington, VA: Association of Analytical Chemists International.

Bacon, C.W., J.K. Porter, and W.P. Norred. 1995. Toxic interaction of fumonisin B_1 and fusaric acid measured by injection into fertile chicken egg. *Mycopathologia* 129:29-35.

Bacon C.W., J.K. Porter, W.P. Norred and U. Leslie. 1996. Production of fusaric acid by *Fusarium* species. *Applied Environmental Microbiology* 62:4039-4043.

Blount, W.P. 1961. Turkey "X" disease. *Journal of British Turkey Federation* 9(2):52,55-58.

CAST (Council for Agricultural Science and Technology). 1989. *Mycotoxins: Economic and Health Risks.* Report No. 116. Ames, IA: Council for Agricultural Science and Technology.

Chang, C.F. and P.B. Hamilton. 1979. Impaired phagocytosis by heterophils from chickens during aflatoxicosis. *Toxicology and Applied Pharmacology* 48:459-466.

Chaouloff, F., D. Laude, D. Merino, B. Serrurrier, and J.L. Elghozi. 1986. Peripheral and central short-term effects of fusaric acid, a DBH inhibitor, on tryptophan and serotonin metabolism in the rat. *Journal of Neural Transmission* 65:219-232.

Chavez-Sanchez, Ma.C., C.A. Martinez, and I. Osorio Moreno. 1994. Pathological effects of feeding young *Oreochromis niloticus* diets supplemented with different levels of aflatoxin B_1. *Aquaculture* 127:49-60.

Chelack, W.S., J. Borsa, R.R. Marquardt, and A.A. Frohlich. 1991. Role of the competitive microbial flora in the radiation-induced enhancement of ochratoxin production by *Aspergillus alutaceus* var. alutaceus. NRRL 3174. *Applied and Environmental Microbiology* 57:2492-2496.

Coulombe, R.A., Jr. 1993. Biological action of mycotoxins. *Journal of Dairy Science* 76:880-891.

Coulombe, R.A., Jr., G.S. Bailey, and J.E. Nixon. 1984. Comparative activation of aflatoxin B_1 to mutagens by isolated hepatocytes from rainbow trout *(Salmo gairdneri)* and coho salmon *(Onchorynchus kisutch). Carcinogenesis* 5:29-33.

Degan, G.H. and H.G. Neumann. 1981. Differences in aflatoxin B_1-susceptibility of rat and mouse are correlated with the capability in vitro to inactivate aflatoxin B_1-epoxide. *Carcinogenesis* (London) 2:299-306.

Doster, R.C., R.O. Sinnhuber, and J.H. Wales. 1972. Acute intraperitoneal toxicity of ochratoxins A and B in rainbow trout *(Salmo gairdneri). Food and Cosmetic Toxicology* 10:85-92.

Dwyer, M.R., L.F. Kubena, R.B. Harvey, K. Mayura, A. B. Sarr, S. Buckley, R.H. Bailey, and T.D. Phillips. 1997. Effects of inorganic adsorbents and cyclopiazonic acid in broiler chickens. *Poultry Science* 76:1141-1149.

Giambrone, J.J., D.L. Ewert, R.D. Wyatt, and C.S. Eidson. 1978. Effect of aflatoxin on the humoral and cell-mediated immune systems of the chicken. *American Journal of Veterinary Research* 39:305-308.

Goel, S., S.D. Lenz, S. Lumlertdacha, R.T. Lovell, R.A. Shelby, M. Li, R.T. Riley, and B.W. Kemppainen. 1994. Sphingolipid levels in catfish consuming *Fusarium moniliforme* corn culture material containing fumonisin. *Aquatic Toxicology* 30:285-294.

Halver, J.E. 1969. Aflatoxicosis and trout hepatoma. In L.A. Goldblatt (Ed.), *Aflatoxin: Scientific Background, Control, and Implications* (pp. 265-306). New York: Academic Press.

Hartley, R.D., B.F. Nesbitt, and J. O'Kelly. 1963. Toxic metabolites of *Aspergillus flavus. Nature* 198:1056-1058.

Hendricks, J.D., R.O. Sinnhuber, J.H. Wales, M.E. Stack, and D.P.H. Hsieh. 1980. Hepatocarcinogenicity of sterigmatocystin and versicolorin A to rainbow trout *(Salmo gairdneri)* embryos. *Journal of National Cancer Institute* 64:1503-1509.

Hunnan, Y.A. and R.E. Bell. 1989. Functions of sphingolipids and sphingolipid breakdown products in cellular regulation. *Science* (Washington, DC) 234:500-507.

Jantrarotai, W. and R.T. Lovell. 1990a. Acute and subchronic toxicity of cyclopiazonic acid to channel catfish. *Journal of Aquatic Animal Health* 2:255-260.

Jantrarotai, W. and R.T. Lovell. 1990b. Subchronic toxicity of dietary aflatoxin B_1 to channel catfish. *Journal of Aquatic Animal Health* 2:248-254.

Jantrarotai, W., R.T. Lovell, and J.M. Grizzle. 1990. Acute toxicity of aflatoxin B_1 to channel catfish. *Journal of Aquatic Animal Health* 2:237-247.

Kubena, L.F., R.B. Harvey, W.E. Huff, D.E. Corrier, T.D. Phillips, and G.E. Rottinghaus. 1990. Efficacy of a hydrated sodium calcium aluminosilicate to reduce the toxicity aflatoxin and T-2 toxin. *Poultry Science* 69:1078-1086.

Kubena. L. F., R.B. Harvey, W.E. Huff, M.H. Elissalde, A.G. Yersin, T.D. Phillips, and G.E. Rottinghaus. 1993. Efficacy of a hydrated sodium calcium aluminosilicate to reduce the toxicity of aflatoxin and diacetoxyscirpenol. *Poultry Science* 72:51-59.

Lacey, J. 1991. Natural occurrence of mycotoxins in growing conserved forage crops. In J.E. Smith and R.S. Henderson (Eds.), *Mycotoxins in Animal Foods* (pp. 365-369). Boca Raton, FL: CRC Press.

Ledoux, D.R. and G.E. Rottinghaus. 1999. In vitro and in vivo testing of adsorbents for detoxifying mycotoxins in contaminated feedstuffs. In T.P. Lyons and K.A. Jacques (Eds.), *Biotechnology in the Feed Industry,* Proceedings of Alltech's Fifteenth Annual Symposium (pp. 369-379). Nottingham, UK: Nottingham University Press.

Ledoux, D.R., G.E. Rottinghaus, A.J. Bermudez, and M. Alonso-Debolt. 1999. Efficacy of a hydrated sodium calcium aluminosilicate to ameliorate the toxic effects of aflatoxin in broiler chicks. *Poultry Science* 78:204-210.

Lee, B.C., J.D. Hendricks and G.S. Bailey. 1991. Toxicity of mycotoxins to fish. In J.E. Smith and R.S. Henderson (Eds.), *Mycotoxins in Animal Foods* (pp. 607-626). Boca Raton, FL: CRC Press.

Lee, D.J., J.H. Wales, J.L. Ayres and R.O. Sinnhuber. 1968. Synergism between cyclopropenoid fatty acids and chemical carcinogens in rainbow trout *(Salmo gairdneri). Cancer Research* 28:2312-2318.

Li, M.H., S.A. Raverty and E.H. Robinson. 1994. Effects of dietary mycotoxins produced by the mold *Fusarium moniliforme* on channel catfish *Ictalurus punctatus. Journal of the World Aquaculture Society* 25:512-516.

Loveland, P.M., J.E. Nixon, and G.S. Bailey. 1984. Glucuronides in rainbow trout *(Salmo gairdneri)* injected with [^{3}H] aflatoxin B_1 and effects on dietary beta-naphthoflavone. *Comparative Biochemistry and Physiology C, Comparative Pharmacology* 78:13-19.

Lumlertdacha, S. and R.T. Lovell. 1995. Fumonisin-contaminated dietary corn reduced survival and antibody production by channel catfish challenged with *Edwardsiella ictaluri. Journal of Aquatic Animal Health* 7:1-8

Lumlertdacha, S., R.T Lovell, R.A. Shelby, S.D. Lenz, and B.W. Kemppainen. 1995. Growth, hematology, and histopathology of channel catfish, *Ictalurus punctatus,* fed toxins from *Fusarium moniliforme. Aquaculture* 130:201-218.

Manning, B.B., 1998. "Fumonisin B_1 from *Fusarium moniliforme* culture material varies in toxicity to channel catfish and rats with purity." Doctoral dissertation. Auburn University, Alabama.

Marquardt, R.R. and A.A. Frohlich. 1992. A review of recent advances in understanding ochratoxicosis. *Journal of Animal Science* 70:3968-3988.

Meronuck, R. and W. Xie. 1999. Mycotoxins in feed. *Feedstuffs* 71(31):123-130.

Norred, W.P. 1993. Fumonisins: Mycotoxins produced by *Fusarium moniliforme. Journal of Toxicology and Environmental Health* 38:309-328.

Norred, W.P., C.W. Bacon, J.K. Porter, and K.A. Voss. 1990. Inhibition of protein synthesis in rat primary hepatocytes by extracts of *Fusarium moniliforme*-contaminated corn. *Food and Chemical Toxicology* 28:89-94.

Norred, W.P., E. Wang, H. Yoo, R.T. Riley, and A.H. Merrill, Jr. 1992. In vitro toxicology of fumonisins and the mechanistic implications. *Mycopathologia* 177:73-78.

Pier, A.C. 1991. The influence of mycotoxins on the immune system. In J.E. Smith and R.S. Henderson (Eds.), *Mycotoxins in Animal Foods* (pp. 489-497). Boca Raton, FL: CRC Press.

Post, G. 1987. Neoplastic diseases of fishes, In *Textbook of Fish Health* (pp. 244-246). Neptune City, NJ: T.F.H. Publications.

Poston, H.A., J.L. Coffin, and G.F. Combs Jr. 1982. Biological effects of dietary T-2 toxin on rainbow trout, *Salmo gairdneri. Aquatic Toxicology* 2:79-88.

Qureshi, M.A. and W.M. Hagler. 1992. Effect of fumonisin B_1 exposure on chicken macrophage functions in vitro. *Poultry Science* 71:104-112.

Riley. R.T., E. Wang, and A.H. Merrill. 1994. Liquid chromatographic determination of sphingosine: Use of free sphinganine to sphingosine ratio as a biomarker for consumption of fumonisins. *Journal of Association of Analytical Chemists International* 77:533-540.

Sargeant, K., J. O' Kelly, R.B.A. Carnaghan, and R. Allcroft. 1961. The assay of a toxic principle in certain groundnut meals. *Veterinary Record* 73:1219-1223.

Sargeant, K., A. Sheridan, J. O' Kelly, and R.B.A. Carnaghan. 1961. Toxicity associated with certain samples of groundnuts. *Nature* 192:1096-1097.

Schoenhard, G.L., J.D. Hendricks, J.E. Nixon, D.J. Lee, J.H. Wales, R.O. Sinnhuber, and N.E. Pawlowski. 1981. Aflatoxicol-induced hepatocellular carcinoma in rainbow trout *(Salmo gairdneri)* and the synergistic effects of cyclopropenoid fatty acids. *Cancer Research* 41:1011-1014.

Smith, T.K. 1992. Recent advances in the understanding of *Fusarium* trichothecene mycotoxicoses. *Journal of Animal Science* 70:3989-3993.

Smith, T.K., E.G. MacMillan, and J.B. Castillo. 1997. Effect of feeding blends of *Fusarium* mycotoxin-contaminated grains containing deoxynivalenol and fusaric acid on growth and feed consumption of immature swine. *Journal of Animal Science* 75:2184-2191.

Smith, T.K. and I.R. Seddon. 1998. Synergism demonstrated between *Fusarium* mycotoxins. *Feedstuffs* 70(25):12-17.

Smith, T.K. and M.G. Sousadias. 1993. Fusaric acid content of swine feedstuffs. *Journal of Agricultural and Food Chemistry* 41:2296-2398.

Steyn, P.S., P.G. Thiel, and D.W. Trinder. 1991. Detection and quantification of mycotoxins by chemical analysis. In J.E. Smith and R.S. Henderson (Eds.), *Mycotoxins in Animal Foods* (pp. 165-222). Boca Raton, FL: CRC Press.

Strength, D.R., D.V. Saradambal, Shoou-Liz Wang, H.H. Daron, and W.P. Schoor. 1982. Glucuronosyl- and sulfo-transferases in fish exposed to environmental carcinogens. *Federation Proceedings* 41(4):1147.

Taylor, D.R. 1999. Mycotoxins binders: What are they and what makes them work? *Feedstuffs* 71(3):41-45

Whitaker, T.B., J.W. Dickens, and F.G. Giesbrecht. 1991. Testing animal feedstuffs for mycotoxins: Sampling, subsampling, and analysis. In J.E Smith and R.S. Henderson (Eds.), *Mycotoxins in Animal Foods* (pp. 153-164). Boca Raton, FL: CRC Press.

Woodward, B., L.G. Young, and A.K. Lun. 1983. Vomitoxin in diets for rainbow trout *(Salmo gairdneri)*. *Aquaculture* 35:93-10.

Wyatt, R.D., P.B. Hamilton, and H.R. Burmeister. 1973. The effects of T-2 toxin in broiler chickens. *Poultry Science* 52:1853-1859.

Yildirim, M., B. Manning, R.T. Lovell, J.M. Grizzle, and G.E. Rottinghaus. 2000. Toxicity of moniliformin and fumonisin B_1 fed singly and in combination in diets for channel catfish. *Journal of World Aquaculture Society* 31:607-616.

Chapter 14

Feed Allowance and Fish Health

Richard T. Lovell
Veronica O. Okwoche
Myung Y. Kim

INTRODUCTION

Winter management of catfish in commercial ponds varies from continual feeding, to following feeding guidelines based upon fish size and temperature (Stickney and Lovell 1977), to no feeding at all. Generally, feeding catfish through the winter months is recommended to prevent weight loss (Lovell and Sirikul 1974) and to maintain health and provide maximum resistance of fish against bacterial infections (MacMillan 1985).

Enteric septicemia (ESC) in channel catfish, caused by the bacterium *Edwardsiella ictaluri,* is an infectious disease of significant economic importance in commercial catfish farming (Plumb 1994). In the southeastern United States, *E. ictaluri* is most virulent during spring when water temperatures are 22 to 25°C (MacMillan 1985; Francis-Floyd et al. 1987; Plumb and Brady 1990). In spring, this favorable environmental condition for an ESC epizootic coincides with a relatively poor nutritional condition of commercially grown channel catfish that do not consume much food during winter. Therefore, winter feeding has been recommended (Dupree and Huner 1984; Lovell 1989) so that the fish will be in the best nutritional condition possible when the season for *E. ictaluri* epizootics arrives. However, few data are available to support the concept that winter feeding increases resistance to infectious diseases in channel catfish. A report by Kim and Lovell (1995) indicates that winter feeding may not be economically practical based upon fish weight change. Information on the effects of winter and early spring feeding on resistance of channel catfish to *E. ictaluri* infection would be valuable in making recommendations on management of channel catfish in commercial ponds in the Southeast.

Little research has been reported regarding the effects of food deprivation on immune responses in fish, although a number of reports are available regarding the effects of nutrient deficiencies on immune responses in channel catfish (Blazer and Wolke 1984; Li and Lovell 1985; Blazer, Ankley, and Finco-Kent 1989; Sheldon and Blazer 1991; Duncan and Lovell 1994; Fracalossi and Lovell 1994; Paripatananont and Lovell 1995). Catfish farmers have observed that during an ESC epizootic, cessation of feeding seems to reduce the number of deaths in the fish (E. H. Robinson, Delta Research and Extension Center, Stoneville, Mississippi, personal communication).

To further evaluate the effects of feeding regimen on resistance to bacterial infection in channel catfish, two similar one-year studies were conducted during a six-month cool weather period to determine the effects of continual feeding, partial feeding, and no feeding on responses of food-size (age two) and small (age one) fish to *E. ictaluri* challenge.

METHODS

Fish and Feeding Protocol

Year One

Age-one (average weight 43 ± 1.0 g) and age-two (average weight 660 ± 16.0 g) channel catfish, obtained from the Alabama Agricultural Experiment Station, were stocked separately in 400 m^2 earthen ponds at the Fisheries Research Unit, Alabama Agricultural Experiment Station, Auburn University on October 5, 1993. Stocking rates were 13,750 fish/ha and 3,750 fish/ha for the age-one and age-two fish, respectively. The experiment began on November 1 and ended on April 30, 1994. Three ponds of fish from each age group (age one and age two) were randomly assigned to each of three overwinter feeding regimens: continual feeding, partial feeding, and no feeding. Fish on the continual-feeding regimen were fed throughout the experimental period when water temperature (1 m depth) was above 6°C. The age-two fish were fed 2.0, 1.75, 1.5, 1.0, 0.5, and 0 percent of body weight every other day when water temperature was >18, 15 to 18, 12 to 15, 9 to 12, 6 to 9, and >6°C, respectively (Stickney and Lovell 1977). The age-one fish were fed 33 percent more than the age-two fish at the prescribed temperatures. Fish on the partial-feeding regimen were fed the same as those on the continual-feeding regimen except during December, January, and February, when they received no feed. The nonfed fish received no feed during the six-month experimental period (November 1 through April 30).

The fed fish received a commercial 26 percent protein, slow sinking, extruded diet during the overwinter period. Fish were fed at 16:00 hours (h) when temperature permitted. Water temperature and dissolved oxygen (DO) were monitored daily throughout the study, and emergency aeration, provided by 0.25 kW lift-type aerators, was used in early spring in ponds if afternoon DO levels dropped to 4 mg/liter or less.

On April 30, when afternoon water temperature had reached approximately 23°C, all fish were removed from the ponds, weighed, and counted. Samples of fish from each pond were then transferred to 1 m^3 circular raceways (age-two fish) or 70-liter aquaria (age-one fish), both of which were supplied with a continual flow of pond water, for bacterial challenge.

Year Two

Fish size in the second experiment was 22 g for year-one fish and 420 g for year-two fish. Starting date was November 1, 1994, and termination date was April 23, 1995. Stocking density, feed, and management were the same as described for year one.

Bacterial Challenge

Maintenance of Fish

Fish in the circular raceways and aquaria continued on the same feeding regimens as they were subjected to in the ponds. Water temperature in the circular raceways was 23 to 25°C in daytime and 1 to 2°C lower at night. Water in the aquaria was maintained at a temperature of 25 ± 2°C. These temperatures represent the optimum temperature for the bacterium *E. ictaluri* (Hawke et al. 1981). The fish were held in the circular raceways or aquaria for five days prior to bacterial challenge.

Preparation of E. ictaluri

A virulent strain of *E. ictaluri* (AL-93-92) isolated from an ESC epizootic was obtained from the Southeastern Cooperative Fish Disease Diagnostic Laboratory, Auburn University, for the challenge. The organism was recycled two times through live channel catfish to enhance the virulence, as described by Vinitnantharat and Plumb (1992). The reisolated pathogen was then cultured in brain heart infusion (BHI) broth (Difco Laboratories, Detroit, Michigan*) for 24 hours at 30°C to yield a cell population of approximately 10^9 cells/ml.

Determining Optimal Cell Concentration for Challenge

Prior to the experimental challenge, the optimum cell concentration that would be used for the experimental challenge was determined. This was obtained by injecting ten age-one fish from the fed treatment with 0.1 ml of bacterial suspensions containing 10^3, 10^4, or 10^5 cells/ml and by injecting ten age-two fish from the fed treatment with 0.1 ml of cell suspension containing 10^6, 10^7, or 10^8 cells/ml and recording mortalities among the fish for ten days. The fish were held in static water in aquaria (age-one fish) or

---

*Use of trade or manufacturer names does not imply endorsement.

circular tanks (age-two fish) with aeration at a temperature of 23 to 25°C. Every other day during the challenge, 10 percent of the water in the holding facilities was changed, and salt (NaCL) was added to each tank or aquaria (200 mg/l) to prevent nitrite toxicity from nitrogenous excretions. The mortality data were used to calculate the LC_{50} values (Plumb and Bowser 1983), which were determined to be $10^{4.25}$ cells/ml of suspension for the year-one and $10^{7.55}$ cells/ml of suspension for the year-two fish for experiment one and to be $10^{3.90}$ cells/ml and $10^{7.85}$ cells/ml for year-one and two fish, respectively, in experiment two. Cell concentrations used for the experimental challenge were 10^4 cells/ml for the year-one and 10^8 cells/ml for the year-two fish in both experiments.

Challenge Conditions

The experimental fish (25 year-two fish and 40 year-one fish from each pond) were anesthetized with 100 mg/l of tricaine methanesulfonate (MS-222, Argent Chemical Laboratories, Redmond, Washington*) and injected intraperitoneally with 0.1 ml of bacterial cell suspension of appropriate concentration. The fish were held in static water in aquaria or circular tanks with continual aeration as described previously. Fish from each pond were held in separate aquaria or tanks. Comparison groups of year-one and year-two fish from each treatment were injected with 0.1 ml of sterile saline and held under similar conditions to serve as controls. Mortality was recorded daily for 16 days. There were no deaths after day 14 postchallenge.

Determination of Antibody Titer

Prior to the challenge in year one and year two, three fish from each replicate pond were assayed and found to be naive to *E. ictaluri* antibodies. Later, antibody titers were determined during year one and year two for five fish from each replicate group that had survived the *E. ictaluri* challenge. Antibody titers were determined by the serum agglutination method of Roberson (1990), using serial 20-fold dilution serums with 0.85 percent NaCl.

Phagocytic Cell Assay

During year one only, head kidneys from 20 year-one fish and 10 year-two fish in each replicate pond were pooled for determination of phagocytic cell activity. Three assays were carried out for each replicate pond. Head kidneys were removed aseptically, placed in phosphate-buffered saline (PBS, pH 7.2, GIBCO BRL, Life Technologies, Inc., Grand Island, New York) and homogenized in a handheld stainless steel tissue grinder with a

---

*Use of trade or manufacturer name does not imply endorsement.

mesh diameter of 0.4 mm. A two-layer discontinuous density gradient of Percoll (Pharmacia LKB Biotechnology, Uppsala, Sweden; see Pharmacia Fine Chemicals, Undated) was used to separate cells. Phagocytic activity was determined by procedures described by Okwoche and Lovell (1997).

EFFECTS OF FEED ALLOWANCE ON GROWTH AND HEALTH

Year One

There was no significant difference in weight gain between the partially fed and continually fed groups in both age groups (see Table 14.1). The unfed fish in both groups showed negative weight change, which was significantly different from the weight changes of the fed groups, which were positive. All fish, including those not fed, appeared healthy and in satisfactory condition at the end of the feeding period.

Unfed year-one fish challenged with *E. ictaluri* began to die one day postchallenge and had all died within five days (see Table 14.1). The fully fed and partially fed fish showed approximately 50 percent mortality at day four, while the unfed fish showed 94 percent mortality at day four. In contrast to the year-one fish, the year-two fish showed higher mortality on the fully and partially fed regimens compared to the unfed regimen (see Table 14.1). On day eight postchallenge, 100 percent of the unfed fish were

TABLE 14.1. Weight Change and Mortality Following *Edwardsiella ictaluri* Challenge by Year-One and Year-Two Channel Catfish Managed by Three Feeding Regimens During Winter

Fish Age Group	Feeding Regimen	Percentage Weight Change	Mortality (%)[1]
Year-One	Unfed	-9a	94a
	Partially fed	50b	50b
	Fully fed	64b	48b
	Pooled SEM	14.1	4.4
Year-Two	Unfed	-10a	23a
	Partially fed	42b	80b
	Fully fed	49b	78b
	Pooled SEM	3.8	3.9

Note: Means in columns within the same age group followed by different letters are significantly different ($P = 0.05$); SEM = standard error of the mean.

[1]Cumulative mortality at 4 days for year-one fish and 12 days for year-two fish.

alive, while only 40 to 50 percent of the fed fish remained alive. By day 12 postchallenge, 67 percent of the unfed fish were alive, while only about 20 percent of the fed fish survived.

Year Two

There was no significant difference in weight gain between the partially fed and continually fed groups of the year-one and year-two fish (see Table 14.2). The unfed fish in both age groups showed significantly lower (negative) weight gains when compared to the fed groups. Although the unfed year-one and year-two fish lost weight during the winter period, they appeared healthy and in satisfactory condition at harvest.

In the year-one fish, mortality rate subsequent to experimental challenge with *E. ictaluri* was significantly higher in the unfed fish, 98.3 percent, when compared to the two fed groups, 52.5 to 55.8 percent (see Table 14.2). In the year-two fish, however, the mortality rate of 9.3 percent in the unfed fish was significantly lower than the 69.3 and 77.3 percent in the partially fed and continually fed groups, respectively. In both age groups there were no significant differences between mortality rates of the partially fed and continually fed fish. Mortality in year-one fish began on day four postchallenge, peaked at days five to seven, and ceased after day nine. In the

TABLE 14.2. Weight Change, Mortality Following *Edwardsiella ictaluri* Challenge, Antibody Production, and Phagocytosis by Year-One and Year-Two Channel Catfish Managed by Three Feeding Regimens During Winter

Fish Age Group	Feeding Regimen	Percentage Weight Change	Mortality (%)[1]	Antibody Titer to *E. ictaluri* (Reciprocal Titer)	Phagocytic Index (Bacterial/ Phagocyte)
Year-One	Unfed	-12a	98.3a	128.8a	2.38a
	Partially fed	92b	55.8b	186.7b	5.06b
	Fully fed	106b	52.5b	288.9c	5.56b
	Pooled SEM	7.6	2.2	7.9	0.09
Year-Two	Unfed	-7a	9.3a	528.0a	3.11a
	Partially fed	38b	69.3b	138.7b	5.69b
	Fully fed	39b	77.3b	142.2b	6.32b
	Pooled SEM	5.6	5.4	4.3	0.21

Note: Means in columns within the same age group followed by different letters are significantly different (*P* - 0.05); SEM = standard error of the mean.

[1]Cumulative mortality at 8 days for year-one fish and 14 days for year-two fish.

year-two fish, mortality in the partially fed and continually fed groups began on day six postchallenge, peaked at days 8 to 11, and ceased on day 14. The unfed year-two fish did not begin to die until day 10 postchallenge and stopped dying on day 12.

No antibodies against *E. ictaluri* antigens were detected in fish prior to the challenge study. Subsequent to challenge, antibody titers in the year-one fish were significantly lower in the unfed fish than in the partially fed or fully fed groups, which showed no difference in antibody production (see Table 14.2). Antibody production for the year-two fish was significantly higher in the unfed fish than in the partially fed and fully fed groups, which were similar. The phagocytic index for unfed year-one and year-two fish was significantly lower than those for the partially fed and fully fed fish in both age groups (see Table 14.2). There was no significant difference in phagocytic index between partially fed and fully fed fish in both age groups.

DISCUSSION

Weight change and challenge data from both experiments agreed closely. The data indicated that continual feeding of year-one and year-two channel catfish during the winter had no significant benefit over discontinued feeding during December, January, and February, with regard to fish weight in the spring. Partial or continual winter feeding resulted in average increases of 64 to 100 percent of initial weight by year-one fish and 38 to 49 percent of initial weight by year-two fish. Weight losses of 7 to 12 percent by the unfed year-one and year-two fish are consistent with other overwinter studies (Lovell and Sirikul 1974; Robinette et al. 1982). Data from these studies show that small channel catfish lose or gain weight in proportion to body size faster than larger fish. In both studies, fish condition was good, and pond mortalities were low in the unfed groups.

Mortality of year-one channel catfish exposed to *E. ictaluri* challenge subsequent to overwinter starvation does not appear to follow the same pattern as that of older fish. Unfed year-one fish showed lower resistance to bacterial infection than the fed fish, but unfed year-two fish had higher resistance than the fed treatments in both experiments. Mortality rates for partial and continual feeding treatments were similar for both age groups. Increased resistance to bacterial infection with winter feeding, as occurred with the young fish, was anticipated inasmuch as research has shown that good nutrition results in increased immune responses in channel catfish (Durve and Lovell 1982; Li and Lovell 1985; Blazer, Ankley, and Finco-Kent 1989; Paripatananont and Lovell 1995). However, the increased resistance to bacterial infection of the starved year-two fish is difficult to explain.

The effects of various nutrients on antibody production and macrophage activity have been reported in channel catfish, but no research has been done on the effects of feeding level. Channel catfish fed diets deficient in vitamin C

(Li and Lovell 1985), vitamin E (Blazer, Ankley, and Finco-Kent 1989), dietary lipids (Sheldon and Blazer 1991), folic acid (Duncan and Lovell 1994), and selenium (Wang 1996) showed reduced phagocytic activity and/or antibody production. Henken, Tigchelaar, and Van Muiswinkle (1987) found that long-term starvation or maintenance feeding of adult African catfish, *Clarias gariepinus,* reduced antibody production. This conclusion fits the data from the year-one fish in this study, in which antibody production was directly related to feeding level, but not data from the year-two fish, in which starvation induced higher antibody titers. The present study shows that phagocytosis of *E. ictaluri* by pronephros macrophages of channel catfish is influenced by feeding level. Phagocytic index was directly related to feeding for both sizes of fish, with starved fish having significantly lower phagocytic activity than fed fish. Reduction in macrophage activity associated with starvation may have influenced the resistance of the year-one fish to *E. ictaluri* challenge, but not that of the year-two fish, which showed increased resistance, with a reduction in macrophage activity.

Increased resistance to infection associated with reduced food intake has been reported in warm-blooded animals. Weindruch and colleagues (1983) found that immune systems of mice subjected to dietary restriction from the time of weaning were better maintained than those in fed control mice. They suggest that the immune systems of the restricted animals may have matured less rapidly and consequently remained more functional for longer periods than in well-fed controls. Fernandes and Good (1984) reported that restricting energy sources for autoimmune strains of mice by reducing the ration of a nutritionally balanced diet by 40 percent impeded or inhibited the development of disease associated with aging. According to Wing and colleagues (1983), fasting has differential influences on immune functions, rather than a uniformly deleterious effect in humans, and nutritional alteration appears to enhance certain effector functions of the host defense system. Kiron and colleagues (1993) reported that the availability of trace elements in various tissues in the host animal, as influenced by dietary intake, may be crucial for immune functions. In warm-blooded animals iron has been reported to play a role in the pathology of bacterial infection by enhancing the multiplication and/or virulence of the invading microorganism (Weinberg 1989); thus, depriving the pathogen of iron may serve as a nonspecific protective mechanism against infection by the host organism (Kluger and Bullen 1987; Weinberg 1989). Bullen, Ward, and Rogers (1991) reported that multiplication and virulence of *Escherichia coli* in humans were enhanced by increased total serum iron concentration as influenced by diet. Effects of starvation on tissue concentration of iron and other trace elements in channel catfish are not known. Starvation alters quantity and composition of serum proteins in fish due to impaired synthesis or proteolysis of serum proteins for energy expenditure and other purposes (Sorvachev 1957; Ellis 1981a, 1981b). Changes in serum protein concentration or composition may be associated with increase or decrease in resistance to bacterial challenge.

This study indicates that starvation increases resistance to *E. ictaluri* infection in channel catfish under some circumstances. In both studies the fish were starved all winter, and resistance was increased in large fish but reduced in small fish. Certain considerations should be addressed before practical implications of these results can be presented. The length of time that channel catfish should be deprived of feed to significantly impact resistance to *E. ictaluri* is important: Would the food-size fish have shown the same increased resistance with a reduction in feed allowance or a shorter period of food deprivation prior to challenge? Another consideration is the possible interaction between fish size and time of food deprivation: Would the small fish, which lost more weight during starvation, have responded adversely to the challenge with a shorter period of food deprivation? Intensity of the bacterial challenge also may influence response of channel catfish to feeding regimen. The experimental challenge in this study was intraperitoneal injection of an LC_{50} concentration of a virulent strain of *Edwardsiella ictaluri;* a less severe challenge, as would occur in a commercial culture, may produce different results. Possibly all the factors, fish size, period and level of feed deprivation, and intensity of challenge, are interrelated in influencing responses of channel catfish to *E. ictaluri* infection.

REFERENCES

Blazer, V.S., G.T. Ankley, and D. Finco-Kent. 1989. Dietary influences on disease resistance factors in channel catfish. *Developmental and Comparative Immunology* 13:43-48.

Blazer, V.S. and R.E. Wolke. 1984. Effect of diet on the immune response of rainbow trout *(Salmo gairdneri). Canadian Journal of Fisheries and Aquatic Sciences* 41:1244-1247.

Bullen, J.J., G.G. Ward, and H.J. Rogers. 1991. The critical role of iron in some clinical infection. *European Journal of Clinical Microbiology and Infectious Diseases* 10:613-617.

Duncan, L.P. and R.T. Lovell. 1994. Influence of vitamin C on the folate requirement of channel catfish, *Ictalurus punctatus,* for growth, hematopoiesis, and resistance to *Edwardsiella ictaluri* infection. *Aquaculture* 127:233-244.

Dupree, H.K. and J.V. Huner. 1984. *Third Report to the Fish Farmers.* Washington, DC: U.S. Fish and Wildlife Service.

Durve, V.S. and RT. Lovell. 1982. Vitamin C and disease resistance in channel catfish *(Ictalurus punctatus). Canadian Journal of Fisheries and Aquatic Sciences* 39:948-951.

Ellis, A.E., 1981a. Nonspecific defense mechanisms in fish and their role in disease processes. *Developments in Biological Standardization* 49:337-352.

Ellis, A.E., 1981b. Stress and the modulation of defense mechanisms in fish. In A.D. Pickering (Ed.), *Stress and Fish* (pp. 147-169). London: Academic Press.

Fernandes, G. and R.A. Good. 1984. Inhibition by diet of lymphoproliferative disease and renal damage in MRL/lp mice. *Proceedings of the National Academy of Sciences of the U.S.A.* 81:6144.

Fracalossi, D.M. and R.T. Lovell. 1994. Dietary lipid sources influence responses of channel catfish *(Ictalurus punctatus)* to challenge with the pathogen *Edwardsiella ictaluri. Aquaculture* 119:287-298.

Francis-Floyd, R., M.H. Beleau, R.R. Waterstate, and P.R. Bowser. 1987. Effect of water temperature on the clinical outcome of infection with *Edwardsiella ictaluri* in channel catfish. *Journal of the American Veterinary Medicine Association* 191:1413-1416.

Hawke, J.P., A.C. McWhorter, A.G. Steigerwalt, and D.J. Brenner. 1981. *Edwardsiella ictaluri* sp. Nov., the causative agent of enteric septicemia of catfish. *International Journal of Systematic Bacteriology* 31:396-400.

Henken, A.M., A.J. Tigchelaar, and W.B. Van Muiswinkle. 1987. Effects of feeding level on antibody production in African catfish, *Clarias gariepinus* Burchell, after injection of *Yersinia ruckeri* O-antigen. *Journal of Fish Diseases* 11:85-88.

Kim, M.K. and R.T. Lovell. 1995. Effect of overwinter feeding regimen on body weight, body composition and resistance to *E. ictaluri* in channel catfish, *Ictalurus punctatus. Aquaculture* 134:237-246.

Kiron, V., A. Gunji, N. Kamoto, S. Sato, Y. Ikeda, and T. Watanabe. 1993. Dietary nutrient dependent variations on natural-killer activity of the leukocytes of rainbow trout. *Gyobyo Kenkyu* 28:71-76.

Kluger, M.J. and J.J. Bullen. 1987. Clinical and physiological aspects. In J.J. Bullen and E. Griffiths (Eds.), *Iron and Infection: Molecular, Physiological and Clinical Aspects* (pp. 243-282). Chichester, England: Wiley.

Li, Y. and R.T. Lovell. 1985. Elevated levels of dietary ascorbic acid increase immune responses in channel catfish. *Journal of Nutrition* 115:123-131.

Lovell, R.T. 1989. *Nutrition and Feeding of Fish.* New York: Van Nostrand Reinhold.

Lovell, R.T. and B. Sirikul. 1974. Winter feeding of channel catfish. *Proceedings of the Southeastern Association of Game and Fish Commissioners* 28:208-216.

MacMillan, J. 1985. Infectious disease. In C. S. Tucker (Ed.), *Channel Catfish Culture* (pp. 434-441). New York: Elsevier Science Publishers.

Okwoche, V.O. and R.T. Lovell. 1997. Cool weather feeding influences responses of channel catfish to *Edwardsiella ictaluri* challenge. *Journal of Aquatic Animal Health* 9:163-171.

Paripatananont, T. and R.T. Lovell. 1995. Responses of channel catfish fed organic and inorganic sources of zinc to *Edwardsiella ictaluri* challenge. *Journal of Aquatic Animal Health* 7:147-154.

Pharmacia Fine Chemicals. Undated. *Percoll: Methodology and Application: Density Marker Beads for Calibration of Gradients of Percoll.* Piscataway, NJ: Pharmacia Fine Chemicals.

Plumb, J.A. 1994. In *Health Maintenance of Cultured Fishes: Principle Microbial Deseases* (pp. 142-148). Boca Raton, FL: CRC Press.

Plumb, J.A. and P.R. Bowser. 1983. *Microbial Fish Disease Laboratory Manual.* Montgomery, AL: Brown Printing Company.

Plumb, J.A. and Y.J. Brady. 1990. Disease of catfish follow seasonal trends. *Highlights of Agricultural Research* 4:37.

Roberson, B.S. 1990. Bacterial agglutination. In J.S. Stolen, T.C. Fletcher, D.P. Anderson, B.S. Roberson, and W.B. Van Muiswinkle (Eds.), *Techniques in Fish Immunology* (pp. 81-86). Fair Haven, NJ: SOS Publications.

Robinette, H.R., S.C. Newton, R.O. Regan, R.R. Stickney, and R.P. Wilson. 1982. Winter feeding of channel catfish in Mississippi, Arkansas, and Texas. *Proceedings of the Southeastern Association of Fish and Wildlife Agencies* 36:162-171.

Sheldon, W.M., Jr. and V.S. Blazer. 1991. Influence of dietary lipid and temperature on bactericidal activity of channel catfish macrophages. *Journal of Aquatic Animal Health* 3:87-93.

Sorvachev, K. 1957. Changes in proteins of carp blood serum during hibernation. *Biochemistry* (Biochimiya) 22:822-827.

Stickney, R.R. and R.T. Lovell. 1977. Nutrition and Feeding of Catfish. *Southeastern Cooperative Series Bulletin* No. 218, Montgomery, AL: Auburn Univeristy.

Vinitnantharat, S. and J.A. Plumb. 1992. Kinetics of the immune response of channel catfish to *Edwardsiella ictaluri. Journal of Aquatic Animal Health* 4:207-214.

Wang, C. 1996. "Composition of Organic and Inorganic Sources of Selenium for Growth, Glutathione Peroxidase Activity, and Immune Responses in Channel Catfish." Doctoral dissertation. Auburn University, Alabama.

Weinberg, E.D. 1989. Cellular regulation of iron assimilation. *Quarterly Review of Biology* 64: 261-290.

Weindruch, R.H., B.H. Devens, H.V. Raff, and R.L. Walford. 1983. Influence of dietary restriction and aging on natural-killer cell activity in mice. *Journal of Immunology* 130:993-998.

Wing, E.J., R.T. Stanko, A. Winkelstein, and S.A. Adibi. 1983. Fasting enhanced immune effector mechanisms in obese subjects. *American Journal of Medicine* 75:91-96.

Chapter 15

Dietary Lipids and Stress Tolerance of Larval Fish

Charles R. Weirich
Robert C. Reigh

INTRODUCTION

Dietary lipids are known to play an important role in animal nutrition (Wilson, Fisher, and Fuqua 1975; Pond, Church, and Pond 1995). They provide energy for growth, maintenance, and reproduction and are sources of essential fatty acids (EFAs) and phospholipids involved in the structure and function of cell membranes. In addition, dietary lipids are involved in the transport of fat-soluble nutrients, such as sterols and vitamins, and in the synthesis of hormones, prostaglandins, and other metabolically active compounds.

Rapid expansion of aquaculture in recent years and the associated increase in the use of formulated feeds has produced a considerable body of research on dietary lipid requirements of fish (Watanabe 1982; Kanazawa 1985; Bell, Henderson, and Sargent 1986; Hepher 1988; Lovell 1989; Sargent, Henderson, and Tocher 1989; Steffens 1989; NRC 1993; Coutteau et al. 1997; Rainuzzo, Reitan, and Olsen 1997). Most of that work has focused on essential fatty acid and phospholipid requirements of juvenile and adult fish, using growth and survival as indicators of diet quality. High survival and sustained growth usually indicate that a diet satisfies the minimum nutritional requirements of the species being fed, but survival and growth alone might not be indicative of a fish's physiological condition (Dhert, Lavens, and Sorgeloos 1992), which can affect its response to stressful conditions. Thus, recent research on lipid requirements has investigated the effects of dietary EFAs and phospholipids on the ability of fish to tolerate stress associated with handling and suboptimal environmental conditions. Much of the work in this area has involved larval fish, particularly marine species, because of the difficulty in raising the larvae of many aquatic species successfully.

This chapter will briefly review the existing literature on the effects of dietary lipids on stress tolerance of larval fishes. The discussion will focus on EFAs and phospholipids because of their importance in moderating stress responses in fish.

ESSENTIAL FATTY ACIDS

An extensive amount of research has been conducted to determine the essential fatty acid requirements of fish; reviews of the lipid requirements of fish have been provided by Castell (1979), Cowey and Sargent (1977), Watanabe (1982), and Bell, Henderson, and Sargent (1986). In general, freshwater fish require dietary sources of polyunsaturated fatty acids (PUFAs) of the n-6 (linoleic acid, 18:2n-6) and n-3 (linolenic acid, 18:3n-3) families for optimal growth (Kanazawa 1985; Henderson and Tocher 1987; Sargent, Henderson, and Tocher 1989; NRC 1993). Marine fish require dietary sources of highly unsaturated fatty acids (HUFAs), particularly eicosapentaenoic acid (EPA, 20:5n-3) and docosahexaenoic acid (DHA, 22:6n-3), because they are unable to elongate and desaturate fatty acids of the n-3 family in sufficient quantity to satisfy minimum requirements (Owen et al. 1975; Cowey et al. 1976; Cowey and Sargent 1977; Kanazawa, Teshima, and Ono 1979; Yamada, Kobayashi, and Yone 1980; Watanabe 1982; Bell, Henderson, and Sargent 1986; NRC 1993). DHA and EPA are important growth promoters in marine fish. They have both structural and metabolic functions and have been associated with such diverse biological processes as pigmentation in larval turbot, *Scophthalmus maximus* (Rainuzzo et al. 1994; Reitan, Rainuzzo, and Olsen 1994), and predation efficiency of larval herring, *Clupea harengus* (Bell et al. 1995).

Although the majority of studies of dietary EFA requirements have been conducted with juvenile fish (Watanabe 1982), in recent years the value of dietary EFAs in larval nutrition has been investigated (Tocher et al. 1985; Falk-Petersen et al. 1986; Koven, Kissil, and Tandler 1989; Koven, et al. 1990; Sargent, Henderson, and Tocher 1989; Sargent, McEvoy, and Bell 1997; Rainuzzo, Reitan, and Olsen 1997; Takeuchi 1997). Most studies have focused on the effect of dietary EFAs on growth and survival of larval marine fishes. Results indicate that larvae possess the same qualitative dietary requirement for n-3 HUFAs (e.g., EPA and DHA) as juvenile marine fish (Sorgeloos, Léger, and Lavens 1988; Sargent, Henderson, and Tocher 1989; Sargent, McEvoy, and Bell 1997; Watanabe 1993; Rainuzzo, Reitan, and Olsen 1997). However, quantitatively, larval marine fish appear to require higher levels of dietary n-3 HUFAs than juvenile fish. The dietary n-3 HUFA requirements of juvenile and larval fish of several marine species are reported to range from 0.5 to 2 percent and 1.2 to 3.9 percent of diet dry weight, respectively (Watanabe 1993; Watanabe and Kiron 1995). Most larval marine fish will not readily accept formulated feeds (Russell 1976). Thus, rearing practices for marine species often incorporate live food organisms, such as *Artemia,* and rotifers, *Brachionus,* that have been enriched with n-3-fortified lipid emulsions to increase their HUFA content (Léger et al. 1986; Bengston, Léger, and Sorgeloos 1991; Sorgeloos et al. 1991; Sorgeloos and Léger 1992).

In addition to the role of essential fatty acids in enhancing growth and survival, the effects of dietary n-3 HUFAs on the tolerance of larval fish to induced stress also have been evaluated. Studies conducted to date (see Table 15.1) share similar experimental methods. Typically, larval fish were fed diets enriched with n-3 HUFAs for a short period of time, usually beginning at, or soon after, the onset of exogenous feeding. Following dietary HUFA treatment, the fish were briefly exposed to a stressor, such as exposure to air or hypersaline conditions, and mortality was recorded (Izquierdo et al. 1989; Watanabe et al. 1989; Dhert, Lavens, and Sorgeloos 1992). With one exception (Ashraf, Bengston, and Simpson 1993), results of these studies demonstrated that larval fish fed diets enriched with n-3 HUFAs exhibited increased resistance to stress when compared with larvae fed non-HUFA-enriched diets. More specifically, results of these experiments identified DHA as the n-3 HUFA that conferred the greatest stress resistance in larval marine fish, since high levels of EPA or other n-3 HUFA were not effective when DHA was limiting.

Low dietary ratios of n-6 to n-3 fatty acids (Sargent, Henderson, and Tocher 1989) and low ratios of palmitoleic acid (18:1n-9) to n-3 fatty acids (Izquierdo et al. 1989) have been hypothesized to play a role in increasing the stress resistance of larval marine fish. However, based on research conducted to date, it appears that a high dietary ratio of DHA to EPA is the most important lipid factor affecting stress resistance. The optimal DHA-to-EPA ratio for larval marine fish has not been determined. However, results of experiments (see Table 15.1) indicate that ratios between 0.5 and 5.6 were sufficient to significantly reduce stress-induced mortality in several species. Kraul and colleagues (1993) recommended a 3:1 ratio to prevent stress-induced mortality in larval dolphinfish, *Coryphaena hippurus.* In contrast, larval striped mullet, *Mugil cephalus* (Ako et al. 1994), and milkfish, *Chanos chanos* (Gapasin et al. 1998), fed HUFA-enriched diets with a DHA-to-EPA ratio of less than 1.0 had significantly lower stress-induced mortality than larvae fed nonenriched diets. It is probable that optimal dietary DHA-to-EPA ratios are species specific and are affected by nondietary factors, such as water quality, the duration and severity of the induced stress, or even by management practices, such as duration of feeding. Dhert and colleagues (1990) determined that HUFA-enriched diets for larval Asian sea bass, *Lates calcarifer,* must be fed at least five days to significantly reduce stress-induced mortality. For most marine fish larvae, the duration of initial feeding ranges from eight days to one month, depending on the species (Sorgeloos et al. 1991; Sorgeloos and Léger 1992; Jones, Kamarudin, and LeVay 1993; Person-Le Ruyet et al. 1993).

TABLE 15.1. Summary of Selected Experiments That Have Evaluated the Effect of Dietary N-3 HUFAs[1] on Tolerance of Larval Fish to Induced Stress

Species	Experimental Design	Results	Source
Red sea bream, *Pagrus major*	Rotifers,[2] enriched with methyl oleate (MO) or selected lipid emulsions to produce varying concentrations of n-3 HUFAs, fed to larvae for 16 days. Larvae then subjected to air exposure for 5 seconds. Mortality recorded.	Mortality of larvae reduced by prior feeding of diets containing increased levels of n-3 HUFAs.	Izquierdo et al. (1989)
	Larvae fed rotifers enriched with MO, EPA,[3] DHA,[4] or a mixture of EPA and DHA for 8 days, then subjected to air exposure for up to 10 seconds. Mortality recorded.	Mortality of larvae fed the DHA-enriched diet significantly lower than that of larvae fed other diets.	Watanabe et al. (1989)
	Larvae fed *Artemia* enriched with ethyl oleate (OA), EPA, or DHA for 11 days, then subjected to air exposure for up to 2 minutes. Mortality recorded.	Mortality of larvae fed DHA-enriched *Artemia* significantly lower than that of larvae fed other diets.	Takeuchi et al. (1991)
	Larvae fed microparticulate diets containing different levels of DHA (0, 1, or 2%) for 20 days, then subjected to air exposure for 30 seconds, increased water temperatures, or low dissolved oxygen concentrations.	Tolerance to stressors significantly improved by inclusion of dietary DHA	Kanazawa (1997)
Asian sea bass, *Lates calcarifer*	Larvae fed *Artemia* diets alone or enriched with HUFAs (EPA and DHA) for 21 days, then exposed to hypersaline conditions (65 g/liter) for up to 120 minutes. Mortality recorded.	Mortality of larvae fed enriched *Artemia* diets significantly lower than larvae fed nonenriched diets.	Dhert et al. (1990)
Yellowtail, *Seriola quinqueradiata*	Larvae fed *Artemia* enriched with OA, EPA, or DHA for 11 days, then subjected to air exposure for up to 2 minutes. Mortality recorded.	Mortality of larvae fed DHA-enriched *Artemia* significantly lower than that of larvae fed other diets.	Toyota et al. (1991)
Rabbitfish, *Siganus guttatus*	*Artemia*, alone or enriched with reference oils to obtain low, medium, and high concentrations of n-3 HUFAs, fed to larvae for five days. Larvae then exposed to hypersaline conditions (70 g/liter) for up to 15 minutes.	Larvae fed *Artemia* enriched with medium and high concentrations significantly lower than that of larvae fed other diets.	Dhert et al. (1992)

Dolphinfish, *Coryphaena hippurus*	Larvae fed *Artemia* and copepod[5] diets containing varying levels of EPA and DHA for up to 10 days, then subjected to air exposure for up to 2 minutes. Mortality recorded.	Mortality of larvae fed diets containing elevated levels of DHA significantly reduced.	Kraul et al. (1993)
Striped mullet, Mugil cephalus	Larvae fed an *Artemia* diet or a combined rotifer/*Artemia* diet alone or enriched with menhaden oil for 20 days, then subjected to air exposure for 15 seconds. Mortality recorded.	Mortality of larvae fed enriched diets significantly lower than larvae fed nonenriched diets.	Ako et al. (1994)
Red Drum, *Sciaenops ocellatus*	Larvae fed a control diet or the control diet supplemented with reference oils to produce varied ratios of DHA to EPA (0.66, 1.25, 2.42, or 3.78) for 18 days, then exposed to hypersaline conditions (70 g/liter) for one hour. Mortality recorded.	Mortality of larvae fed a diet containing a DHA/EPA ratio of 3.78 significantly reduced compared to larvae fed other diets.	Brinkmeyer and Holt (1998)
Milkfish, *Chanos chanos*	Larvae fed a rotifer/*Artemia* diet alone or enriched with HUFAs or HUFAs and vitamin C for 25 days, then exposed to hypersaline conditions (65 g/liter) for up to 90 minutes. Mortality recorded.	Mortality of larvae fed enriched diets significantly lower than larvae fed the nonenriched diet.	Gapasin et al. (1998)
Inland silverside, *Menidia beryllina*	Larvae fed *Artemia* enriched with low, medium, or high levels of HUFAs for 3 weeks, then exposed to hypersaline water (75 g/liter) for 120 minutes. Mortality recorded.	Mortality not influenced by dietary HUFA level.	Ashraf et al. (1993)

1 Highly unsaturated fatty acid(s)
2 *Brachionus plicatilis*
3 Eicosapentanoic acid
4 Docosahexanoic acid
5 *Euterpina acutifrons*

The specific mechanism by which DHA confers increased stress resistance in larval marine fish is unclear. DHA is present at high concentrations in cell membranes, especially neural tissues (Tocher and Harvie 1988; Bell and Dick 1991). It has been speculated that a low DHA-to-EPA ratio in the phospholipid fraction of cell membranes may reduce membrane fluidity (Watanabe 1993) and affect the movement of substances into and out of cells. Bell, Henderson, and Sargent (1986) and Dhert, Lavens, and Sorgeloos (1992) suggested that this effect would be most acute in surface tissues, such as gill epithelia, that are involved in osmoregulation and respiration. DHA deficiencies in gill tissue could reduce fish performance and contribute to the deleterious effects of stress.

It is noteworthy that stress resistance of larval inland silverside, *Menidia beryllina,* the only freshwater species studied (see Table 15.1), was not significantly improved by ingestion of HUFA-enriched diets (Ashraf, Bengston, and Simpson 1993). This could have resulted from relatively low dietary requirements for DHA and EPA in this species, or a DHA-to-EPA ratio in the HUFA-enriched diet (0.45) that was not high enough to affect stress resistance.

PHOSPHOLIPIDS

Phospholipids, or lecithins, comprise up to 50 percent of the lipids in biological membranes, where they serve as structural components of the lipid bilayer in cell membranes and regulate membrane fluidity. The biological membranes of fish, similar to those of other animals, contain several categories of phospholipids, primarily phosphatidylcholine (PC) and phosphatidylethanolamine (PE), with lesser amounts of phosphotidylserine, phosphatidylinositol, cardiolipin, and sphingomyelin (Sargent, Henderson, and Tocher 1989). The phospholipid composition of membranes varies among tissues (Anderson 1970; Hazel 1979, 1984; Bell, Simpson, and Sargent 1983) and can be altered in fish by physiological processes, such as homeoviscous regulation, that restructure the lipid composition of membranes in response to changes in environmental temperature (Hazel 1984). In rainbow trout, Oncorhynchus mykiss; common carp, Cyprinus carpio; and goldfish, Carrassius auratus, PE levels in membranes increase and PC levels decrease in response to decreased water temperatures (Miller, Hill, and Smith 1976; Hazel 1979; Wodtke 1981). The opposite effect (decreased PE and increased PC) occurs in rainbow trout during acclimation to increased water temperatures (Hazel and Carpenter 1985).

Although de novo synthesis of phospholipids occurs in fish (Henderson and Tocher 1987; Sargent et al. 1993), inclusion of phospholipids in the diets of juvenile and larval fish of several species has been shown to increase performance (Coutteau et al. 1997). Moreover, unlike dietary HUFA supplementation, beneficial effects of phospholipid supplementation have been observed in both freshwater and marine species. Geurden, Radünz-Neto, and Bergot (1995) reported that larvae of common carp fed a phospholipid-enriched (2 percent egg lecithin) diet experienced significantly better growth and survival than larvae fed phospholipid-free diets. Similar effects were observed by Kanazawa and colleagues (1981) and Kanazawa, Teshima, Kobayashi, et al. (1983) in ayu, *Plecoglossus altivelis,* larvae fed diets containing 3 percent egg lecithin or soybean lecithin. Growth and survival of larval red sea bream, *Pagrus major;* rock bream, *Oplegnathus fasciatus;* and Japanese flounder, *Paralichthys olivaceus,* have been improved by feeding phospholipid-supplemented diets containing up to 7 percent soybean lecithin (Kanazawa, Teshima, Inamori, et al. 1983; Kanazawa 1993).

High levels of lecithin supplementation can be impractical in some diet formulations, and phospholipid content and composition can vary considerably among and within lecithin sources. Most experiments investigating the effects of dietary phospholipids on fish performance have used egg and soybean lecithins with variable phospholipid content. However, the few studies conducted with purified phospholipid fractions indicate that phosphatidylcholine and phosphatidylinositol are the primary growth-promoting phospholipids in lecithin (Coutteau et al. 1997). It is notable that phosphatidylcholine also has been identified as the lecithin fraction that confers increased stress resistance in shrimp, *Penaeus japonicus* and *P. vannamei* (Camara 1994). Additional research is needed to identify the best types and sources of phospholipids for larval fish, and the dietary levels required for optimal growth and health.

Only two studies have been conducted to evaluate the effect of phospholipids on stress tolerance of larval fish (see Table 15.2). Results from both experiments indicated that dietary phospholipid supplementation can effectively mitigate stress. The specific mechanism by which dietary phospholipid supplementation confers stress resistance in larval fish and the class(es) of phospholipids responsible for stress-reducing effects have not been determined, but it is thought that beneficial effects are related to changes in the structure and/or function of biological membranes.

TABLE 15.2. Summary of Selected Experiments That Have Evaluated the Effect of Dietary Phospholipids on Tolerance of Larval Fish to Induced Stress

Species	**Experimental Design**	**Results**	**Source**
Rock bream, *Oplenganthus fasciatus*	Larvae fed microparticulate diets containing different levels of soybean lecithin (SBL; 0, 3, 5, or 7%) for 28 days, then subjected to air exposure for 30 seconds. Mortality recorded.	Mortality of larvae significantly reduced by inclusion of dietary SBL.	Kanazawa (1993)
Red sea bream, *Pagrus major*	Larvae fed microparticulate diets containing different levels of soybean lecithin (SBL; 0, 3, or 5%) for 20 days, then subjected to air exposure for 30 seconds, increased water temperatures, or low dissolved oxygen concentrations.	Tolerance of larvae to stressors significantly improved by inclusion of dietary SBL.	Kanazawa (1997)

CONCLUSIONS

Research indicates that the stress tolerance of larval fish can be improved by feeding diets with proper levels of highly unsaturated fatty acids, primarily docosahexaenoic acid (22:6n-3) and eicosapentaenoic acid (20:5n-3), and phospholipids, especially phosphatidylcholine and phosphatidylinositol. Such dietary manipulations could be used to increase the success of larval fish culture. However, research is needed to evaluate the effect of dietary lipids on stress resistance in a greater number of fish species, especially freshwater species. Investigations also are needed to identify the mechanisms responsible for lipid-induced stress resistance in larval fish and to determine the specific lipid types responsible for stress-reducing effects.

REFERENCES

Ako, H., C.S. Tamaru, P. Bass, and C-S. Lee. 1994. Enhancing the resistance to physical stress in larvae of *Mugil cephalus* by the feeding of enriched *Artemia nauplii. Aquaculture* 122:81-90.

Anderson, T.R. 1970. Temperature adaptation and the phospholipids of membranes in goldfish *(Carrassius auratus). Comparative Biochemistry and Physiology* 33:663-687.

Ashraf, M., D.A. Bengston, and K.L. Simpson. 1993. Effects of fatty acid enrichment on survival, growth, and salinity-stress-test performance of inland silversides. *The Progressive Fish-Culturist* 55:280-283.

Bell, M.V., R.S. Batty, J.R. Dick, K. Fretwell, J.C. Navarro, and J.R. Sargent. 1995. Dietary deficiency of docosahexaenoic acid impairs vision at low light intensities in juvenile herring (*Clupea harengus* L.). *Lipids* 30:443-449.

Bell, M.V. and J.R. Dick. 1991. Molecular species composition of the major diacylglycerophospholipids from muscle, liver, retina, and brain of cod *(Gadus morhua). Lipids* 26:565-573.

Bell, M.V., R.J. Henderson, and J.R. Sargent. 1986. The role of polyunsaturated fatty acids in fish. *Comparative Biochemistry and Physiology* 83B:711-719.

Bell, M.V., C.M.F. Simpson, and J.R. Sargent. 1983. N-3 and n-6 polyunsaturated fatty acids in the phosphoglycerides of salt-secreting epithelia from two marine fish species. *Lipids* 18:720-726.

Bengston, D.A., P. Léger, and P. Sorgeloos. 1991. Use of *Artemia* as a food source in aquaculture. In R.A. Browne, P. Sorgeloos, and C.N.A. Trotman (Eds.), *Handbook on Artemia Biology* (pp. 255-285). Boca Raton, FL: CRC Press.

Brinkmeyer, R.L. and G.J. Holt. 1998. Highly unsaturated fatty acids in diets for red drum *(Sciaenops ocellatus)* larvae. *Aquaculture* 161:253-268.

Camara, M.R. 1994. "Dietary phosphatidylcholine requirements of *Penaeus japonicus* Bate and *Penaeus vannamei* Boone (Crustacea, Decapoda, Penaeidae)." Doctoral dissertation. University of Ghent, Belgium.

Castell, J.O. 1979. Review of lipid requirements of finfish. *Proceedings of the World Symposium on Finfish Nutrition and Fish Feed Techniques* 1:20-23.

Coutteau, P., I. Geurden, M.R. Camara, P. Bergot, and P. Sorgeloos. 1997. Review on the dietary effects of phospholipids in fish and crustacean larviculture. *Aquaculture* 155:149-164

Cowey, C.B., J.M. Owen, J.W. Adron, and C. Middleton. 1976. Studies on the nutrition of marine flatfish: The effect of dietary fatty acids on the growth and fatty acid composition of turbot *(Scopthalmus maximus). British Journal of Nutrition* 36:479-486.

Cowey, C.B. and J.R. Sargent. 1977. Lipid nutrition in fish. *Comparative Biochemistry and Physiology* 57B:269-273.

Dhert, P., P. Lavens, M. Duray, and P. Sorgeloos. 1990. Improved larval survival at metamorphosis of Asian sea bass *(Lates calcarifer)* using ω3-HUFA-enriched live food. *Aquaculture* 90:63-74.

Dhert, P., P. Lavens, and P. Sorgeloos. 1992. A simple test for quality evaluation of cultured fry of marine fish. *Medinal Faculteit Landbouwwetenschappen Universität* Gent 57:2135-2142.

Falk-Petersen, S., I.B. Falk-Petersen, J.R. Sargent, and T. Haua. 1986. Lipid class and fatty acid composition of eggs from Atlantic halibut *(Hippoglossus hippoglosus). Aquaculture* 52:207-211.

Gapasin, R.S.J., R. Bombeo, P. Lavens, P. Sorgeloos, and H. Nelis. 1998. Enrichment of live food with essential fatty acids and vitamin C: Effects on milkfish *(Chanos chanos)* larval performance. *Aquaculture* 162:269-286.

Geurden, I., J. Radünz-Neto, and P. Bergot. 1995. Essentiality of dietary phospholipids for carp *(Cyprinus carpio)* larvae. *Aquaculture* 131:303-314.

Hazel, J.R. 1979. The influence of temperature adaptation on the composition of neutral lipid fraction of rainbow trout *(Salmo gairdneri)* liver. *Journal of Experimental Zoology* 207:33-42.

Hazel, J.R. 1984. Effects of temperature on the structure and metabolism of cell membranes in fish. *American Journal of Physiology* 246:R460-470.

Hazel, J.R. and R. Carpenter. 1985. Rapid changes in the phospholipid composition of gill membranes during thermal acclimation of the rainbow trout, *Salmo gairdneri. Journal of Comparative Physiology* 155B:597-602.

Henderson, R.J. and D.R. Tocher. 1987. The lipid composition and biochemistry of freshwater fish. *Progressive Lipid Research* 26:281-347.

Hepher, B. 1988. *Nutrition of Pond Fishes.* Melbourne, Australia: Cambridge University Press.

Izquierdo, M.S., T. Watanabe, T. Takeuchi, T. Arakawa, and C. Kitajima. 1989. Requirement of larval red sea bream *Pagrus major* for essential fatty acids. *Nippon Suisan Gakkaishi* 55:859-867.

Jones, D.A., M.S. Kamarudin, and L. LeVay. 1993. The potential for replacement of live feeds in larval culture. *Journal of the World Aquaculture Society* 24:199-210.

Kanazawa, A. 1985. Essential fatty acid and lipid requirements of fish. In C.B. Cowey, A.M. Mackie, and J.G. Bell (Eds.), *Nutrition and Feeding in Fish* (pp. 281-298). London, England: Academic Press.

Kanazawa, A. 1993. Essential phospholipids of fish and crustaceans. In S.J. Kaushik and P. Luquet (Eds.), *Fish Nutrition in Practice* (pp. 519-530). Paris: Les Colloques 61, INRA.

Kanazawa. A. 1997. Effects of docosahexanoic acid and phospholipids on stress tolerance of fish. *Aquaculture* 155:129-134.

Kanazawa, A., S. Teshima, S. Inamori, and H. Matsubara. 1981. Effects of phospholipids on growth, survival rate, and incidence of malformation in the larval ayu. *Kagoshima University* (Japan) *Fisheries Research Report* 30:301-309.

Kanazawa, A., S. Teshima, S. Inamori, and H. Matsubara. 1983. Effects of dietary phospholipids on growth of the larval red sea bream and knife jaw. *Kagoshima University* (Japan) *Fisheries Research Report* 32:109-114.

Kanazawa, A., S. Teshima, T. Kobayshi, M. Takae, T. Iwashita, and R. Uehara. 1983. Necessity of dietary phospholipids for growth of the larval ayu. *Kagoshima University* (Japan) *Fisheries Research* 32:115-120.

Kanazawa, A., S. Teshima, and K. Ono. 1979. Relationship between essential fatty acid requirements of aquatic animals and the capacity for bioconversion of linolenic acid to highly unsaturated fatty acids. *Comparative Biochemistry and Physiology* 63B:295-298.

Koven, W.M., G.W. Kissil, and A. Tandler. 1989. Lipid and n-3 requirements of *Sparus aurata* larvae during starvation and feeding. *Aquaculture* 79:185-191.

Koven, W.M., A. Tandler, G.W. Kissil, D. Sklan, O. Friezlander, and M. Harel. 1990. The effect of dietary n-3 polyunsaturated fatty acids on growth, survival and swim bladder development in *Sparus aurata* larvae. *Aquaculture* 91:131-141.

Kraul, S., K. Brittain, R. Cantrell, T. Nagao, H. Ako, A. Ogasawara, and H. Kitagawa. 1993. Nutritional factors affecting stress resistance in the larval mahimahi *Coryphaena hippurus*. *Journal of the World Aquaculture Society* 24:186-193.

Léger, P., D.A. Bengston, K.L. Simpson, and P. Sorgeloos. 1986. The use and nutritional value of *Artemia* as a food source. *Oceanography and Marine Biology: An Annual Review* 24:521-623.

Lovell, R.T. 1989. *Nutrition and Feeding of Fish.* New York: Van Nostrand Rheinhold.

Miller, N.G.A., M.W. Hill, and M.W. Smith. 1976. Positional and species analysis of membrane phospholipids extracted from goldfish adapted to different environmental temperatures. *Biochimica et Biophysica Acta* 55:644-654.

NRC (National Research Council). 1993. *Nutrient Requirements of Fish.* Washington, DC: National Academy Press.

Owen, J.M., J.W. Adron, C. Middleton, and C.B. Cowey. 1975. Elongation and desaturation of dietary fatty acids in turbot *(Scopthalmus maximus)* and rainbow trout *(Salmo gairdneri)*. *Lipids* 10:528-531.

Person-Le Ruyet, J., J.C. Alexandre, L. Thébaud, and C. Mugnier. 1993. Marine fish larvae feeding: Formulated diets or live prey? *Journal of the World Aquaculture Society* 24:211-224.

Pond, W.G., D.C. Church, and K.R. Pond. 1995. *Basic Animal Nutrition and Feeding.* New York: John Wiley and Sons.

Rainuzzo, J.R., K.I. Reitan, L. Jorgensen, and Y. Olsen. 1994. Lipid composition in turbot larvae fed live feed cultured by emulsion of different lipid classes. *Comparative Biochemistry and Physiology* 107A:699-710.

Rainuzzo, J.R., K.I. Reitan, and Y. Olsen. 1997. The significance of lipids at early stages of marine fish: A review. *Aquaculture* 155:103-115.

Reitan, K.I., J.R. Rainuzzo, and Y. Olsen. 1994. Influence of lipid composition of live feed on growth, survival, and pigmentation of turbot *(Scopthalmus maximus)* larvae. *Aquaculture International* 2:33-48.

Russell, F.S. 1976. *The Eggs and Planktonic Stages of British Marine Fish.* London, England: Academic Press.

Sargent, J.R., J.G. Bell, M.V. Bell, R.J. Henderson, and D.R. Tocher. 1993. The metabolism of phospholipids and polyunsaturated fatty acids in fish. In B. Lahlou and P. Vitiello (Eds.), *Aquaculture: Fundamental and Applied Research, Coastal and Estuarine Studies,* 43 (pp. 103-124). Washington, DC: American Geophysical Union.

Sargent, J.R., R.J. Henderson, and D.R. Tocher. 1989. The lipids. In J.E. Halver (Ed.), *Fish Nutrition* (pp. 153-219). San Diego, CA: Academic Press.

Sargent, J.R., L.A. McEvoy, and J.G. Bell. 1997. Requirements, presentation, and sources of polyunsaturated fatty acids in marine fish larval feeds. *Aquaculture* 155:117-127.

Sorgeloos, P., P. Lavens, P. Léger, W. Tackert, and D. Versichele. 1991. State of the art in larviculture of fish and shellfish. In P. Lavens, P. Sorgeloos, E. Jaspers, and F. Ollevier (Eds.), *Larvi '91—Fish and Crustacean Larviculture Symposium, European Aquaculture Society Special Publication* 15 (pp. 3-5). Oostende, Belgium: European Aquaculture Society.

Sorgeloos, P. and P. Léger. 1992. Improved larviculture outputs of marine fish, shrimp, and prawn. *Journal of the World Aquaculture Society* 23:251-264.

Sorgeloos, P., P. Léger, and P. Lavens. 1988. Improved larval rearing of European and Asian sea bass, sea bream, mahi-mahi, siganid and milkfish using enrichment diets for *Brachionus* and *Artemia. World Aquaculture Magazine* 19(4):78-79.

Steffens, W. 1989. *Principles of Fish Nutrition.* Chichester, England: Ellis Horwood Limited.

Takeuchi, T. 1997. Essential fatty acid requirements of aquatic animals with emphasis on fish larvae and fingerlings. *Reviews in Fisheries Science* 5:1-25.

Takeuchi, T., M. Toyota, and T. Watanabe. 1991. Dietary value to larval red sea bream of *Artemia nauplii* enriched with EPA and DHA. In *Abstracts of the Annual Meeting of the Japanese Society of Scientific Fisheries* (p. 21). Tokyo, Japan.

Tocher, D.R., A.J. Fraser, J.R. Sargent, and J.C. Gamble. 1985. Fatty acid composition of phospholipids and neutral lipids during embryonic and early larval development in Atlantic herring (*Clupea harengus* L.). *Lipids* 20:69-74.

Tocher, D.R. and D.G. Harvie. 1988. Fatty acid compositions of the major phosphoglycerides from fish neural tissues; n-3 and n-6 polyunsaturated fatty acids in rainbow trout *(Salmo gairdneri)* and cod *(Gadus morhua)* brains and retina. *Lipids* 24:585-588.

Toyota, M., T. Takeuchi, and T. Watanabe. 1991. Dietary value to larval yellowtail of *Artemia nauplii* enriched with EPA and DHA. In *Abstracts of the Annual Meeting of the Japanese Society of Scientific Fisheries* (p. 25). Tokyo, Japan.

Watanabe, T. 1982. Lipid nutrition in fish. *Comparative Biochemistry and Physiology* 73B:3-15.

Watanabe, T. 1993. Importance of docosahexanoic acid in marine larval fish. *Journal of the World Aquaculture Society* 24:152-161.

Watanabe, T., M.S. Izquierdo, T. Takeuchi, S. Satoh, and C. Kitajima. 1989. Comparison between eicosapentanoic and docosahexanoic acids in terms of essential fatty acid efficacy in larval red sea bream. *Nippon Suisan Gakkaishi* 55:1635-1640.

Watanabe, T. and V. Kiron. 1995. Red sea bream *(Pagrus major)*. In W.R. Bromage and R.J. Roberts (Eds.), *Broodstock Management and Egg and Larval Quality* (pp. 398-413). London, England: Blackwell Science.

Wilson, E.D., K.H. Fisher, and M.E. Fuqua. 1975. *Principles of Nutrition.* New York: John Wiley and Sons.

Wodtke, E. 1981. Temperature adaptation of biological membranes: Effects of acclimation temperature on the unsaturation of the main neutral and charged phospholipids in mitochondrial membranes of the carp, *Cyprinus carpio. Biochimica et Biophysica Acta* 64:698-709.

Yamada, K., K. Kobayashi, and Y. Yone. 1980. Conversion of linolenic acid to ω-3 highly unsaturated fatty acids in marine fishes and rainbow trout. *Bulletin of the Japanese Society of Scientific Fisheries* 46:1231-1233.

Chapter 16

Modulation of Environmental Requirements of Finfish Through Nutrition

Joseph R. Tomasso
Delbert M. Gatlin III
Charles R. Weirich

INTRODUCTION

To approach optimal production, aquaculturists typically attempt to modify the culture environment to meet the environmental requirements of the species being cultured. Environmental modification may be as simple as aeration of culture ponds or as complicated as temperature, photoperiod, and dissolved ion manipulation in recirculation systems (Tomasso 1993; Tomasso 1997). An alternate approach is to modify the organism to better match the available environment. This can be done genetically or nutritionally. A genetic approach could involve either the use of classic selection and hybridization techniques or simply selecting existing stocks with a trait of value to an aquaculturist (Tomasso and Carmichael 1991). A nutritional approach would involve the use of dietary additives or supplements that would enhance the ability of the animal to deal with some aspect of a nonoptimal environment.

Here we present examples of how dietary additives or supplements may help cultured fishes deal with three specific nonoptimal aspects of their environment—low temperature, low salinity, and high nitrite concentrations. It is not our purpose to present a complete literature review; rather, we offer these examples as evidence that nutritional modification of environmental requirements of cultured fishes may provide an alternative to environmental and genetic modifications.

MODULATION OF TOLERANCE TO LOW TEMPERATURE

Red drum, *Sciaenops ocellatus,* are found along the Gulf of Mexico and south- and mid-Atlantic coasts of the United States. Due to overfishing, most of the red drum fisheries were closed during the 1980s. This led to the development of a red drum aquaculture industry along the Gulf and Atlantic coasts of the southeastern United States. Originally, production was primarily in ponds, and fish were left in ponds during the winter. Surprisingly, these practices resulted in fish kills due to low winter temperatures. Apparently, wild red drum are able to find warm-water refuges, but fish confined in shallow culture ponds were exposed to lethal temperatures.

In an attempt to increase the tolerance of red drum to low temperatures, Craig, Neill, and Gatlin (1995) investigated the influence of dietary lipid manipulation. Juvenile fish were fed diets containing fatty acids from a variety of lipids (menhaden oil, corn oil, coconut oil, and saturated menhaden oil). After being fed their respective diets for six weeks, the fish were subjected to cold tolerance assays in which the temperature was reduced 1°C per day from 26 to 19°C and then 0.5°C per day until all fish died. Median-lethal temperatures (the temperature at which 50 percent of the exposed animals died) were significantly affected by dietary lipids, with fish fed diets containing menhaden oil having the lowest median-lethal temperature (5.1°C in brackish water, 3.9° C in seawater), and fish fed diets containing coconut oil having the highest median-lethal temperature (8.6°C in brackish water, 9.4°C in seawater).

Fish fed the diet containing menhaden oil had more highly unsaturated fatty acids available than fish fed the other diets. During cold acclimation, poikilotherms may increase the concentration of highly unsaturated fatty acids in cell membranes to allow retention of membrane fluidity during exposure to low temperatures (Hazel 1979; 1984).

MODULATION OF TOLERANCE TO LOW SALINITY

Red drum are euryhaline fish that can survive in water with dissolved solid concentrations from <1g/liter to hyperstrength seawater (Crocker et al. 1981). However, growth may be inhibited in dilute water due to the need to partition energy away from growth to ionoregulation (Bryan, Ham, and Neill 1988). In an eight-week growth study in freshwater, red drum fed a diet supplemented with 2 percent sodium chloride had significantly better weight gain and feed efficiency than fish fed an unsupplemented diet (Gatlin et al. 1992). In contrast, similar studies with channel catfish, *Ictalurus punctatus;* rainbow trout, *Oncorhynchus mykiss;* and Atlantic salmon, *Salmo salar,* found no advantage to salt supplementation of diets (Shaw et al. 1975; Morgan and Andrews 1979; Salman and Eddy 1988).

Feeding salt to freshwater-acclimated fishes also has been shown to increase survival when fish are abruptly transferred to saltwater. Feeding diets supplemented with 10 g/liter sodium chloride to two tilapia species and a hybrid tilapia resulted in increased survival when the fish were abruptly transferred to seawater (Al-Amoudi 1987). Two to three weeks of feeding the salt-supplemented diets appeared to produce the maximum effect. Similar results were observed in Atlantic salmon parr after being fed rations supplemented with synthetic sea salts (Basulto 1976). Best results were obtained after feeding a diet supplemented with 12 g/liter sea salt for 107 days. Feeding salts to freshwater fishes probably aids in saltwater acclimation by increasing plasma sodium and chloride concentrations, which, in turn, stimulates salt excretory mechanisms normally used in seawater environments. This arrangement allows the fish to begin the physiological seawater acclimation process before the fish are actually exposed to seawater.

MODULATION OF TOLERANCE TO NITRITE EXPOSURE

In high-density aquaculture systems, nitrite may reach toxic levels due to an imbalance in the nitrification processes. Nitrite enters the blood of fish by active transport and causes, among other things, elevated levels of methemoglobin (Tomasso 1994). Channel catfish fed high levels of vitamin C (up to 7,950 mg/kg diet) and exposed to nitrite for 22 hours (h) had lower methemoglobin levels than fish fed control levels of vitamin C (63 mg/kg diet) and exposed to nitrite in a similar manner (Wise, Tomasso, and Brandt 1988). However, the protective effect of vitamin C was apparent only if the fish were fed 48 h or less before the beginning of nitrite exposure, apparently due to the rapid clearance of this water-soluble vitamin from the fish. A similar protective effect of vitamin C to nitrite-induced methemoglobinemia was observed in steelhead trout, *Oncorhynchus mykiss* (Blanco and Meade 1980).

CONCLUSIONS

Modification of environmental requirements of cultured fishes through the use of dietary additives and supplements appears to show promise. The greatest potential may be in situations where the species is only marginally out of its environmental range. For example, the few degrees in low-temperature tolerance gained by red drum may represent the difference between survival and death in many southern U.S. aquaculture operations.

REFERENCES

Al-Amoudi, M.M. 1987. The effect of high salt diet on the direct transfer of *Oreochromis mossambicus, O. spilurus,* and *O. aureus/O. niloticus* hybrids to sea water. *Aquaculture* 64:333-338.

Basulto, S. 1976. Induced saltwater tolerance in connection with inorganic salts in the feeding of Atlantic salmon *(Salmo salar). Aquaculture* 8:45-55.

Blanco, O. and T. Meade. 1980. Effect of dietary ascorbic acid on the susceptibility of steelhead trout *(Salmo gairdneri)* to nitrite toxicity. *Revista Biologia Tropical* 28:91-107.

Bryan, J.D., K.D. Ham, and W.H. Neill. 1988. Biophysical model of osmoregulation and its metabolic cost in red drum. *Contributions in Marine Science* 30(supplement):169-182.

Craig, S.R., W.H. Neill, and D.M. Gatlin III. 1995. Effects of dietary lipid and salinity on growth, body composition, and cold tolerance of juvenile red drum *(Sciaenops ocellatus). Fish Physiology and Biochemistry* 14:49-61.

Crocker, P.A., C.R. Arnold, J.A. DeBoer, and G.J. Holt. 1981. Preliminary evaluation of survival and growth of juvenile red drum *(Sciaenops ocellata)* in fresh and salt water. *Proceedings of the World Mariculture Society* 12:122-134.

Gatlin, D.M. III, D.S. MacKenzie, S.R. Craig, and W.H. Neill. 1992. Effects of dietary sodium chloride on red drum juveniles in waters of various salinities. *Progressive Fish-Culturist* 54:220-227.

Hazel, J.R. 1979. The influence of temperature adaptation on the composition of neutral lipid fraction of rainbow trout *(Salmo gairdneri)* liver. *Journal of Experimental Zoology* 207:33-42.

Hazel, J.R. 1984. Effects of temperature on the structure and metabolism of cell membranes in fish. *American Journal of Physiology* 246:R460-470.

Morgan, W.M. and J.W. Andrews. 1979. Channel catfish: The absence of an effect of dietary salt on growth. *Progressive Fish-Culturist* 41:155-156.

Salman, N.A. and F.B. Eddy. 1988. Effect of dietary sodium chloride on growth, food intake and conversion efficiency in rainbow trout (*Salmo gairdneri* Richardson). *Aquaculture* 70:131-144.

Shaw, H.M., R.L. Sanders, H.C. Hall, and E.B. Henderson. 1975. Effect of dietary sodium chloride on growth of Atlantic salmon *(Salmo salar). Journal of the Fisheries Research Board of Canada* 32:1813-1819.

Tomasso, J.R. 1993. Environmental requirements and environmental diseases of freshwater and estuarine temperate fishes. In M. Stoskopf (Ed.), *Fish Medicine* (pp. 78-89). Philadelphia: W.B. Saunders.

Tomasso, J.R. 1994. Toxicity of nitrogenous wastes to aquaculture animals. *Reviews in Fisheries Science* 2:291-314.

Tomasso, J.R. 1997. Environmental requirements and noninfectious diseases. In R.M. Harrell (Ed.), *Striped Bass and Other* Morone *Culture* (pp. 253-270). New York: Elsevier Science Publishers.

Tomasso, J.R. and G.J. Carmichael. 1991. Differential resistance of nitrite among strains and strain crosses of channel catfish. *Journal of Aquatic Animal Health* 3:51-54.

Wise, D.J., J.R. Tomasso, and T.M. Brandt. 1988. Ascorbic acid inhibition of nitrite-induced methemoglobinemia in channel catfish. *Progressive Fish-Culturist* 50:77-80.

Chapter 17

Vaccines: Prevention of Diseases in Aquatic Animals

Phillip H. Klesius
Craig A. Shoemaker
Joyce J. Evans
Chhorn Lim

INTRODUCTION

Aquaculture constitutes a vital and rapidly growing segment of world agriculture. The U.S. Department of Agriculture (1998) reported that 20 percent of fish and seafood consumed in the United States is farm raised. The U.S. aquaculture industry is a recognized segment of the livestock production, larger than lamb, veal, and mutton combined (USDA 1998). Advances in U.S. aquaculture technologies and the strong public demand for safe and wholesome fish products are expected to accelerate industry growth in the next millennium. Catfish, tilapia, salmon, hybrid striped bass, and rainbow trout production make up significant segments of U.S. aquaculture. Catfish continues to lead the U.S. aquaculture production at 535 to 545 million pounds (USDA 1998). In 1989, 1.6 billion fingerlings were cultured and susceptible to infectious agents that we estimate may claim 30 to 50 percent of these fish before they reach food size. Successful vaccination of these young fish against both enteric septicemia (ESC; *Edwardsiella ictaluri*) and *Flavobacterium columnare,* two primary infections of catfish, would undoubtedly enhance their total survival to food-size fish.

INFECTIOUS DISEASES IN AQUACULTURE SYSTEMS

The majority of cultured fish are intensively grown by stocking large numbers of animals in limited space, promoting stressful conditions. This causes increased risk of the rapid spread and severity of infectious diseases. Numerous bacterial, viral, fungal, and parasitic infections cause disease in farm-raised fish (Plumb 1999). The type of aquaculture system, variability of water quality and temperature in different seasons and production cycles, species of fish cultured, sources and handling of fish, nutrition and feeding, and production and health management practices all combine to determine the kinds of infectious organisms and the periods of greatest risk of disease outbreaks in cultured fish. "All in and all out" types of production systems, such as are practiced in the poultry industry, require preventive health management programs that include vaccination to reach optimal production. In contrast, many aquaculture production systems require manipulation in which all sizes of fish are grown together, and frequent sorting and movement is necessary. These culture practices seem to favor the more frequent occurrence of different kinds of infection and different degrees of severity during the various production phases. Cultured fish appear to be more susceptible to bacterial, viral, and fungal infections. Nutrition, injury, stress, parasitism, and immune status influence the susceptiblity of fish to different kinds of infectious agents. For example, nutritionally deprived fish have a weaker immune system and become more susceptible to infection and disease than fish maintained on diets that favor growth and health (Klesius, Lim, and Shoemaker 1999).

Contact of susceptible animals with infectious agents may result in severe infection, morbidity, and mortality. Infection may also result with little or no external signs. Subclinical infection results in reduced performance, which causes economic loss that is less obvious than that caused by death from severe infection. Animals grown in germ-free environments were found to attain weight gains as much as 10 percent more than animals in contact with common environmental microbes (Lev and Forbes 1959). The reason for this appeared to be stimulation of the immune system of animals exposed to environmental microbes. Immunostimulation is growth suppressive due to actions of immune cells that secrete cytokines on nonimmune cells. On the other hand, viral infections cause immunosuppression and enhanced susceptibility to other infectious agents. Assessment of the immune status of cultured fish is not easily determined and thus not usually done in the production of fish. There is a need to develop methods to accurately and rapidly estimate the immune status and the degree of stress of cultured fish.

Marine or freshwater recirculating aquaculture systems are unique in terms of the factors predisposing to infections, severity and recurrence of diseases, status of fish resistance, and species of infectious agents. Common and recurring infectious agents are often different between systems and geographic locations in the world. In the most serious circumstances, infectious diseases are responsible for the abandonment of aquaculture enterprises due to complete loss of profitability. The treatment of infectious disease is only partly effective in reducing the loss of profitability, and chemotherapy appears to be less effective in reuse water systems. Further, treatment of sick and dying aquatic animals requires the use of chemicals and antibiotics that are often harmful to the environment and to the ensurance of safe food and water. Regulations and laws are mandating major reductions in chemical and antibiotic treatments of disease in aquatic animals. If chemical and antibiotic treatment strategies are becoming more and more a nonviable means for management of infectious diseases of aquatic animals, what is the alternative strategy?

Management of aquatic animal health by vaccination is critical to continued profitable aquatic animal production. Vaccination is a proven cost-effective health maintenance tool that has enjoyed notable success. For example, in Norway, the introduction of vaccines against cold-water vibriosis aided significantly in decreasing antibiotic sales (Markestad and Graves 1997). In a period of four years, the estimated amount of antibiotics used per kilogram of salmon cultured was reduced from 231 g to 6.4 g. The vaccines effectively prevented major losses in farmed salmonids due to cold-water vibriosis without the continuous use of antibiotics. The diminished use of antibiotics in aquaculture significantly reduces the health risks of antibiotic-resistant infectious agents in the environment and food chain. Vaccination is being recognized as the most cost-effective means by which economic losses and animal suffering from infectious diseases can be prevented in aquaculture.

IMMUNE CAPACITY OF AQUATIC ANIMALS

Immunity is the inherited capacity to recognize and respond in a defensive way against infectious agents. Recognition of infectious agents, expansion of cells for defense against the infectious agents, and coordination of cells of the immune system involved in self-defense by regulatory substances are all important to active (specific) immunity. Shrimp (and other crustaceans and mollusks) and fish differ greatly in their ability and the degree to which they can carry out these functions of the immune system. The capacity to recognize, expand the specific recognition, express specific recognition, and coordinate the specific expression of defense is much more primitive in shrimp than in fish. Shrimp possess restricted capacities to mount a humoral and cellular defense response to infectious diseases. Lectins, agglutinins, lysozymes, and serine proteases are important humoral defense molecules. The defense cells are nongranular, semigranular, and granular hemocytes. The defense system of shrimp lacks specific immunity and memory, both of which are important for active acquired immunity stimulated by vaccines. Regulation of the hemocytes in defensive mechanisms appears to be accomplished by lectin-receptor interactions, rather than by cytokines, as in higher animals. Beta-glucans have been used with limited success to try to enhance protection of the treatment of infectious diseases, especially viral infections. Genetic enhancement of nonspecific defense mechanisms, and better feeding, water quality, and disease management appear to be the future of disease control in intensive shrimp culture. Viral interference may also be a possible mechanism to investigate for the development of control strategies to prevent viral infections in shrimp.

Based on the current state of knowledge, it is not advisable to employ vaccines in shrimp aquaculture. Any reported beneficial results are poorly documented and most likely due to happenstance, rather than the stimulation of actively acquired immunity. Crustaceans and mollusks lack the capacity for specific immunity against infectious agents and the use of a vaccine does not provide a specific protective immunity (Klesius and Shoemaker 1997a; Soderhall and Thornqvist 1997).

ACTIVE AND PASSIVE IMMUNIZATION

Active immunization is the current farm-raised animal industry standard. Passive immunization is the administration of immune antibodies to a susceptible animal and is not currently used. Active immunization stimulates long-lasting and revokable protection against any given antigenic form of an infectious agent. A monovalent vaccine provides active immunity against a specific antigenic form of infectious agent, but it may not afford protection against infectious agents with different antigenic makeups. Infectious agents may exist in many different antigenic makeups or possess the genetic ability to change to new antigenic makeups. This situation is often overcome by the use of a polyvalent vaccine that stimulates active immunity against the more commonly occurring antigenic makeups of the infectious agent. Polyvalent vaccines may also be composed of different species of infectious agents to provide active immunity against the common infectious species of the aquaculture system.

VACCINE STRATEGIES

The vaccine strategy is focused on the protection of fish against specific infections that cause problems in the particular type of aquaculture system. Formulation of a vaccine strategy first must be made compatible to the production cycle of the fish species, then the type of aquaculture system and its management. Salmonids, catfish, tilapia, hybrid striped bass, sea bream, and turbot are raised under different aquaculture and management systems. The successful vaccine strategy recognizes these differences. Among fish species, differences in the inherited capacity of their immune systems appear to exist, and the successful vaccine may have to be slightly different for each species. We believe that prevention of infectious disease in reuse water aquaculture systems may enjoy the best results by vaccination. Successful vaccine strategies are based on the benefits and adverse effects of vaccination against a certain type of infectious agent, age of fish when vaccinated, the choice of vaccine type, the vaccination method, choice of type of water and temperature for vaccination, and the need for a booster dose and adjuvants for optimal vaccination. The particulars that influence the outcome of, and solutions to, successful vaccination strategies are summarized in

Table 17.1. The efficacy of the vaccination needs to be high enough to protect the greatest numbers of fish against the maximum load of infectious agents. Vaccination should be performed prior to the time of greatest risk of disease outbreak(s). The protection needs to last the entire period of disease risk and should extend from fry to food-size fish.

Evidence for antibiotic therapy causing vaccination failure indicates that certain antibiotics are immunosuppressive. These types of antibiotics are aminoglycosides, cephalosporins, imidazoles, rifampicin, sulfonamides, and tetracyclines (Finch 1980; Nagi et al. 1984). The degree of their immunosuppressive effects is unknown, but certainly with live vaccines their effects could be the cause of failures. More research is needed in drug therapy and immunization in fish.

Advances are being made in the understanding of the complex interactions between genetics and immune responses. Vaccination can be enhanced by breeding of improved fish stocks capable of uniformly responding to different types of vaccine and management strategies. The genetic influence of fish responses to antifurunculosis vaccine was demonstrated in salmon (Stromsheim et al. 1994). Some lines had greater numbers of fish that had enhanced vaccine responses than other lines. Preliminary evidence has shown the families of channel catfish have different responses to the modified live *Edwardsiella ictaluri* vaccine (Wolters, W., ARS Catfish Genetics Laboratory, Stoneville, MS, personal communication).

Vaccine Types

The vaccine types that may be employed in fish are shown in Table 17.2. The type of vaccine available is a major factor in the success of a disease

TABLE 17.1. Particulars That Influence the Outcomes of, and Solutions to, Optimal Vaccination

Particular	Solution
1. Genetic variability in immune responses within species and families	Identification of polygenic traits and selective breeding for uniform enhanced vaccine responders and fewer nonresponders.
2. Age variability	Determine the earliest age for particular species for the vaccine type and route of administration.
3. Nutritional variability	Formulation of diets that favor enhanced immune responses.
4. Stress variability	Identify and minimize stressors in production cycles that suppress the immune system (e.g., temperature, crowding, handling, transport, and stress from vaccination).
5. Microbial load variability	Lessen the microbial load and infectious agents that suppress the immune system for successful vaccination.
6. Chemical and antibiotic treatments that suppress the immune system	Prior to vaccination, stop antibiotic and chemical treatments that may interfere with optimal vaccination.

TABLE 17.2. Types, Formulation, and Commercially Licensed Vaccines for Fish

Type	Alternative Name(s)	Formulation	Licensed
Killed	Inactivated Bacterin (bacterial) Autogenous (single isolate from a facility)	Whole bacterium, virus, parasite, or fungus	Yes Bacterial only
Toxoid	Antitoxin	Neutralized toxin	No
Subunit (can be killed or live)	Recombinant Gene deleted Vector	Portion of pathogen (single protein)	No
DNA	Genetic Nucleic acid	Naked DNA	No
Modified Live	Attenuated	Whole bacterium, virus, parasite, or fungus	No

preventive strategy. Each type of vaccine offers both advantages and disadvantages. Regulations that govern the licensing and how a vaccine may be employed often determine what types of vaccines are available to the industry, and these regulations differ among the countries of the world (Birnbaum 1997; Lee 1997). Safety of the vaccine is the first consideration. The vaccine to be licensed must not cause the disease, initiate adverse reactions, or be responsible for disease transmision to the environment and other animals. The effectiveness of the vaccine in the prevention of infectious diseases under defined conditions is the next consideration. The vaccine is licensed for species of fish, age of the vaccinates, routes of administration, and dosage with and without booster and adjuvant.

Killed or Inactivated Vaccine

The regulatory issues and other considerations have pushed the development and employment of fish vaccines that are of the killed type. The infectious agent is killed or inactivated by chemical, heat, or irradiation treatment. Safety is the advantage of the killed vaccine. The disadvantages of killed vaccines are that protection is limited to those mediated by antibody and a booster dose and adjuvants are often needed to provide effective long-lasting protection. Inexpensive manufacturing costs and easy development and approval for licensing make killed vaccines the most common commercial vaccine for fish.

Toxoid Vaccine

Toxoid is also an inactivated or killed form of vaccine. Chemical treatment of a toxin that renders it inactive or in a toxoid form is used in its manufacture. No toxoid vaccines are commercially licensed for aquaculture.

Toxoid vaccines are advantageous because of safety and effectiveness in the neutralization of toxicity by antibody. High manufacturing costs and the need for a booster are the disadvantages of the toxoid vaccine.

Subunit Vaccine

Subunit vaccines are a more recent development in vaccine technology, and none are licensed for aquaculture. Subunit vaccines may be killed or inserted into a live organism. The protective antigen of the infectious agent is identified, separated from the infectious agent, and then purified to make a subunit vaccine. In a recombinant form of a subunit vaccine, the gene that codes for the purified protective antigen is identified, cloned, and inserted into an avirulent vector. The avirulent vector may be grown to produce the subunit vaccine in vitro or used as the vaccine directly. Among the disadvantages of a subunit vaccine is that the immune responses are restricted to the antigen selected, which may only provide partial protection. Another disadvantage is that the antibody-mediated protection stimulated by immunization with an incomplete structured protein often fails to provide effective protection of long duration. Safety (killed subunit vaccine), ease, and cost of manufacturing are advantages. The development of subunit vaccines for aquaculture is a favored approach by many investigators.

DNA Vaccine

The newest approach to improve the safety, efficacy, and duration of killed and live vaccines is by employing DNA vaccination. The injection of naked DNA into an animal host and its subsequent ability to cause antigen expression on the surface of host cells to stimulate both antibody and cell-mediated immunities is not completely understood (Donnelly et al. 1997). DNA vaccines consist of a bacterial plasmid with a strong viral promoter, the gene of interest (coding for protective antigen), and a polyadenylation/transcriptional terminal sequence. The plasmid is grown in bacteria or other suitable microbes, purified, dissolved in a suitable delivery solution, and then intramuscularly injected into the host. The DNA is taken up by host muscle cells where the encoded protein antigen is made and expressed on the surfaces of the muscle cells. Leong and colleagues (1997) provide an excellent review of and insight into the production and delivery of DNA vaccines in fish. The ability to stimulate both antibody and cell-mediated immunities of long duration is a major advantage of the DNA-based vaccine. The expression of protein antigens only by the antigen-presenting muscle cells is a limitation of the DNA vaccine. However, the proper structure and composition of the expressed protein antigen is maintained, as opposed to protein antigens produced for a subunit vaccine. Questions of safety, the difficulties of manufacturing and licensing, and the costs of DNA vaccines are among the disadvantages for employment in aquaculture. Due to the newest and preliminary progress in DNA-based vaccine development and delivery, it is difficult to forecast the future of this type of immunization in aquaculture.

Modified Live Vaccine

Modified live or attenuated vaccines have attracted great attention because they are the most successful vaccines in the livestock and poultry industries. Infectious agents are converted from a virulent (pathogenic) to an avirulent (nonpathogenic) form by numerous passages through cell cultures, increasing concentration of certain antibiotics or foreign host animals. Avirulent forms of infectious agents also exist in nature and may be selected and isolated by the application of molecular techniques. The modified live vaccine maintains its antigenic makeup but lacks the ability to cause pathology. Modified live vaccine can also be produced by either gene insertion or deletion to disrupt or knock out virulent genes of the infectious agent. Questions of safety and certain handling and storage problems are disadvantages that are outweighed by the advantages of modified live vaccines. Administration by many routes in one dose, stimulation of both antibody and cell-mediated immunities of long duration, lack of need for adjuvants, and inexpensive manufacturing costs are among the advantages.

It is worthy of note that modified, live, nonattenuated infectious agents have been made by genetic alterations for employment as aquaculture vaccines (Vaughan, Smith, and Foster 1993; Lawrence, Cooper, and Thune 1997). The major and limiting disadvantage of this type of nutritionally modified organism-based vaccine is the retention of its virulent properties, making high immunizing doses unsafe. At a low immunizing dose, the vaccine agent usually is not retained in the host long enough to stimulate complete protection of long duration. No vaccines of this type are licensed or employed in aquaculture. The major advantage, disadvantage, cost, and route of administration for each type of vaccine is summarized in Table 17.3.

TABLE 17.3. Major Advantages, Disadvantages, Costs, and Routes of Administration for Different Types of Vaccines

Type	Advantage	Disadvantage	Costs	Route of Administration
Killed	Safety	Poor to moderate efficacy	Low	Injection, immersion, and feeding
Subunit	Safety	Poor to moderate efficacy	High	Injection, immersion, and feeding
DNA	Excellent efficacy	Safety	High	Injection
Live modified	Excellent efficacy	Safety	Low	Injection, immersion, and feeding

Commercially Licensed Vaccines for Fish

Table 17.4 shows the licensed vaccines employed in fish. Licensed fish vaccines now include killed or inactivated and modified live vaccines. Most vaccines are bacterins because they are composed of killed bacteria. Enteric redmouth disease of rainbow trout, *Oncorhynchus mykiss,* caused by *Yersinia ruckeri* is successfully prevented by the employment of a formalin-killed *Y. ruckeri* vaccine administered by immersion, spray, injection, or feeding routes (Furones, Rogers, and Munn 1993; Toranzo and Barja 1993; Stevenson 1997). The vaccine is used to protect Atlantic salmon *(Salmo salar),* chinook salmon *(Oncorhynchus tshwyscha),* and other salmonids from this disease. Two antigenic makeups (serovar one and two) are known. The antiyersiniosis vaccine is composed of serovar one. This suggests that a degree of cross protection among these two antigenic makeups is afforded by this monovalent vaccine. Mortalities do occur in the vaccinates, but they are significantly reduced. The other economic benefits are enhanced feed conversions and reduced employment of antibiotics and medicated diets (Rogers 1991).

Polyvalent vaccines against *Vibrio anguillarum,* serovar one and two, *V. salmonicida,* and *Aeromonas salmonicida* for prevention of vibriosis, cold-water vibriosis, and furunculosis diseases of salmonids were developed and licensed. The vaccines were first developed as monovalent vaccines. These vaccines are the only polyvalent types employed in aquaculture

TABLE 17.4. Commercially Licensed Vaccines for Fish

Vaccine	Type	Disease	Species	Route(s)
Yersina ruckeri	Killed	Enteric redmouth	Rainbow trout	Immersion
Vibrio anguillarum	Killed	Vibriosis	Salmonids	Injection and immersion
Vibrio salmonicida	Killed	Cold-water vibriosis	Salmonids	Injection and immersion
Aeromonas salmonicida	Killed	Furunculosis	Salmonids	Injection and immersion
Polyvalent *V. anguillarum, V. salmonicida,* and *A. salmonicida*	Killed	Cold-water vibriosis and furunculosis	Salmonids	Oil adjuvant Injection
Edwardsiella ictaluri	Killed	Enteric septicemia	Channel catfish	Injection and immersion
	Modified live	Enteric septicemia	Channel catfish	Immersion
Flexibacter columnaris	Killed	Columnaris	Channel catfish	Injection

and have a high efficacy for each of the diseases. These vaccines are a formalin-killed combination of all three infectious agents that are administered as injectable oil-and-water adjuvant formulations. Granulomatous side effects of the oil adjuvant vaccine cause major problems due to their extensive negative effect on carcass quality and, hence, the loss of profit. Vinitnantharat, Gravningen, and Greger (1999) showed that efficacy of the polyvalent vaccine was near 100 percent against four *Vibrio* pathogens. Excellent reviews of the antivibriosis and furunculosis vaccines were written by Toranzo, Santos, and Barja (1997) and Ellis (1997). Advances in vibriosis and furunculosis vaccines are discussed, as well as the advantages and disadvantages in the employment of these vaccines. The vaccines need to be emulsified in oil and water and delivered by intraperitoneal injection to provide efficacy and protection of long duration. The vaccines have successfully prevented catastrophic disease of cultured Atlantic salmon and reduced the employment of antibiotics by marine netpen farmers since 1993. The predictable and satisfactory efficacy of these salmonid vaccines justifies their employment as a preventive strategy that is cost-effective.

One killed vaccine (bacterin) is currently licensed for immunization against enteric septicemia of catfish (ESC). This vaccine is a coated product and is administered by injection, immersion, or feeding. The efficacy of this vaccine is not well documented in extensive field trials, and the employment of this vaccine to prevent ESC is currently very limited in the catfish industry. An antibody type of immune response is stimulated by this coated bacterin. However, passive antibody studies have shown no protection against *E. ictaluri* infection (Klesius and Sealey 1995; Shoemaker and Klesius 1997). Mechanisms of immunity against *E. ictaluri* infection are reviewed by Klesius (1992).

The recent development and employment of a modified live vaccine against ESC was accomplished (Klesius and Shoemaker 1999). Many of the advantages of a modified live vaccine for prevention of an endemic and commonly occurring disease of major species of intensively cultured fish can be seen by employment of this anti-ESC vaccine. Control of *E. ictaluri* infection by feeding antibiotic-medicated diet is not currently used by catfish producers. The reasons are the additional expense and limited effectiveness in already sick and dying fish. Furthermore, oxytetracycline and ormethoprim-sulfamethoxine resistance has been documented. Fry to food-size catfish are susceptible to ESC, which is rapidly spread in intensive culture conditions. Our strategy was to develop a modified live vaccine that would have efficacy in fry seven to ten days old (Shoemaker, Klesius, and Bricker 1999), provide life-long protection, stimulate cell-mediated immunity, and require a single immunization by an immersion administration. Fry are stocked into fingerling ponds at 10 to 12 days of age, and this appeared to be the easiest and least expensive time for vaccination. Immersion of large numbers of catfish fry (50,000 to 100,000) into a vaccine bath for two minutes has been shown to be efficacious (see Table 17.5). The short immersion exposure to a modified live vaccine allowed immunization of the fry with minimum stress. The attenuated *E. ictaluri* infection persists in the fry for a sufficient time to stimulate protective specific cell-mediated responses. The safety of the modified live vaccine was documented in both experimental and field trials involving more than 5 million catfish in 1997 and 1998. The results of a field trial are presented in Table 17.6. The anti-ESC vaccine successfully protected large numbers of channel catfish fingerlings from clinical signs of disease and mortality due to *E. ictaluri* infection during a period of nine months of culture.

The modified live vaccine is composed of a rifampicin-resistant mutant that is no longer virulent. The loss of virulence was shown to be due to alteration of the lipopolysaccharide composition of the mutant's cell surface. Replication of the nonvirulent mutant in the vaccinate is necessary to stimulate a strong cell-mediated protective response (Shoemaker, Klesius, and Plumb 1997; Shoemaker and Klesius 1997). A question remains as to the extent of protection against various antigenic strains of *E. ictaluri* using this monovalent vaccine. Research has documented that *E. ictaluri* strains are not homogeneous in antigenic makeup, but heterogeneous (Klesius and Shoemaker 1997b). It is highly likely that different strains exist throughout the many farms in southeastern states producing the majority of catfish. However, more effective protection against different strains was observed in the vaccinates challenged 60 days after immunization with the modified live vaccine. Research has also shown that the genetic makeup of a catfish line affects the efficacy of the vaccine. The duration of the vaccine efficacy was

TABLE 17.5. Efficacy of Modified Live *Edwardsiella ictaluri* Vaccine in Channel Catfish

Vaccine Trial	**Number Vaccinated**	**Relative Percent Survival (RPS)**	**Percent Control Mortality**
1	60	98.3	100.0
2	125	96.6	94.6
3	60	96.8	94.4
4	60	93.9	26.4
5	60	78.9	79.3

TABLE 17.6. Field Trial Results of Modified Live *Edwardsiella ictaluri* Vaccine in Channel Catfish

Treatment	**Number of Fish**	**Survival (%)**	**Enteric Septicemia Signs or Mortality**
Immunized at 21 days posthatch	225,000	214,000 (95%)	No
Nonimmunized at 21 days posthatch	225,000	105,000 (47%)	Yes

found to be at least eight months without booster immunization. The licensing, manufacturing, and employment of this modified live vaccine is expected to reduce losses in profit due to ESC mortality and sickness by 50 to 70 percent. The vaccine is expected to protect approximately 1.6 to 2 billion fingerlings produced in 1999 against the economic impact of ESC. Klesius and Shoemaker (1999) recently reviewed the safety and efficacy of this modified live ESC vaccine.

A commercial and licensed catfish vaccine against columnaris disease, caused by *Flavobacterium columnare,* is available. This columnaris vaccine is a killed bacterin type that may be administered by immersion, injection, and feeding. However, the extent of its employment and efficacy in the field is not well documented in the literature. Efficacious vaccines against *Flavobacterium* species have proven difficult to produce and employ. Bernardet (1997) provides an excellent review on recent research aimed at the development of these vaccines for fish. A combination vaccine against *F. columnare* and *E. ictaluri* diseases is needed for the catfish industry, and development and employment of such a vaccine should be an area of intense research. Columnaris and ESC are the two major diseases of farm-raised catfish; they frequently occur together, or columnaris disease follows ESC.

IMMUNIZATION ROUTES EMPLOYED IN FISH VACCINATION

Immersion

A comparison of route of administration, stress, labor costs, adjuvant, and booster requirements for fish vaccines is presented in Table 17.7. Fish are immersed in the vaccine, which has been diluted in the water containing the fish (for example, the hatchery water) for short periods of time (two to ten minutes). The immersion route has favorable characteristics, especially for batch immunization of young fish. Larger fish are usually not immunized by immersion. A major advantage of this route is the short duration and reduced stress on the fish to be vaccinated.

TABLE 17.7. Comparison of Administration, Stress, Labor Costs, Adjuvant, and Booster Requirements

Characteristics	Immersion	Injection	Feeding
Administration	Easy	Moderate	Easy
Number of fish vaccinated/unit of time	Large	Small	Large
Degree of stress on vaccinates	Light	Intensive	Light
Labor costs/fish	Low	High	Low
Adjuvant	No	Yes	No
Ease of booster administration	Moderate	Difficult	Easy

Injection

The currently employed route of administrating some vaccines is by injection, especially in salmonids. Bacterins require adjuvants and boostered immunizations for optimal efficacy. The size of the fish to be vaccinated limits the injection route to larger fish (greater than 5 g). Other known limitations are that adjuvants cause adhesions, transient reduced feeding and performance, and increased health risks to the workers performing the vaccination. Injection administration is usually by the intraperitoneal route, often using a vaccine injection gun. Anesthetization is usually necessary for fish to be injected.

Feeding

Vaccination by feeding is currently the least frequently employed means of vaccine administration. The technology needed for effective administration of vaccines by feeding is lacking and there remains a need to overcome the problems in manufacturing, handling, and storage of oral vaccines. In addition, improvements are needed in protecting antigens from digestion in the digestive tract, and in the uptake and presentation of these antigens to stimulate protective immunity in the vaccinates. Some success in feeding vaccines has been reported, but further research on this means of vaccine administration is necessary.

Spray

Spray vaccination is another method of immunizing fish. This means of administration has been employed in vaccinating salmonids. The immunization of large numbers of fish regardless of size, the short time period, the reduced stress on the vaccinates, and the relative low costs are among the advantages. The major disadvantage is the relative low efficacy achieved by spraying vaccines. Additional research is needed to enhance the efficacy of this means of vaccine delivery before it is more widely employed in aquaculture.

REASONS FOR VACCINE FAILURES

Table 17.8 lists some of the reasons for vaccine failures. Of these, improper handling and administration, host and environmental factors, overwhelming infection, inadequate duration of immunity, and antigenic differences between vaccine and field strains are most germane to fish vaccinology. Another reason for vaccine failure is the ability of some antigens from infectious agents to misdirect or hinder the induction of active immunization. Immunization with the vaccine containing the secreted P 57 antigen of *Renibacterium salmoninarum* resulted in the failure to develop a

TABLE 17.8. Reasons for Vaccine Failures

1. Insufficient time to develop immunity
2. Improperly handled and administrated
3. Host factors
4. Overwhelming challenge exposure by pathogen(s)
5. Inadequate duration of immunity
6. Antigenic differences between vaccine and field strains
7. Multiple vaccine interference
8. Inappropriate or poor environmental conditions at time of vaccination

protective immune response against bacterial kidney disease (Kaattari and Piganelli 1997). This phenomenon is most likely an explanation for the failure of other vaccines, especially viral vaccines in fish. The ability of parasites to alter their antigenic composition in the various life forms is another explanation for the failure of antiparasitic vaccines.

AGE OF IMMUNIZATION

The literature on immunity in fish larvae was reviewed by Zapata and colleagues (1997). The development of the immune system begins early during embryogenesis. The cell-mediated immune capacity most likely develops earlier than antibody immune capacity. They concluded that exact age of immunological maturity was difficult to determine, especially in relation to when fry can be immunized. This earliest age of immunization was found to vary with the species of fish and the vaccine. The majority of vaccine trials in young fish was based on their size rather than their age. It is commonly believed that vaccination of fish is best after 21 days posthatch. Shoemaker, Klesius, and Bricker (1999) found that channel catfish fry at 12 days posthatch were successfully immunized with the modified live *Edwardsiella ictaluri* vaccine. More recent studies indicate that channel catfish fry were immunizable as young as seven to ten days posthatch, as previously indicated in this chapter. *In ovo* vaccination is becoming an industry standard for poultry and no evidence for tolerance in active protective immunity is documented.

IMMUNOSTIMULANTS AND ADJUVANTS

The use of immunostimulants may be a promising approach to enhancing the defense mechanisms of fish (Anderson 1992) and possibly shrimp (Klesius and Shoemaker 1997a). An immunostimulant is a substance that is usually recognized and leads to the activation of cells of the immune system (fish) and defense system (shrimp). The activated cells have enhanced nonspecific capacities against infectious agents. In fish, these cells are monocytes and in shrimp the cells are hemocytes. Beta-glucans are an example of immunostimulants that have been used to enhance fish vaccine efficacy and activate shrimp hemocytes in vitro. The source and chemical extraction process strongly influence the polymer size, aqueous solubility, and activities of the beta-glucan. The source and composition of selected beta-glucans include those of barley of 1,3-1,4 glucan linkage and yeasts of 1,3-1,6 glucan linkage.

Adjuvants were created to make antigens optimally immunogenic. An adjuvant commonly employed is an oil adjuvant such as found in polyvalent vibriosis vaccine for salmonids. The types of oil used include mineral and squalene. The use of this adjuvant formulation causes severe pathology that is more severe than aluminum salt or beta-glucan adjuvant vaccines. The degree of adjuvant-induced pathology was found to vary among different adjuvant, commercial vaccines in Atlantic salmon. The adjuvant vaccines were found to depress the growth of fish.

There is an immediate need to improve adjuvant formulation so as to limit economic losses due to carcass lesions and poor performance.

CONCLUSION

A parallel to the development and employment of fish vaccines is the successful management of disease in poultry by vaccination. Fish raised under commercial, intensive conditions are very vulnerable to environmental exposure to a number of existing or emerging pathogens. Vaccination of fish needs to become an integral part of fish health management programs. Currently, few vaccines are commercially available and active immunization with bacterin vaccines is all that is employed. Newer vaccines that include modified live, subunit, and DNA vaccines are needed. Enhanced delivery methods including immersion, feeding, and spray routes are needed. Effective injectable vaccines that do not require either adjuvants or boosters need to be developed. Alternatively, adjuvants that do not produce adverse reactions should be devised and employed. Ideal vaccines, regardless of their type of administration, should be able to stimulate both antibody and cell-mediated immunity against infectious agents. There exists a great need to develop polyvalent combined vaccines against the common and frequently occurring infectious agents within an aquaculture system. Diseases could be prevented by a single vaccination of the youngest fish for the duration of their greatest risk to infectious agents in the production cycles. Finally, the genetics and basics of immunity in fish need to be better understood, so that lines of fish can be developed that are more efficient in their response to the protection afforded by vaccination.

Table 17.9 provides a list of selected diseases for which vaccines are needed in fish. Vaccines could provide effective protection against these bacterial, viral, and parasitic diseases with their commercial development and employment. Recent research progress toward the development of vaccines against these diseases has been reviewed (Gudding et al. 1997). It is expected that some of these vaccines are currently being reviewed for licensing and manufacturing in the next five years. The development and employment of new and more efficacious vaccines is needed to sustain the growth of the aquaculture industry in the next century. This effort will require greater partnerships between governments, universities, private institutes, and industries involved in vaccine research. Fish vaccine devclopment requires an informed understanding of the various kinds of aquaculture systems, the species of fish cultured, the nature of the fish immune system for

TABLE 17.9. Selected Diseases Lacking Commercially Licensed Vaccines

Vaccine	Disease	Species
Renibacterium salmoninarum	Bacterial kidney	Salmonids
Piscirickettsia salmonis	Rickettsial septicemia	Salmonids
Pasteurella piscida	Pasteurellosis	Salmonids
Aeromonas sp.	Motile aeromonas septicemia	Salmonids, catfish, and other species
Enterococcus seriolicida	Enterococcal septicemia	Marine species
Edwardsiella tarda	Edwardsiellosis	Various species
Streptococcus sp. *(iniae)*	Streptococcosis	Tilapia, rainbow trout, and hybrid bass
Flavobacterium psychrophilum	Cold-water disease	Salmonids
Flavobacterium psychrophilum	Rainbow trout fry syndrome	Rainbow trout
Infectious pancreatic necrosis virus	Infectious pancreatic necrosis	Salmonids
Hemorrhagic septicemia virus	Viral hemorrhagic septicemia	Salmonids
Proliferative gill parasite	Proliferative gill	Channel catfish
Ichthyophthirius multifiliis	White spot (Ich)	Various species
Saprolegnia sp.	Winter-kill disease	Channel catfish

different species of fish, and the needs of the customers or fish producers. Fish vaccinology offers unique challenges that differ from those of the other animal industries. Basic knowledge learned from warm-blooded animal vaccinology research may be expected to benefit the advancement of fish vaccinology.

REFERENCES

Anderson, D.P. 1992. Immunostimulants, adjuvants, and vaccine carriers in fish: Applications to aquaculture. In M. Faisal and F.M. Hetrick (Eds.), *Annual Review of Fish Diseases* Volume 2 (pp. 281-307). New York: Pergamon Press.

Bernardet, J.-F. 1997. Immunization with bacterial antigens: Flavobacterium and Flexibacter infections. In R. Gudding, A. Lillehaug, P.J. Midtlyng, and F. Brown (Eds.), *Fish Vaccinology: Developments in Biological Standardization* Volume 90 (pp. 179-188). Karger, Switzerland: Basel.

Birnbaum, N.G. 1997. Regulation of veterinary biological products for fish in the United States. In R. Gudding, A. Lillehaug, P.J. Midtlyng, and F. Brown (Eds.), *Fish Vaccinology: Developments in Biological Standardization* Volume 90 (pp. 335-340). Karger, Switzerland: Basel.

Donnelly, J.J., J.B. Ulmer, J.W. Shiver, and M.A. Liu. 1997. DNA vaccines. *Annual Reviews of Immunology* 15:617-648.

Ellis, A.E. 1997. Immunization with bacterial antigens: Furunculosis. In R. Gudding, A. Lillehaug, P.J. Midtlyng, and F. Brown (Eds.), *Fish Vaccinology: Developments in Biological Standardization* Volume 90 (pp. 107-116). Karger, Switzerland: Basel.

Finch, R. 1980. Immunomodulating effects on antimicrobial agents. *Journal of Antimicrobial Chemotherapy* 6:691-699.

Furones, M.D., C.J. Rogers, and C.B. Munn. 1993. *Yersinia ruckeri* the caustive agent of enteric redmouth disease (ERM) in fish. In M. Faisal and F.M. Hetrick (Eds.), *Annual Review of Fish Diseases* Volume 3 (pp. 105-125). New York: Pergamon Press.

Gudding, R., A. Lillehaug, P.J. Midtlyng, and F. Brown. 1997. *Fish Vaccinology. Developments in Biological Standardization.* Volume 90. Karger, Switzerland: Basel.

Kaattari, S.L. and J.D. Piganelli. 1997. Immunization with bacterial antigens: Bacterial kidney disease. In R. Gudding, A. Lillehaug, P.J. Midtlyng, and F. Brown (Eds.), *Fish Vaccinology: Developments in Biological Standardization* Volume 90 (pp. 145-152). Karger, Switzerland: Basel.

Klesius, P.H. 1992. Immune system of channel catfish: An overture on immunity to *Edwardsiella ictaluri.* In M. Faisal and F. M. Hetrick (Eds.), *Annual Review of Fish Diseases.* Volume 2 (pp. 325-338). New York: Pergamon Press.

Klesius, P., C. Lim, and C. Shoemaker. 1999. Effect of feed deprivation on innate resistance and antibody response to *Flavobacterium columnare* in channel catfish, *Ictalurus punctatus. Bulletin European Association of Fish Pathologists* 19(4): 156-158.

Klesius, P.H. and W.M. Sealey. 1995. Chacteristics of serum antibody in enteric septicemia of catfish. *Journal of Aquatic Animal Health* 7:205-210.

Klesius, P.H. and C.A. Shoemaker. 1997a. Enhancement of disease resistance in shrimp: A review. In Alston, D.E., B.W. Green, and H.C. Clifford III (Eds.), *Focusing on Sustainable Shrimp and Tilapia Farming* (pp. 31-35). Central American Symposium on Aquaculture Proceedings 4. Tegucigalpa, Honduras: World Aquaculture Society, Latin American Chapter.

Klesius, P.H. and C.A. Shoemaker. 1997b. Heterologous isolates challenge of channel catfish, *Ictalurus punctatus,* immune to *Edwardsiella ictaluri. Aquaculture* 157:147-155.

Klesius, P.H. and C.A. Shoemaker. 1999. Development and use of modified live *Edwardsiella ictaluri* vaccine against enteric septicemia of catfish. In R.D. Schultz (Ed.), *Advances in Veterinary Medicine.* Volume 41 (pp. 523-537). San Diego, CA: Academic Press.

Lawrence, M.L., R.K. Cooper, and R.L. Thune. 1997. Attenuation, persistence and vaccine potential of an *Edwardsiella ictaluri* purA mutant. *Infection and Immunity* 65:4642-4651.

Lee, A. 1997. European regulations relevant to the marketing and use of fish vaccines. In R. Gudding, A. Lillehaug, P.J. Midtlyng, and F. Brown (Eds.), *Fish Vaccinology: Developments in Biological Standardization.* Volume 90 (pp. 341-346). Karger, Switzerland: Basel.

Leong, J.C., E. Anderson, L.M. Bootland, P.-W. Chiou, M. Johnson, C. Kim, D. Mourich, and G. Trobridge. 1997. Fish vaccine antigens produced or delivered by recombinant DNA technologies. In R. Gudding, A. Lillehaug, P.J. Midtlyng, and F. Brown (Eds.), *Fish Vaccinology: Developments in Biological Standardization.* Volume 90 (pp. 267-277). Karger, Switzerland: Basel.

Lev, M. And M. Forbes. 1959. Growth response to dietary penicillin of germ-free chicks with a defined intestinal flora. *British Journal of Nutrition* 13:78-84.

Markestad, A. and K. Graves. 1997. Reduction of antibacterial drug use in Norwegian fish farming due to vaccination. In R. Gudding, A. Lillehaug, P.J. Midtlyng, and F. Brown (Eds.), *Fish Vaccinology: Developments in Biological Standardization.* Volume 90 (pp. 365-369). Karger, Switzerland: Basel.

Nagi, S., A. N. Sahin, G. Wagner, and J. Williams. 1984. Adverse effects of antibiotics on the development of gut-associated lymphoid tissues and the serum immunoglobulins in chickens. *American Journal of Veterinary Research* 45:1425-1429.

Plumb, J. A. 1999. *Health Maintenance and Microbial Diseases of Cultured Fishes.* Ames, IA: Iowa State University Press.

Rogers, C.J. 1991. The usage of vaccination and antimicrobial agents for control of *Yersinia ruckeri. Journal of Fish Diseases* 14:291-301.

Shoemaker, C.A. and P.H. Klesius. 1997. Protective immunity against enteric septicemia in channel catfish, *Ictalurus punctatus* (Rafinesque), following controlled exposure to *Edwardsiella ictaluri. Journal of Fish Diseases* 20:101-108.

Shoemaker, C.A., P.H. Klesius, and J.M. Bricker. 1999. Efficacy of a modified live *Edwardsiella ictaluri* vaccine in channel catfish as young as seven days post hatch. *Aquaculture* 176:189-193.

Shoemaker, C.A., P.H. Klesius, and J.A. Plumb. 1997. Killing of *Edwardsiella ictaluri* by macrophages from channel catfish immune and susceptible to enteric septicemia of catfish. *Veterinary Immunology and Immunopathology* 58:181-190.

Soderhall, K., and P.-O. Thornqvist. 1997. Crustacean immunity—A short review. In R. Gudding, A. Lillehaug, P.J. Midtlyng, and F. Brown (Eds.), *Fish Vaccinology: Developments in Biological Standardization.* Volume 90 (pp. 45-51). Karger, Switzerland: Basel.

Stevenson, R.M.W. 1997. Immunization with bacterial antigens: Yersiniosis. In R. Gudding, A. Lillehaug, P.J. Midtlyng, and F. Brown (Eds.), *Fish Vaccinology: Developments in Biological Standardization.* Volume 90 (pp. 117-124). Karger, Switzerland: Basel.

Stromsheim, A., D.M. Eide, K.T. Fjalestad, H.J.S. Larsen, and K.H. Roed. 1994. Genetic variation in humoral immune response in Atlantic salmon (*Salmo salar* L.) against *Aeromonas salmonocida* O-layer. *Veterinary Immunology and Immunopathology* 4:341-352.

Toranzo, A.E. and J.L. Barja. 1993. Virulence factors of bacteria pathogenic for coldwater fish. In M. Faisal and F.M. Hetrick (Eds.), *Annual Review of Fish Diseases.* Volume 3 (pp. 5-36). New York: Pergamon Press.

Toranzo, A.E., Y. Santos, and J.L. Barja. 1997. Immunization with bacterial antigens: Vibrio infections. In R. Gudding, A. Lillehaug, P.J. Midtlyng, and F. Brown (Eds.), *Fish Vaccinology: Developments in Biological Standardization.* Volume 90 (pp. 93-105). Karger, Switzerland: Basel.

USDA (U.S. Department of Agriculture). 1998. *Aquaculture Outlook.* Document Number ERS-LPD-AQS-7. Washington, DC: USDA.

Vaughan, L.M., P.R. Smith, and T.J. Foster. 1993. An aromotic-dependent mutant of the fish pathogen *Aeromonas salmonicida* is attenuated in fish and is effective as a live vaccine against the salmonid disease furunculosis. *Infection and Immunity* 61:2172-2181.

Vinitnantharat, S., K. Gravningen, and E. Greger. 1999. Fish vaccines. In R.D. Schultz (Ed.), *Advances in Veterinary Medicine.* Volume 41 (pp. 539-550). San Diego, CA: Academic Press.

Zapata, A.G, M. Torroba, A. Varas, E. Jimenez. 1997. Immunity in fish larvae. In R. Gudding, A. Lillehaug, P.J. Midtlyng, and F. Brown (Eds.), *Fish Vaccinology: Developments in Biological Standardization.* Volume 90 (pp. 23-32). Karger, Switzerland: Basel.

Index

Page numbers followed by the letter "f" indicate figures; those followed by the letter "t" indicate tables.

Order Your Own Copy of This Important Book for Your Personal Library!

NUTRITION AND FISH HEALTH

_______in hardbound at $89.95 (ISBN: 1-56022-887-3)

COST OF BOOKS______

OUTSIDE USA/CANADA/ MEXICO: ADD 20%____

POSTAGE & HANDLING________
(US: $4.00 for first book & $1.50 for each additional book) Outside US: $5.00 for first book & $2.00 for each additional book)

SUBTOTAL________

in Canada: add 7% GST____

STATE TAX____
(NY, OH & MIN residents, please add appropriate local sales tax)

FINAL TOTAL____
(If paying in Canadian funds, convert using the current exchange rate, UNESCO coupons welcome.)

❑ **BILL ME LATER:** ($5 service charge will be added)
(Bill-me option is good on US/Canada/Mexico orders only; not good to jobbers, wholesalers, or subscription agencies.)

❑ Check here if billing address is different from shipping address and attach purchase order and billing address information.

Signature________________________

❑ **PAYMENT ENCLOSED:** $________________

❑ **PLEASE CHARGE TO MY CREDIT CARD.**

❑ Visa ❑ MasterCard ❑ AmEx ❑ Discover ❑ Diner's Club ❑ Eurocard ❑ JCB

Account # ________________________

Exp. Date________________________

Signature________________________

Prices in US dollars and subject to change without notice.

NAME________________________

INSTITUTION________________________

ADDRESS________________________

CITY________________________

STATE/ZIP________________________

COUNTRY________________ COUNTY (NY residents only)________________

TEL________________ FAX________________

E-MAIL________________

May we use your e-mail address for confirmations and other types of information? ❑ Yes ❑ No
We appreciate receiving your e-mail address and fax number. Haworth would like to e-mail or fax special discount offers to you, as a preferred customer. **We will never share, rent, or exchange your e-mail address or fax number.** We regard such actions as an invasion of your privacy.

Order From Your Local Bookstore or Directly From

The Haworth Press, Inc.

10 Alice Street, Binghamton, New York 13904-1580 • USA
TELEPHONE: 1-800-HAWORTH (1-800-429-6784) / Outside US/Canada: (607) 722-5857
FAX: 1-800-895-0582 / Outside US/Canada: (607) 722-6362
E-mail: getinfo@haworthpressinc.com

PLEASE PHOTOCOPY THIS FORM FOR YOUR PERSONAL USE.

www.HaworthPress.com

BOF00

Milton Keynes UK
Ingram Content Group UK Ltd.
UKHW020015071024
449327UK00031B/2796